Controversies in Hepatocellular Carcinoma

Nancy S. Reau • Amit G. Singal
Editors

Controversies in Hepatocellular Carcinoma

Springer

Editors
Nancy S. Reau
Section of Hepatology
Rush University Medical Center
Chicago, IL, USA

Amit G. Singal
Department of Medicine
University of Texas Southwestern Medical Center
Dallas, TX, USA

ISBN 978-3-031-90224-6 ISBN 978-3-031-90225-3 (eBook)
https://doi.org/10.1007/978-3-031-90225-3

This Springer imprint is published by the registered company Springer Nature Switzerland AG
The registered company address is: Gewerbestrasse 11, 6330 Cham, Switzerland

I would like to dedicate this book to my loving wife, Purva, and amazing daughters, Avanya and Amaya, who tolerate me being on my computer for too many hours of the day, including the time as I work on this book. I would also like to dedicate this book to my mom and dad who spent unmeasurable effort guiding me through life, who taught me innumerable life lessons, and whom I could never thank enough.

–Amit G. Singal, MD, MS

I would like to dedicate this book to my family, who tolerate proofreading, late hours, and constant requests for "better" words: my wonderful spouse, Rajeev, who now knows nearly as much about liver disease as his orthopedic passion, and my three children, Aria, Krishan, and Alina, who are considering any career other than medicine.

–Nancy S. Reau, MD

We also want to recognize the patients, scientists, and clinicians, without whom there would be no innovation, no controversy, and no improvement in therapeutics for our patients with liver cancer.

Introduction

Controversies in Hepatocellular Carcinoma has been written for clinicians who care for patients with hepatocellular carcinoma (HCC) as well as health services researchers conducting work in this area. Although HCC receives less attention in the lay media compared with other common cancers, it is one of the leading causes of cancer-related death worldwide as well as one of the leading causes of death in patients with cirrhosis. While the burden has historically been greatest in East Asia and Africa, the incidence of HCC is increasing in many parts of the Western world in parallel with an increasing prevalence of metabolic dysfunction-associated steatotic liver disease. HCC remains a challenging disease for many providers given the need to understand and manage both liver disease and cancer, underlying the need for education efforts regarding optimal management.

Despite improvements over time, HCC continues to be a highly fatal cancer with a 5-year survival that has remained below 20%. In the United States, HCC has remained one of the few solid tumors whose mortality has continued to worsen over time. Fortunately, we have seen tremendous advances in both screening and therapeutics over the past several years, which have the possibility to change this disturbing trend.

In this book, we have tackled some of the emerging controversial topics and questions that come up in HCC clinical practice and research. We have selected 33 questions ranging from screening and diagnosis to therapeutics and survivorship. Authors were tasked with defending or refuting a controversial question, thus the title of each chapter is a controversy not a factual statement. Notably, some of the questions tackle topics with emerging data, whereas others tackle important clinical questions but in data-free zones. We have tried to make the content relevant to providers across multiple disciplines, including hepatologists, medical oncologists, surgeons, interventional radiologists, and radiation oncologists, reflecting the multidisciplinary team that is needed to care for these patients in clinical practice.

If there is a particular topic or question of interest, each chapter is a stand-alone section and can be read as such. However, there are others that may enjoy reading the entire book to gain knowledge across all covered topic areas. Overall, we hope the book is enjoyable and helps with management of this patient population.

Dallas, TX, USA — Amit G. Singal
Chicago, IL, USA — Nancy S. Reau

Financial Disclosers

Nancy S. Reau has served as a consultant or on advisory boards for Abbott, Arbutus, Gilead Sciences, Salix, and VIR. Research support (paid to Rush) has been received from Gilead, AbbVie and Salix.

Amit G. Singal has served as a consultant or on advisory boards for Genentech, AstraZeneca, Eisai, Exelixis, Bayer, Merck, Elevar, Boston Scientific, Sirtex, FujiFilm Medical Sciences, Exact Sciences, Helio Genomics, Roche, Glycotest, Abbott, ImCare, Curve Biosciences, DELFI, Mursla, and Universal Dx.

Acknowledgments

We would like to thank our coauthors, not only for their contributions to this book but for their expertise and dedication to this field.

Contents

About the Editors and Contributors

About the Editors

Nancy S. Reau, MD, is Professor of Internal Medicine, Richard B. Capps Chair of Hepatology, Associate Director of Solid Organ Transplantation, and Section Chief of Hepatology at Rush University Medical Center in Chicago, Illinois. She received her medical degree from the Ohio State University College of Medicine in Columbus, where she completed a residency and fellowship in gastroenterology/hepatology followed by a second fellowship in advanced transplant hepatology at Johns Hopkins Hospital in Baltimore, Maryland. Her primary research interests focus on viral hepatitis and complications of chronic liver disease.

Professor Reau has authored or coauthored more than 100 peer-reviewed articles. She is the previous editor-in-chief of *Clinical Liver Disease* and was an author of the American Association for the Study of Liver Diseases/Infectious Diseases Society of America hepatitis C guidance document.

Amit G. Singal, MD, MS, is a Professor of Medicine and Chief of Hepatology at UT Southwestern Medical Center in Dallas, Texas. He is an expert in hepatocellular carcinoma (HCC) management and serves as the Medical Director of the UT Southwestern Liver Tumor Program. Dr. Singal received his medical degree from the University of Michigan Medical School, his residency in internal medicine at the University of Washington School of Medicine, and then completed fellowships in gastroenterology and transplant hepatology at the University of Michigan. Dr. Singal is board certified in gastroenterology and transplant hepatology.

Dr. Singal's research focuses on the prevention, screening, and treatment of HCC, including evaluation of interventions to improve early tumor detection and reduce HCC-related mortality. He has published more than 300 peer-reviewed articles and received several grants, including from the National Institutes of Health, Cancer Prevention Research Institute of Texas, and the US Department of Defense. For his work, he has been awarded the David Bruton Professorship in Clinical Cancer Research, Dedman Family Scholar in Clinical Care, and Blue Faery Award for Excellence in Liver Cancer Research.

Contributors

Joseph C. Ahn Division of Gastroenterology and Hepatology, Mayo Clinic, Rochester, MN, USA

Naim Alkhouri Department of Hepatology, Arizona Liver Health, Chandler, AZ, USA

Sukeshi Patel Arora Division of Hematology/Oncology, Mays Cancer Center, University of Texas (UT) Health San Antonio, San Antonio, TX, USA

Bulent Arslan Rush University Medical Center, Chicago, USA

Ashwini Arvind Division of Gastroenterology, University of Texas at Southwestern (UTSW) Medical Center, Dallas, TX, USA

Anthony Bejjani Department of Medicine, Division of Hematology/Oncology, Geffen School of Medicine at University of California Los Angeles, Jonsson Comprehensive Cancer Center, Los Angeles, CA, USA

Stefan Brancel Henry Ford Hospital, Detroit, MI, USA

Edie Y. Chan Department of Surgery, Rush University Medical Center, Chicago, IL, USA

Daniel T. Chang Department of Radiation Oncology, Stanford University, Stanford, CA, USA

Victoria Chernyak Department of Radiology, University of Washington, Seattle, WA, USA

Guilherme M. Cunha Liver Imaging Group, Department of Radiology, UC San Diego, San Diego, CA, USA

Sheila Eswaran Department of Medicine, Rush University Medical Center, Chicago, IL, USA

Jordan Feld Toronto General Hospital Research Institute (TGHRI), Toronto Centre for Liver Disease, University Health Network, Toronto, ON, Canada

Richard S. Finn Department of Medicine, Division of Hematology/Oncology, Geffen School of Medicine at University of California Los Angeles, Jonsson Comprehensive Cancer Center, Los Angeles, CA, USA

Michele Finotti Department of Surgical Oncological and Gastroenterological Sciences, Padua University, Padua, Italy

Kathryn J. Fowler Diagnostic Radiology, Scripps Medical Group, San Diego, CA, USA

Miles W. Grunvald Colon and Rectal Surgery Fellow, University of Chicago Medical Center, Chicago, IL, USA

Dharani Guttikonda Endeavor Health Glenview, Chicago, IL, USA

Alexander A. Harris Radiation Oncology Consultants Ltd., Ascension Alexian Brothers Medical Center, Elk Grove, IL, USA

Ruben Hernaez Section of Gastroenterology, Michael E. DeBakey Veterans Affairs Medical Center, Houston, TX, USA
Center for Innovations in Quality, Effectiveness and Safety (IQuESt), Michael E. DeBakey Veterans Affairs Medical Center, Houston, TX, USA
Section of Gastroenterology and Hepatology, Department of Medicine, Baylor College of Medicine, Houston, TX, USA

George N. Ioannou Director of Hepatology, Veterans Affairs Puget Sound Health Care System, University of Washington, Seattle, WA, USA

Shikha Jain Division of Hematology and Oncology, University of Illinois Cancer Center, Chicago, IL, USA

Michelle R. Ju Division of Surgical Oncology, Department of Surgery, UT Southwestern Medical Center, Dallas, TX, USA

Fasiha Kanwal Section of Gastroenterology, Michael E. DeBakey Veterans Affairs Medical Center, Houston, TX, USA
Center for Innovations in Quality, Effectiveness and Safety (IQuESt), Michael E. DeBakey Veterans Affairs Medical Center, Houston, TX, USA
Section of Gastroenterology and Hepatology, Department of Medicine, Baylor College of Medicine, Houston, TX, USA

Yuko Kono Division of Gastroenterology and Hepatology, Department of Medicine, University of California San Diego, San Diego, CA, USA
Department of Radiology, University of California San Diego, San Diego, CA, USA

Laura Kulik Medicine (Gastroenterology and Hepatology), Radiology, Surgery (Organ Transplantation), Northwestern University, Chicago, IL, USA

Allison J. Kwong Stanford Transplant Outcomes Research Center, Stanford University, Stanford, CA, USA

Vennis Lourdusamy Bayhealth Medical Group, Dover, DE, USA

Raoel Maan Department of Gastroenterology and Hepatology, Erasmus MC University Medical Center, Rotterdam, The Netherlands

Jorge A. Marrero Division of Gastroenterology and Hepatology, University of Pennsylvania, Philadelphia, PA, USA

William R. Masch Division of Abdominal Imaging, Department of Radiology, University of Michigan, Ann Arbor, MI, USA

Neil Mehta University of California, San Francisco, CA, USA

Mishal Mendiratta-Lala Division of Abdominal Imaging, Department of Radiology, University of Michigan, Ann Arbor, MI, USA

K. V. Narayanan Menon Medical City Dallas Transplant Institute, Dallas, TX, USA

Alexandra Morris Henry Ford Hospital, Detroit, MI, USA

Arvind R. Murali Orlando Health Liver Center, Orlando Health, University of Iowa, Iowa City, IA, USA

Pierre Nahon Liver Unit, APHP, Bobigny, France

Mindie H. Nguyen Division of Gastroenterology and Hepatology, Stanford University Medical Center, Palo Alto, CA, USA
Department of Epidemiology and Population Health, Stanford University Medical Center, Palo Alto, CA, USA
Stanford Cancer Institute, Stanford University Medical Center, Palo Alto, CA, USA

Neehar D. Parikh Division of Gastroenterology and Hepatology, University of Michigan, Ann Arbor, MI, USA

Anjana Pillai Division of Gastroenterology and Hepatology, University of Illinois Chicago, Chicago, IL, USA

Ashwin Rao Section of Gastroenterology and Hepatology, Baylor College of Medicine, Houston, TX, USA

Nicole E. Rich Department of Medicine, University of Texas Southwestern Medical Center, Dallas, TX, USA

Ian A. Rowe Leeds Institute for Medical Research, University of Leeds, Leeds, England, UK
Leeds Liver Unit, Leeds Teaching Hospitals NHS Trust, Leeds, England, UK

Reena Salgia Division of Gastroenterology and Hepatology, Henry Ford Hospital, Detroit, MI, USA

Myron Schwartz Icahn School of Medicine at Mount Sinai, New York, NY, USA

Scott Schwartz Henry Ford Hospital, Detroit, MI, USA

Ranya Selim Division of Gastroenterology and Hepatology, Henry Ford Hospital, Detroit, MI, USA

Olivier Seror Radiology Unit, APHP, Bobigny, France

Tracey G. Simon Division of Gastroenterology, Massachusetts General Hospital, Harvard Medical School, Boston, MA, USA

Amit G. Singal Department of Medicine, University of Texas Southwestern Medical Center, Dallas, TX, USA

Claude B. Sirlin Department of Radiology, Columbia University Irving Medical Center, New York, NY, USA

William Small Jr Department of Radiation Oncology, Stritch School of Medicine, Loyola University, Maywood, IL, USA

Jordan Tasse Rush University Medical Center, Chicago, USA

Emily Harding-Theobald Division of Gastroenterology, University of Colorado, Aurora, CO, USA

Tarita O. Thomas Department of Radiation Oncology, Feinberg School of Medicine, Northwestern University, Chicago, IL, USA

Adriaan J. van der Meer Department of Gastroenterology and Hepatology, Erasmus MC University Medical Center, Rotterdam, The Netherlands

Alessandro Vitale Department of Surgical Oncological and Gastroenterological Sciences, Padua University, Padua, Italy

Jennifer Wang Division of Gastroenterology and Hepatology, University of Illinois Chicago, Chicago, IL, USA

Mike T. Wei Division of Gastroenterology and Hepatology, Stanford University Medical Center, Palo Alto, CA, USA

Stephanie R. Wilson Department of Radiology, University of Calgary, Calgary, AB, Canada
Department of Medicine, University of Calgary, Calgary, AB, Canada

Ju Dong Yang Division of Digestive and Liver Diseases, Department of Medicine, Comprehensive Transplant Center, Samuel Oschin Comprehensive Cancer Institute, Cedars Sinai Medical Center, Los Angeles, CA, USA

Adam C. Yopp Division of Surgical Oncology, Department of Surgery, UT Southwestern Medical Center, Dallas, TX, USA

Hooman Farhang Zangneh Toronto General Hospital Research Institute (TGHRI), Toronto Centre for Liver Disease, University Health Network, Toronto, ON, Canada

Jinyu Zhang Division of Gastroenterology and Hepatology, Henry Ford Hospital, Detroit, MI, USA

Part I

Risk Assessment and Chemoprevention

Should All Patients with Hepatitis B Virus Undergo Hepatocellular Carcinoma Surveillance?

1

Mike T. Wei and Mindie H. Nguyen

Hepatocellular carcinoma (HCC) surveillance should be performed in all patients with hepatitis B virus (HBV) cirrhosis as well as other high-risk groups, which can be defined using risk scores. There are limited data for populations in whom surveillance is not recommended (e.g., immune-tolerant or inactive-carrier state), and shared decision-making with patients can be performed.

Main Points

1. There have been several HCC risk scores in patients with HBV infection that incorporate risk factors including older age, male sex, family history of HCC, higher DNA levels, and the presence of cirrhosis.
2. Risk scores are likely useful to identify high- and low-risk patients, although validation in multiracial cohorts including a combination of treated and untreated patients is still required.
3. Data for HCC risk are sparse for populations outside surveillance recommendations, including immune-tolerant and inactive carrier patients. These studies are difficult given issues with classification of patients into distinct phenotypes, particularly over extended periods of time. In light of limited data, shared decision-making can be performed.

M. T. Wei
Division of Gastroenterology and Hepatology, Stanford University Medical Center, Palo Alto, CA, USA
e-mail: mtwei@stanford.edu

M. H. Nguyen (✉)
Division of Gastroenterology and Hepatology, Stanford University Medical Center, Palo Alto, CA, USA

Department of Epidemiology and Population Health, Stanford University Medical Center, Palo Alto, CA, USA

Stanford Cancer Institute, Stanford University Medical Center, Palo Alto, CA, USA
e-mail: mindiehn@stanford.edu

As of 2015, the World Health Organization estimated that 257 million people had chronic hepatitis B (CHB). However, only 27 million people knew they had the infection, and 4.5 million people were on treatment [1]. A modeling study in 2016 gave a higher estimate of 292 million people affected globally with HBV infection [2]. Within the United States, the estimated HBV burden is between 0.84 million and 1.59 million individuals, with a disease awareness rate of only 15% among affected people [3, 4]. A more recent study using a large real-world nationwide data set of 138 million Americans with private health insurance (2005–2015) showed that even among people with better health insurance, the HBV diagnosis rate is only about 15%. Further, the overall treatment rate was dismally low at 5%, and only 30–50% for patients with cirrhosis and HCC [5].

CHB can lead to cirrhosis, HCC, liver failure, and death. Because there is no cure for CHB, management of CHB entails long-term viral suppression of the disease [6]. While viral suppression decreases the risk of development of HCC, this risk is not completely eliminated. [7–10] In Asian studies, current CHB treatment reduces the risk of HCC by around 30% among cirrhotic patients, and around 80% among noncirrhotic patients [11, 12]. Currently, liver cancer is the fourth most common cause of cancer death, estimated at 782,000 deaths in 2018 [13]. The incidence of HCC is higher in men, with a male-to-female ratio of 2.5–3:1 [14]. The risk overall is different depending on the region studied. In North America and Europe, the incidence rate of HCC is 0.02 per 100 person-years among inactive HBV carriers, 0.3 for CHB patients without cirrhosis, and 2.2 for patients with compensated cirrhosis. In East Asia, the incidence is 0.2 for inactive carriers, 0.6 for CHB patients without cirrhosis, and 3.7 for patients with compensated cirrhosis [15].

Surveillance is important as it has been found to improve survival. [16–18] In the one available randomized controlled trial, Zhang et al. [19] in 2004 showed that even though only 58.2% of patients in the HCC surveillance arm completed all screening opportunities, patients receiving HCC surveillance

N. S. Reau, A. G. Singal (eds.), *Controversies in Hepatocellular Carcinoma*, https://doi.org/10.1007/978-3-031-90225-3_1

had earlier detection of lower-stage HCC and had a reduction in HCC mortality of 37%. Several guidelines exist regarding screening for HCC among patients with CHB. In general, HCC surveillance is recommended if the risk of HCC is 0.2% or greater per year [20]. While the American Association for the Study of Liver Diseases (AASLD), European Association for the Study of the Liver (EASL), and Asian Pacific Association for the Study of the Liver (APASL) agree that patients with cirrhosis should receive surveillance for HCC, the guidelines are less clear for noncirrhotic patients (Table 1.1) [21–23]. Notably, APASL does not comment beyond surveillance for cirrhotic patients, choosing to allow each country to tailor recommendations based on its economic situation [23]. While society guidelines provide practice recommendations, updates in research may further shape and improve guidelines and practice [24].

Overall, risk factors for HCC include increased age, male sex, family history of HCC, increased HBV DNA levels, higher alanine aminotransferase (ALT), and presence of cirrhosis. A more extensive list of risk factors is shown in Table 1.2 [7, 25]. With information gathered regarding HCC risk factors among patients with CHB, several risk scores have been generated to determine high-risk individuals that may warrant regular surveillance (Tables 1.1, 1.2, and 1.3). Risk scores may be instrumental in identifying patients with higher risk of HCC development and shaping future guidelines regarding patients requiring regular surveillance.

Risk Scores Developed for Untreated Patients

The 3 most-cited risk scores have been the Guide With Age, Gender, Hepatitis B Virus DNA, Core Promoter Mutations, and Cirrhosis (GAG-HCC); Chinese University–Hepatitis B Virus DNA, Cirrhosis (CU-HCC); and Risk Estimation for Hepatocellular Carcinoma in Chronic Hepatitis B (REACH-B). These have been developed using untreated Asian cohorts.

Guide with Age, Gender, Hepatitis B Virus DNA, Core Promoter Mutations, and Cirrhosis (GAG-HCC)

In 2009, Yuen et al. [26] developed an HBV risk score (Guide with age, gender, HBV DNA, core promoter mutations, and cirrhosis; GAG-HCC) to determine the risk of developing HCC 5 or 10 years after presentation. Yuen and colleagues [26] studied a cohort of 820 untreated patients and developed the score using age, sex, HBV DNA level, presence of core promotor mutations, and cirrhosis. The risk score achieved an area under the receiver operating characteristic (AUROC) of 0.88 (95% CI, 0.82–0.93) and 0.89 (95% CI, 0.85–0.93) for 5 and 10 years, respectively. As hospitals may have difficulty obtaining core promotor mutations for the risk score, the group developed a version without core promotor mutations. This score achieved an AUROC of 0.87 (95% CI, 0.82–0.93) and 0.88 (95% CI, 0.82–0.92) for 5 years and 10 years, respectively. At a cutoff of 100 for 5 years, this version of the risk score had a sensitivity of 67.8% (95% CI, 62.4–99.3%) and a specificity of 88.1% (95% CI, 59.0–91.5%).

Chinese University–Hepatitis B Virus DNA, Cirrhosis

In 2010, Wong et al. [27] studied 1005 untreated patients as the training cohort and 424 for the validation cohort. Chinese University–Hepatitis B Virus DNA, Cirrhosis (CU-HCC), based on age, albumin, bilirubin, HBV DNA, and presence of cirrhosis, achieved a 5-year AUROC of 0.76 (95% CI, 0.66–0.86) as well as a 10-year AUROC of 0.78 (95% CI:

Table 1.1 Hepatocellular carcinoma screening guidelines for patients with chronic hepatitis B virus

AASLD[21]	EASL[22]	APASL[23]
Cirrhosis	Cirrhosis	Cirrhosis
Asian/black men ≥ 40 year	HBV carriers with active hepatitis	
Asian women ≥ 50 year	Family history of HCC	
First-degree family member with HCC		
HDV positive		

Abbreviations: *AASLD* American association for the study of liver diseases, *APASL* Asian Pacific association for the study of the liver, *EASL* European association for the study of the liver, *HBV* hepatitis B virus, *HCC* hepatocellular carcinoma, *HDV* hepatitis D virus

Table 1.2 Risk factors for hepatocellular carcinoma for patients with chronic hepatitis B virus

Patient factors	Comorbidities	Hepatitis B characteristics
Male sex	Obesity	Genotype (C > B)
Increased age	Diabetes	Hepatitis E antigen positive
Family history of HCC	Alcohol	Hepatitis B viral load
Asian or African	Concurrent hepatitis C or D	Level of HBsAg
Presence of cirrhosis	Smoking	

Abbreviations: *HBsAg* hepatitis B virus surface antigen, *HCC* hepatocellular carcinoma

Table 1.3 HCC risk scores

	Author/country/ single- vs multicenter/year	Risk variables	Population used to develop risk score	AUROC[a]	Sensitivity	Specificity
GAG-HCC [26]	Yuen/Hong Kong/ single center/2009	Age, sex, HBV DNA level, cirrhosis	820 untreated patients	5-year: 0.87 (95% CI, 0.82–0.93); 10-year: 0.88 (95% CI, 0.82–0.92)	5-year cutoff: 100; 10-year cutoff: 82; 5-year: 67.8% (95% CI, 62.4–99.3%); 10-year: 86.6% (69.7–96.3%)	5-year: 88.1% (95% CI, 59.0–91.5%); 10-year: 75.6% (95% CI, 65.5–90.3%)
CU-HCC [27]	Wong/Hong Kong/ single center/2010	Age, albumin, bilirubin, HBV DNA, cirrhosis	Development cohort: 1005 untreated patients; validation cohort: 424 untreated patients	5-year: 0.76 (95% CI, 0.66–0.86); 10-year: 0.78 (95% CI, 0.71–0.86)	Cutoff: 5 5-year: 78.3% (95% CI, 74.3–82.2%); 10-year: 81.0% (95% CI, 77.2–84.7%)	Cutoff: 5 5-year: 72.8% (95% CI, 68.5–77.2%); 10-year: 75.7% (95% CI, 72.0–79.3%)
REACH-B [28]	Yang/multinational[b] Asian/ multicenter/2011	Age, sex, ALT, HBeAg seropositivity, HBV DNA level	Development cohort: 3584 untreated patients without cirrhosis; validation cohort: 1505 untreated patients	Validation cohort without cirrhosis: 3-year: 0.902 (95% CI, 0.884–0.918); 5-year 0.783 (0.759–0.806), and 10-year: 0.806 (95% CI, 0.783–0.828)	c	c
REVEAL-HBV [34]	Lee/Taiwan/ multicenter/2013	Age, sex, ALT level, HBeAg status, HBV DNA level, hepatitis B surface antigen level, genotype	3342 untreated patients: Two-thirds allocated to development cohort, one-third to validation cohort	5-year: 0.84; 10-year: 0.86; 15-year: 0.87	c	c
LSM-HCC [35]	Wong/Hong Kong/ single center/2014	Age, albumin, HBV DNA level, and liver stiffness measurement	Development cohort: 1035; validation cohort: 520	3-year: 0.89 (95% CI, 0.84–0.95); 5-year: 0.83 (95% CI, 0.71–0.94)	Cutoff: 11 3-year: 100% (95% CI, 67.6–100%) 5-year: 92.3% (95% CI, 66.7–98.6%)	Cutoff: 11 3-year: 70.7% (95% CI, 66.6–74.5%) 5-year: 71.2% (95% CI, 67.1–75.0%)
PAGE-B [11]	Papatheodoridis/ multinational[b] Europe/ multicenter/2016	Age, sex, and platelet count	Development cohort: 1325; validation cohort: 490	C statistic: 5-year: 0.82	Cutoff: 10 5-year: 100% Cutoff: 17 5-year: 76.0%	Cutoff: 10 5-year: 19.6% Cutoff: 17 5-year: 77.3%
APA-B [38]	Chen/Taiwan/ multicenter/2017	Age, platelet count, and AFP level following 12 mo of treatment	Development cohort: 883; validation cohort: 442	3-year: 0.892 (95% CI, 0.828–0.956); 5-year: 0.862 (95% CI, 0.795–0.929)	Cutoff: 6 5-year: 90.3%	Cutoff: 6 5-year: 79.1%
CAMD [29]	Hsu/Taiwan and Hong Kong/ multicenter/2018	Age, sex, presence of cirrhosis and diabetes	Development cohort: 23,851; validation cohort: 19,321	C statistic: 1-year: 0.74 (95% CI, 0.71–0.77); 3 y: 0.75 (95% CI, 0.72–0.77)	c	c
REAL-B [39]	Yang/ multinational[b]/ multicenter/2020	Age, sex, alcohol use, presence of diabetes and cirrhosis, platelet count, AFP	Development cohort: 5365; validation cohort: 2683	3-year: 0.83 (95% CI, 0.78–0.87); 5-year: 0.81 (95% CI, 0.77–0.85); 10-year 0.81 (95% CI, 0.78–0.84)	c	c

Abbreviations: *AFP* alpha fetoprotein, *ALT* alanine transaminase, *APA-B* age, platelets, and alpha fetoprotein, *AUROC* area under the receiver operating characteristic, *CAMD* cirrhosis, patient age, male sex, and diabetes, *CU-HCC* Chinese University–Hepatitis B virus DNA, cirrhosis, *GAG-HCC* guide with age, gender, hepatitis B virus DNA, core promoter mutations, and cirrhosis, *HBeAg* hepatitis B E antigen, *HBV* hepatitis B virus, *LSM-HCC* liver stiffness measurements–hepatocellular carcinoma, *PAGE-B* platelets, age, and gender, *REACH-B* risk estimation for hepatocellular carcinoma in chronic hepatitis B, *REAL-B* real-world effectiveness from the asia pacific rim liver consortium for HBV, *REVEAL-HBV* risk evaluation of viral load elevation and associated liver disease/cancer in hepatitis B virus

[a]If validation cohort is available, AUROC or C statistic is based off validation cohort

[b]Multinational is defined as more than 2 countries involved

[c]Sensitivity/specificity not available

0.71–0.86). With a cutoff score of 5, CU-HCC was able to achieve a 5-year sensitivity of 78.3% (95% CI, 74.3–82.2%) and a corresponding specificity of 72.8% (95% CI, 68.5–77.2%).

Risk Estimation for Hepatocellular Carcinoma in Chronic Hepatitis B (REACH-B)

In 2011, Yang et al. [28] developed the REACH-B score based on age, sex, ALT, hepatitis B E antigen (HBeAg) seropositivity, and HBV DNA level. Unlike GAG-HCC and CU-HCC, REACH-B did not include patients with cirrhosis. With this 17-point risk score, developed using 3584 untreated patients, Yang and colleagues [28] were able to achieve an AUROC of 0.902 (95% CI, 0.884–0.918), 0.783 (95% CI, 0.759–0.806), and 0.806 (95% CI, 0.783–0.828) at 3 years, 5 years, and 10 years, respectively, for the validation cohort.

However, GAG-HCC, CU-HCC, and REACH-B have been limited in applicability, especially regarding patients with different ethnicities and for patients undergoing treatment [29–32]. In 2015, Papatheodoridis et al. [30] examined 1756 adult White individuals with CHB from various European nations (Greece, Italy, Spain, the Netherlands, Turkey) who had received entecavir (ETV) or tenofovir disoproxil fumarate (TDF) for 6 months or longer. The population included patients with and without cirrhosis. For the different subgroups tested, the C statistic was 0.60–0.76 across the 3 risk scores. In addition, in 2015, Arends et al. [31] used the European network of excellence for Vigilance against Viral Resistance (VIRGIL) to study 891 CHB patients treated for 3 months or longer with ETV between 2005 and May 2013 in 11 European referral centers. Among White individuals, GAG-HCC, CU-HCC, and REACH-B achieved a *C* statistic ranging from 0.54 to 0.74. However, the 3 risk scores performed favorably when applied to Asian patients receiving treatment. In 2013, Wong et al. [32] evaluated 1531 patients with CHB in Hong Kong who were treated with ETV for 12 months or longer, between December 2005 and August 2012. At baseline, the AUROC was 0.80 (95% CI, 0.75–0.86) for CU-HCC, 0.76 (95% CI, 0.70–0.82) for GAG-HCC, and 0.71 (95% CI, 0.62–0.81) for REACH-B. The on-treatment risk scores remained rather consistent. At 3 years of entecavir therapy, the AUROC was 0.95 (0.91–0.99) for CU-HCC, 0.95 (0.93–0.98) for GAG-HCC, and 0.97 (0.95–0.99) for REACH-B. In the meantime, several other scores have been developed, which we will summarize later.

In 2010, Yang et al. [33] developed 3 models to determine risk of HCC, based on untreated patients from the Risk Evaluation of Viral Load Elevation and Associated Liver Disease/Cancer in Hepatitis B Virus (REVEAL-HBV) cohort. All models included variables of age, sex, family history of HCC, alcohol consumption, and ALT. Model 2 additionally had variables for HBeAg status and HBV DNA level, while model 3 further included HBV genotype. These models achieved an AUROC of 0.82 to 0.87 for the 5-year and 10-year prediction.

Risk Evaluation of Viral Load Elevation and Associated Liver Disease/Cancer in Hepatitis B Virus (REVEAL-HBV)

In 2013, Lee et al. [34] developed a risk model using the Risk Evaluation of Viral Load Elevation and Associated Liver Disease/Cancer in Hepatitis B Virus (REVEAL-HBV) cohort, based on age, sex, ALT, HBeAg status, HBV DNA level, hepatitis B surface antigen level, and genotype. Similar to the prior scores, this was also developed in an entirely Asian and untreated cohort. In the validation cohort, the risk score had an AUROC of 0.84 (5-year), 0.86 (10-year), and 0.87 (15-year).

Liver Stiffness Measurements–Hepatocellular Carcinoma (LSM-HCC)

Wong et al. [35] in 2014 created a scoring system (Liver Stiffness Measurements–Hepatocellular Carcinoma; LSM-HCC) using age, albumin, HBV DNA level, and liver stiffness measurement with transient elastography based on patients seen at the Prince of Wales Hospital in Hong Kong. In this study, 38% of the development cohort and 32% of the validation cohort received antiviral therapy. The LSM-HCC score had an AUROC of 0.89 (95% CI, 0.84–0.95) at 3 years and 0.83 (95% CI, 0.71–0.94) at 5 years.

Studies Assessing the Effect of Antiviral Therapy on Risk of Hepatocellular Carcinoma

Most current scoring systems have not focused on patients receiving treatment, even though treatment has been found to decrease the risk of HCC. A series of different studies have been performed using historical controls to assess the effect of antiviral therapy on the risk of HCC [12]. In 2013, Wong et al. [36] compared 1466 CHB patients who received ETV for 12 months or longer with 424 historical controls. The historical controls were recruited between December 1997 and July 2000 at the Princess of Wales Hospital and followed for an average of 114 months. Wong and colleagues [36] found no statistically significant difference in the 5-year cumulative probability of HCC between treated (6.6%, 95% CI, 5.1–8.1%) and untreated (6.5%, 95% CI, 5.3–7.7%) patients. While there was no statistically significant difference in the cumulative probability of HCC for noncirrhotic untreated and treated patients, among cirrhotic patients, treated patients (13.8%, 95% CI, 11.3–16.3%) had a statistically significantly decreased risk of HCC compared to untreated patients (26.4%, 95% CI, 20.7–32.1%; $P = 0.036$).

While other studies have replicated the findings by Wong et al. [12] a recent large study of 591 pairs of propensity score–matched untreated and TDF-treated patients from the United States has shown significant HCC risk reduction both in cirrhotic and noncirrhotic patients (77% and 73% risk reduction, respectively) [37]. Given the risk reduction provided by antiviral therapy, there may be additional nuance to guidelines regarding surveillance in the event treatment is started and viral suppression is achieved. Several risk scores have sought to evaluate patients receiving antiviral therapy.

Risk Scores Developed for Patients Treated with Antiviral Therapy

Platelets, Age, and Gender (PAGE-B)

In 2016, Papatheodoridis et al. [11] developed the Platelets, Age, and Gender (PAGE-B) score based on adult patients who had received ETV or TDF for 12 months or longer, from centers in Italy, Greece, Spain, the Netherlands, and Turkey. The study included patients who were treatment-experienced and those who were naive. Patients with decompensated cirrhosis, HCC diagnosed before ETV/TDF, concurrent hepatitis C/D or HIV, or prior liver transplant were excluded. PAGE-B, which is based on age, sex, and platelet count, was developed with a cohort of 23,851 patients and validated with a cohort of 19,321. The 5-year *C* statistic for the validation cohort was 0.82 [11].

Age, Platelets, and Alpha Fetoprotein (APA-B)

Chen et al. [38] in 2017 developed the Age, Platelets, and Alpha Fetoprotein (APA-B) risk score, which was developed using 1325 treatment-naive patients receiving ETV for 12 months or longer between January 2007 and August 2012 in Taiwan. APA-B is based on age, platelet count, and alpha fetoprotein (AFP) level following 12 months of treatment. Patients were excluded if they had other forms of liver disease, had baseline HCC, or were within 12 months of ETV, or had decompensated cirrhosis. In the validation group, APA-B was able to achieve an AUROC of 0.892 (95% CI, 0.828–0.956) at 3 years and 0.862 (95% CI, 0.795–0.929) at 5 years.

Cirrhosis, Patient Age, Male Sex, and Diabetes (CAMD)

Hsu et al. [29] in 2018 created the Cirrhosis, Patient Age, Male Sex, and Diabetes (CAMD) score to predict HCC among patients with CHB on ETV or TDF. Hsu and colleagues [29] used national claims databases in Taiwan (National Health Insurance Research Database) and the Hong Kong Hospital Authority Database to identify 23,851 eligible Taiwanese patients taking ETV or TDF for 3 months or longer between January 2008 through December 2013 to construct the model. Exclusions included malignancy, decompensated cirrhosis, renal failure, other viral hepatitis, or patients who took lamivudine, adefovir, or telbivudine for 3 months or longer, or developed HCC or died within 3 months of antiviral therapy. The CAMD score is based on age, sex, presence of cirrhosis, and diabetes. A total of 19,321 Hong Kong patients served as the validation cohort, and achieved a C statistic of 0.74 (95% CI, 0.71–0.77) at 1 year and 0.75 (95% CI, 0.72–0.77) at 3 years.

Real-World Effectiveness from the Asia Pacific Rim Liver Consortium for HBV (REAL-B)

The Real-World Effectiveness From the Asia Pacific Rim Liver Consortium for HBV (REAL-B) risk score was developed and validated using patient data collected from the REAL-B registry, which included 8048 Asian patients treated with oral nucleot(s)ides from 25 US and Asia Pacific centers. In this study, patients with less than 12 months' follow-up and those with immunosuppressive treatment or viral coinfection were excluded. The REAL-B score ranges from 0 to 13 and is based on age, sex, alcohol, presence of diabetes mellitus and cirrhosis, platelet count, and AFP. The AUROCs for 3-, 5-, and 10-year prediction of HCC development in the validation cohort were 0.83 (95% CI, 0.78–0.87), 0.81 (95% CI, 0.77–0.85), and 0.81 (95% CI, 0.78–0.84), respectively. In this study cohort, the REAL-B risk score performed statistically significantly better than PAGE-B and other risk scores developed for untreated (REACH-B, GAG-HCC) or partially treated patients (CU-HCC) [39].

Can We Use Risk Scores to Guide Care?

The development of different risk scores helps to demonstrate the importance of different risk factors (especially cirrhosis), as well as the nuances of HCC risk for patients without cirrhosis. Overall, most of these risk scores seek to allocate patients into different risk categories. For studies in which an optimal risk score cutoff has been identified (GAG-HCC, APA-B, PAGE-B, CU-HCC, LSM-HCC), there is variable sensitivity and specificity. PAGE-B (5-year 100%) and LSM-HCC (3-year 100%, 5-year 92.3%) were able to maintain high sensitivity, making these scores potentially useful scoring systems to rule out patients with low long-term risk of HCC.

However, our current risk scores in general have been studied using either Asian patients or European patients, not both, and as such lack universal applicability. In addition, treatment is known to reduce HCC risk among cirrhotic patients and may be a useful variable to have as well. Studies so far have been either focused primarily on untreated or treated patients and tend not to include treatment as an actual variable. Further, while risk scores provide risk categorization, how they translate to actual practice remains less clear.

While the obvious strategy is that there should be more frequent screening among patients with high risk and less for patients with low risk, where these cutoffs lie remains less clear. It is hoped that, in the future, with a well-validated risk score, this development may be a realistic scenario for CHB care.

Much work on risk scores has already been accomplished. This large body of work provides significant data on existing risk factors for CHB. The existing data also help to point out ways for further development. It is possible in the not too distant future that there may be risk scores developed with multiple ethnicities in mind as well as treatment. With such a well-validated risk score, we may be able to accomplish the difficult task of guiding HCC screening among CHB patients without cirrhosis.

What Should We Do About Immune-Tolerant and Inactive Chronic Hepatitis B Patients?

The "immune-tolerant" phase, also known as HBeAg-positive chronic HBV infection, is characterized by positive HBeAg, high HBV DNA, and ALT within the upper limit of normal (ULN) [40]. In a retrospective study by Kim et al. [41] reviewing the outcomes of 413 untreated immune-tolerant patients to 1497 immune-active patients on nucleot(s)ide analogs, on multivariate analysis, the immune-tolerant group was found to have a higher risk of HCC (hazard ratio 2.54; 95% CI, 1.54–4.18) than the immune-active group. However, in a recent editorial evaluating the study by Kim and colleagues, Zhou and Terrault [42] question whether members of the cohort were considered immune tolerant throughout the study period, and whether there was an issue of misclassification in some of the patients. The definition of immune-tolerant patients was based on only one normal ALT measurement and not on the persistence of normal ALT levels over time, as defined by various treatment guidelines [21, 40].

Separately, the "inactive carrier" phase, also known as HBeAg-negative chronic HBV infection, is characterized by negative HBeAg, positive anti-HBe, HBV DNA of less than 2000 IU/mL, and ALT within ULN. In one study, 4376 HBeAg-negative, HBsAg-positive patients with ALT less than 2 times the ULN in Taiwan were followed for a mean period of 13.4 ± 5.2 years. During this period, liver cirrhosis occurred in 2.7% and HCC in 0.5%. Of note, the highest ALT during the follow-up period (MaxALT) remained less than 2 times the ULN among 80.8%, while 5.9% of patients had ALT greater than 5 times the ULN. On multivariate analysis evaluating risk factors for HCC development, independent risk factors included liver cirrhosis, age at entry, and MaxALT [43]. CHB patients receiving nucleot(s)ide analogs have been found to have higher risk of developing HCC compared to patients with inactive CHB [44]. In another study by Chen et al. [45] following 62 patients who had achieved spontaneous HBeAg seroconversion with persistently normal ALT ($\leq$ 36 U/L) for a median of 18.1 years, none of the patients developed cirrhosis or HCC, regardless of HBV level (< 5 log 10 or $\geq$ 5 log 10).

Overall, the data for HCC development remain sparse for immune-tolerant (HBeAg-positive chronic HBV infection) and inactive carrier (HBeAg-negative chronic HBV) patients. In addition, accurate classification of patients into the correct clinical phenotypes can be difficult and, in many cases, requires long-term observation [21, 40]. In fact, after 1 year of observation, about half of a large cohort of about 6100 patients from the United States and Taiwan did not meet the criteria for any of the defined clinical phases for CHB and these patients had higher HCC incidence than inactive patients [46]. Further, the data are also sparse for groups currently not recommended for HCC surveillance by the AASLD guidelines (men aged < 40 year and women aged < 50 year) [21]. Reasons for this paucity in data may be because larger studies and longer follow-up would be required to produce reliable data for this lower-risk population.

Therefore, the current practice guidelines with their age cutoffs as well as the limitation of our current knowledge regarding the natural history including HCC risk among lower-risk groups should be discussed and considered with patients, along with potential benefits and harms of HCC surveillance, including for individuals not meeting current guideline criteria.

Conclusion

Given the significant risk cirrhosis poses to HCC, all cirrhotic patients with CHB should receive HCC screening. However, screening guidance for CHB patients without cirrhosis remains less clear. Risk scores may in the future become a reliable way to help stratify patients to determine the need for screening. However, at this point more studies are needed to develop and validate clinical risk scores to determine risk of HCC and, as a result, need for screening. It is hoped that, with additional studies, risk scores may help identify a more nuanced and cost-effective way of helping to efficiently determine patients who may need regular screening. For the time being, informed discussion should be made with patients, realizing the pitfalls of current practice guidelines, the risk of missed opportunity for early HCC diagnosis and curative treatment, balanced with the risk, cost, and inconvenience of long-term HCC surveillance among lower-risk patients.

References

1. World Health Organization. Hepatitis B. Volume; 2017.
2. Polaris Observatory Collaborators. Global prevalence, treatment, and prevention of hepatitis B virus infection in 2016: a modelling

study. Lancet Gastroenterol Hepatol. 2018;3(6):383–403. https://doi.org/10.1016/S2468-1253(18)30056-6.
3. Le MH, Yeo YH, Cheung R, Henry L, Lok AS, Nguyen MH. Chronic hepatitis B prevalence among foreign-born and U.S.-born adults in the United States, 1999-2016. Hepatology. 2020;71(2):431–43. https://doi.org/10.1002/hep.30831.
4. Lim J, Nguyen MH, Kim WR, Gish R, Perumalswami P, Jacobson IM. Prevalence of chronic hepatitis B virus infection in the United States. Am J Gastroenterol. 2020;115(9):1429–38. https://doi.org/10.14309/ajg.0000000000000651.
5. Ogawa E, Yeo YH, Dang N, et al. Diagnosis rates of chronic hepatitis B in privately insured patients in the United States. JAMA Netw Open. 2020;3(4):e201844. https://doi.org/10.1001/jamanetworkopen.2020.1844.
6. Wei MT, Le AK, Chang MS, et al. Antiviral therapy and the development of osteopenia/osteoporosis among Asians with chronic hepatitis B. J Med Virol. 2019;91(7):1288–94. https://doi.org/10.1002/jmv.25433.
7. Wong VWS, Janssen HLA. Can we use HCC risk scores to individualize surveillance in chronic hepatitis B infection? J Hepatol. 2015;63(3):722–32. https://doi.org/10.1016/j.jhep.2015.05.019.
8. Papatheodoridis GV, Lampertico P, Manolakopoulos S, Lok A. Incidence of hepatocellular carcinoma in chronic hepatitis B patients receiving nucleos(t)ide therapy: a systematic review. J Hepatol. 2010;53(2):348–56. https://doi.org/10.1016/j.jhep.2010.02.035.
9. Thursz M, Brown A. Can antiviral therapy of chronic hepatitis B prevent the development of hepatocellular carcinoma? Gut. 2011;60(8):1025–6. https://doi.org/10.1136/gut.2010.236521.
10. Sung JJY, Tsoi KKF, Wong VWS, Li KCT, Chan HLY. Meta-analysis: treatment of hepatitis B infection reduces risk of hepatocellular carcinoma. Aliment Pharmacol Ther. 2008;28(9):1067–77. https://doi.org/10.1111/j.1365-2036.2008.03816.x.
11. Papatheodoridis G, Dalekos G, Sypsa V, et al. PAGE-B predicts the risk of developing hepatocellular carcinoma in Caucasians with chronic hepatitis B on 5-year antiviral therapy. J Hepatol. 2016;64(4):800–6. https://doi.org/10.1016/j.jhep.2015.11.035.
12. Papatheodoridis GV, Chan HLY, Hansen BE, Janssen HLA, Lampertico P. Risk of hepatocellular carcinoma in chronic hepatitis B: assessment and modification with current antiviral therapy. J Hepatol. 2015;62(4):956–67. https://doi.org/10.1016/j.jhep.2015.01.002.
13. World Health Organization. Cancer; 2018.
14. Hemming AW, Berumen J, Mekeel K. Hepatitis B and hepatocellular carcinoma. Clin Liver Dis. 2016;20(4):703–20. https://doi.org/10.1016/j.cld.2016.06.007.
15. El-Serag HB, Kanwal F. Epidemiology of hepatocellular carcinoma in the United States: where are we? Where do we go? Hepatology. 2014;60(5):1767–75. https://doi.org/10.1002/hep.27222.
16. Wong GLH, Wong VWS, Tan GM, et al. Surveillance programme for hepatocellular carcinoma improves the survival of patients with chronic viral hepatitis. Liver Int. 2008;28(1):79–87. https://doi.org/10.1111/j.1478-3231.2007.01576.x.
17. Chan SL, Mo FKF, Wong VWS, et al. Use of antiviral therapy in surveillance: impact on outcome of hepatitis B-related hepatocellular carcinoma. Liver Int. 2012;32(2):271–8. https://doi.org/10.1111/j.1478-3231.2011.02634.x.
18. Santi V, Trevisani F, Gramenzi A, et al. Semiannual surveillance is superior to annual surveillance for the detection of early hepatocellular carcinoma and patient survival. J Hepatol. 2010;53(2):291–7. https://doi.org/10.1016/j.jhep.2010.03.010.
19. Zhang BH, Yang BH, Tang ZY. Randomized controlled trial of screening for hepatocellular carcinoma. J Cancer Res Clin Oncol. 2004;130(7):417–22. https://doi.org/10.1007/s00432-004-0552-0.
20. Bruix J, Sherman M, American Association for the Study of Liver Diseases. Management of hepatocellular carcinoma: an update. Hepatology. 2011;53(3):1020–2. https://doi.org/10.1002/hep.24199.
21. Terrault NA, Lok AS, McMahon BJ, et al. Update on prevention, diagnosis, and treatment of chronic hepatitis B: AASLD 2018 hepatitis B guidance. Hepatology. 2018;67(4):1560–99. https://doi.org/10.1002/hep.29800.
22. European Association for the Study of the Liver; European Organisation for Research and Treatment of Cancer. EASL-EORTC clinical practice guidelines: management of hepatocellular carcinoma. J Hepatol. 2012;56(4):908–43. https://doi.org/10.1016/j.jhep.2011.12.001.
23. Sarin SK, Kumar M, Lau GK, et al. Asian-Pacific clinical practice guidelines on the management of hepatitis B: a 2015 update. Hepatol Int. 2016;10(1):1–98. https://doi.org/10.1007/s12072-015-9675-4.
24. Nguyen MH, Wong G, Gane E, Kao JH, Dusheiko G. Hepatitis B virus: advances in prevention, diagnosis, and therapy. Clin Microbiol Rev. 2020;33(2):e00046–19. https://doi.org/10.1128/CMR.00046-19.
25. Trépo C, Chan HLY, Lok A. Hepatitis B virus infection. Lancet. 2014;384(9959):2053–63. https://doi.org/10.1016/S0140-6736(14)60220-8.
26. Yuen MF, Tanaka Y, Fong DYT, et al. Independent risk factors and predictive score for the development of hepatocellular carcinoma in chronic hepatitis B. J Hepatol. 2009;50(1):80–8. https://doi.org/10.1016/j.jhep.2008.07.023.
27. Wong VWS, Chan SL, Mo F, et al. Clinical scoring system to predict hepatocellular carcinoma in chronic hepatitis B carriers. J Clin Oncol. 2010;28(10):1660–5. https://doi.org/10.1200/JCO.2009.26.2675.
28. Yang HI, Yuen MF, Chan HL, et al. Risk estimation for hepatocellular carcinoma in chronic hepatitis B (REACH-B): development and validation of a predictive score. Lancet Oncol. 2011;12(6):568–74. https://doi.org/10.1016/S1470-2045(11)70077-8.
29. Hsu YC, Yip TCF, Ho HJ, et al. Development of a scoring system to predict hepatocellular carcinoma in Asians on antivirals for chronic hepatitis B. J Hepatol. 2018;69(2):278–85. https://doi.org/10.1016/j.jhep.2018.02.032.
30. Papatheodoridis GV, Dalekos GN, Yurdaydin C, et al. Incidence and predictors of hepatocellular carcinoma in Caucasian chronic hepatitis B patients receiving entecavir or tenofovir. J Hepatol. 2015;62(2):363–70. https://doi.org/10.1016/j.jhep.2014.08.045.
31. Arends P, Sonneveld MJ, Zoutendijk R, et al. Entecavir treatment does not eliminate the risk of hepatocellular carcinoma in chronic hepatitis B: limited role for risk scores in Caucasians. Gut. 2015;64(8):1289–95. https://doi.org/10.1136/gutjnl-2014-307023.
32. Wong GLH, Chan HLY, Chan HY, et al. Accuracy of risk scores for patients with chronic hepatitis B receiving entecavir treatment. Gastroenterology. 2013;144(5):933–44. https://doi.org/10.1053/j.gastro.2013.02.002.
33. Yang HI, Sherman M, Su J, et al. Nomograms for risk of hepatocellular carcinoma in patients with chronic hepatitis B virus infection. J Clin Oncol. 2010;28(14):2437–44. https://doi.org/10.1200/JCO.2009.27.4456.
34. Lee MH, Yang HI, Liu J, et al. Prediction models of long-term cirrhosis and hepatocellular carcinoma risk in chronic hepatitis B patients: risk scores integrating host and virus profiles. Hepatology. 2013;58(2):546–54. https://doi.org/10.1002/hep.26385.
35. Wong GLH, Chan HLY, Wong CKY, et al. Liver stiffness-based optimization of hepatocellular carcinoma risk score in patients with chronic hepatitis B. J Hepatol. 2014;60(2):339–45. https://doi.org/10.1016/j.jhep.2013.09.029.
36. Wong GLH, Chan HLY, Mak CWH, et al. Entecavir treatment reduces hepatic events and deaths in chronic hepatitis B patients with liver cirrhosis. Hepatology. 2013;58(5):1537–47. https://doi.org/10.1002/hep.26301.
37. Nguyen MH, Yang HI, Le A, et al. Reduced incidence of hepatocellular carcinoma in cirrhotic and noncirrhotic patients with chronic hepatitis B treated with tenofovir – a propensity score–matched

study. J Infect Dis. 2019;219(1):10–8. https://doi.org/10.1093/infdis/jiy391.
38. Chen CH, Lee CM, Lai HC, et al. Prediction model of hepatocellular carcinoma risk in Asian patients with chronic hepatitis B treated with entecavir. Oncotarget. 2017;8(54):92431–41. https://doi.org/10.18632/oncotarget.21369.
39. Yang HI, Yeh ML, Wong GL, et al. Real-world effectiveness from the Asia Pacific rim liver consortium for HBV risk score for the prediction of hepatocellular carcinoma in chronic hepatitis B patients treated with oral antiviral therapy. J Infect Dis. 2020;221(3):389–99. https://doi.org/10.1093/infdis/jiz477.
40. European Association for the Study of the Liver. EASL 2017 clinical practice guidelines on the management of hepatitis B virus infection. J Hepatol. 2017;67(2):370–98. https://doi.org/10.1016/j.jhep.2017.03.021.
41. Kim GA, Lim YS, Han S, et al. High risk of hepatocellular carcinoma and death in patients with immune-tolerant-phase chronic hepatitis B. Gut. 2018;67(5):945–52. https://doi.org/10.1136/gutjnl-2017-314904.
42. Zhou K, Terrault N. Immune tolerant HBV and HCC: time to revise our tolerance levels for therapy? AME Med J. 2018;3:27. https://doi.org/10.21037/amj.2018.02.01.
43. Tai DI, Lin SM, Sheen IS, Chu CM, Lin DY, Liaw YF. Long-term outcome of hepatitis B e antigen-negative hepatitis B surface antigen carriers in relation to changes of alanine aminotransferase levels over time. Hepatology. 2009;49(6):1859–67. https://doi.org/10.1002/hep.22878.
44. Cho YY, Lee JH, Chang Y, et al. Comparison of overall survival between antiviral-induced viral suppression and inactive phase chronic hepatitis B patients. J Viral Hepat. 2018;25(10):1161–71. https://doi.org/10.1111/jvh.12927.
45. Chen YC, Huang SF, Chu CM, Liaw YF. Serial HBV DNA levels in patients with persistently normal transaminase over 10 years following spontaneous HBeAg seroconversion. J Viral Hepat. 2012;19(2):138–46. https://doi.org/10.1111/j.1365-2893.2011.01450.x.
46. Li X, Yeo Y, Le A, et al. High proportion and long-term outcomes of patients with chronic hepatitis B have indeterminate clinical phenotype as per AASLD 2018 guidelines. Hepatology. 2018;68:1213A.

Should Hepatocellular Carcinoma Surveillance Be Restricted to Those with Cirrhosis Among Patients with Hepatitis C?

2

Jordan Feld and Hooman Farhang Zangneh

Hepatocellular carcinoma (HCC) surveillance should be restricted to those with cirrhosis after sustained virological response (SVR) given low HCC risk in the absence of cirrhosis.

Main Points

1. Patients with cirrhosis continue to have a high risk of HCC after SVR and should continue HCC surveillance, whereas those without cirrhosis are at low risk and do not warrant surveillance.
2. However, it can be challenging to accurately identify those with vs without cirrhosis, particularly distinguishing bridging (F3) vs confirmed (F4) fibrosis.
3. Pretreatment fibrosis-4 index (FIB-4), at a threshold of 3.25, may be a readily available and accurate means to identify patients with cirrhosis who warrant continued HCC surveillance after SVR.

HCC is one of the most devastating complications of chronic hepatitis C virus (HCV) infection. It is the fifth most common cancer worldwide, [1] and has an overall dismal prognosis, except for those diagnosed at a very early stage who may receive potentially curative therapies [2]. For patients with established cirrhosis, the risk of HCC is high, even after virological cure; therefore, all guidelines recommend long-term surveillance with imaging with or without alpha fetoprotein (AFP) [3, 4] However, while the HCC risk is clearly lower in those without cirrhosis, the thresholds to define who should receive surveillance remain controversial, particularly after cure of HCV infection [5]. Balancing the potential benefits of an early cancer diagnosis with the risks and costs associated with widespread long-term screening is challenging, and as with many issues in medicine, highlights the tension between making decisions for an individual patient against the priorities of a health care system [6]. Using emerging data on the incidence of HCC before and after HCV cure, the controversies around surveillance from its effectiveness to its cost-effectiveness in different populations are explored.

Hepatocellular Carcinoma Surveillance Effectiveness

Remarkably, there are no randomized controlled trials of HCC surveillance in patients with HCV-related cirrhosis [3, 5]. Indeed, aside from a randomized study of limited quality in hepatitis B virus carriers showing a 37% reduction in mortality with surveillance, [7] there is no high-level evidence supporting HCC surveillance. Hence, there remains at least some controversy surrounding whether surveillance truly leads to a reduction in mortality (as proof of efficacy) in this population [8]. Attempts to conduct prospective randomized trials to confirm the benefit of surveillance have failed because of poor enrollment and because many consider it unethical to perform such studies given the clear benefits for individuals with early diagnosis [2, 9]. Furthermore, lower-quality data demonstrating the favorable effects of surveillance are abundant [5]. A recent meta-analysis concluded that HCC surveillance improved the life expectancy of patients with cirrhosis, even after controlling for lead-time bias, compared with no surveillance [10]. As a consequence, all international guidelines recommend surveillance for patients with HCV cirrhosis, even after sustained virological response. Where there is less unanimous agreement is on the

J. Feld (✉) · H. F. Zangneh
Toronto General Hospital Research Institute (TGHRI), Toronto Centre for Liver Disease, University Health Network, Toronto, ON, Canada
e-mail: Jordan.Feld@uhn.ca; Hooman.f.zangneh@mail.utoronto.ca

N. S. Reau, A. G. Singal (eds.), *Controversies in Hepatocellular Carcinoma*, https://doi.org/10.1007/978-3-031-90225-3_2

role of surveillance in patients with advanced fibrosis but without clear evidence of cirrhosis (F3), particularly after the achievement of SVR [3, 4, 8].

Hepatocellular Carcinoma Surveillance Logistics

Ultrasound (US) is noninvasive, relatively inexpensive, and lacks the risk of contrast or radiation exposure, making it the preferred test for HCC surveillance. However, its performance is operator-dependent and suboptimal in obese patients and in cirrhotic livers [2]. Computed tomography or magnetic resonance imaging may have high accuracy for HCC diagnosis, but the cost and potential harms limit their use for surveillance [2].

The utility of AFP as an adjunct to US has been challenged because of low sensitivity for early-stage HCC (32–49%) and poor specificity related to elevation in active hepatitis and some non-HCC malignancies [11]. However, a recent meta-analysis from Tzartzeva et al. [12] showed that the addition of AFP to US increased the sensitivity to 63% (95% CI, 48–75%), leading both the American Association for the Study of Liver Diseases (AASLD) and the European Association for the Study of the Liver (EASL) to recommend US with or without AFP as the preferred surveillance approach [3, 4]. Notably, AFP may be more useful in the post-SVR setting where issues with low specificity due to AFP elevation from active hepatitis are of less concern; however, this has not been rigorously evaluated.

The current consensus for surveillance interval, as determined by the HCC growth rates rather than cost-effectiveness, is on biannual surveillance, regardless of a higher or lower HCC risk in a specific patient group [3, 6]. Unfortunately, despite the accepted benefits, less than 20% of patients with HCV cirrhosis undergo surveillance [13].

Sustained Virological Response

Defined as undetectable serum HCV RNA assessed 12 weeks (24 after interferon [IFN]-based therapy) after completion of antiviral therapy, [14] SVR is associated with a reduction in all-cause and liver-related mortality, the survival benefits being the consequence of a reduction both in extrahepatic and end-stage liver complications, including HCC [15–17]. Although there is a considerable decrease in HCC risk post SVR, it is not fully eliminated. [18–21] As before treatment, the risk of HCC after SVR is highly related to fibrosis stage. Those with minimal fibrosis (F0-F2) have negligible HCC rates in the absence of other liver diseases and do not require surveillance. In patients with clear evidence of cirrhosis, even after SVR, the incidence of HCC remains consistently above 1%/year with a much higher incidence reported in patients with other risk factors (advanced age, male sex, and diabetes) [20–22]. For individuals with advanced fibrosis but without clear evidence of cirrhosis, most studies report HCC incidence rates well below cost-effectiveness thresholds for surveillance. The major challenge in patients with F3 fibrosis is determining whether they actually might already have cirrhosis.

Although there were initial concerns that patients treated with direct-acting antivirals (DAAs) had a higher incidence of HCC than those cured with IFN-based therapy, multiple studies have now confirmed that this association is due to differences in the patient populations [23]. DAAs are very well tolerated even in patients with cirrhosis; therefore, many more cirrhotic patients, even those with decompensated cirrhosis, achieve SVR with DAAs than with IFN. After controlling for baseline factors, the post-SVR HCC incidence does not differ between DAA- and IFN-treated patients [24]. However, the observation that greater numbers of patients with more advanced liver disease and with other risk factors for HCC will achieve SVR makes the decision of whether to recommend HCC surveillance in the post-SVR setting more important.

Regression of Hepatitis C Virus–Induced Cirrhosis Post Sustained Virological Response

Although fibrosis was initially considered an irreversible condition, long-term follow-up studies of patients with many liver conditions have now clearly shown that fibrosis may regress when the inciting cause of liver injury is removed. [14] Remarkably, even in patients with documented cirrhosis, HCV cure can lead to such regression of fibrosis that repeat biopsies post treatment show less than or equal to F2 fibrosis in some but not all patients (10–49%) [19, 25]. Cirrhosis regression studies are difficult to interpret because of differences in the time lag between pre-SVR and post-SVR biopsies, sampling error, and interobserver variability, leading to wide ranges in regression estimates [18, 26]. Noninvasive measures of fibrosis have largely replaced liver biopsy in pretreatment assessment. Unfortunately, all noninvasive tools are influenced by hepatic inflammation so that improvement in fibrosis scores is almost universal post SVR [27]. Whether this truly indicates regression of fibrosis is difficult to discern and small studies have shown that even in patients with improvement in liver stiffness measurements to below the cirrhotic range post SVR, many patients still have cirrhosis on liver biopsy [18].

Furthermore, it is not clear, even with biopsy-proven regression of cirrhosis, whether regression reduces the risk of HCC [18]. Although the hallmark of cirrhosis is collagen deposition, other changes occur in the cirrhotic liver.

Changes in blood flow, intrahepatic immunity, and overall architecture may be as or more important than the amount of collagen in terms of HCC risk. Therefore, changes in fibrosis assessment post SVR must be interpreted carefully, and at least based on current data, pretreatment fibrosis assessment should be used to guide post-SVR HCC surveillance decisions [28]. Although the post-SVR resolution of liver fibrosis and liver remodeling may lead to a decline in HCC risk over time to a level at which HCC surveillance may no longer be warranted, as time passes after SVR, patients age and may acquire diabetes, obesity, or alcohol use, potentially limiting the expected decrease in HCC risk with SVR [28].

International Guidelines Recommendations

Although there is general agreement across international guidelines, some disagreement persists as to who should undergo HCC surveillance both before and after SVR. The most recent AASLD HCC guidelines recommend limiting surveillance to those with documented cirrhosis before treatment, acknowledging that the benefits of surveillance are uncertain in those with F3 fibrosis [3]. EASL has similar recommendations [4]. Previously, the AASLD/Infectious Diseases Society of America HCV guidance document recommended surveillance for all those with advanced fibrosis (F3/F4) before treatment, [14] but recent updates now limit the recommendation to those with cirrhosis on pretreatment assessment [29]. Highlights of these recommendations are reviewed and compared in Table 2.1.

The Controversy of Bridging Fibrosis

A major limitation of all tools used to assess liver fibrosis, including liver biopsy, is the ability to distinguish between F3 and F4. Results indicating F3 fibrosis raise concerns for clinicians that cirrhosis may already be present but missed because of sampling error or simple imprecision of the tools used. In truth, fibrosis progression is likely not homogeneous throughout the liver and it is thus possible that parts of the liver will show evidence of cirrhosis, whereas other areas may not yet have progressed to that stage [26]. To address these concerns, most guidelines have erred on the side of caution and recommended HCC surveillance in patients with advanced fibrosis to ensure that none with cirrhosis was missed. [4] While this approach may be reasonable, it would be preferable if tools were available to more accurately risk-stratify individuals to avoid unnecessary surveillance, which is not only costly but may lead to false-positive diagnoses and is inconvenient and potentially anxiety-provoking for patients. Furthermore, the situation may differ in the DAA era. With the low efficacy of IFN-based treatment and thus the high risk of liver disease progression, guidelines have advocated that all those with advanced fibrosis (F3 or F4) undergo HCC surveillance. With the development of highly effective DAAs, almost all patients will be cured shortly after diagnosis and thus those with F3 will not progress (Fig. 2.1).

Notably, in studies relying on liver biopsy, the risk of HCC post SVR is indeed much lower in those with F3 than F4 fibrosis, as van der Meer and colleagues [38] found a cumulative 8-year incidence of 1.8% in those with F3

Table 2.1 Recommendations for surveillance in patients with hepatitis C virus–related risk for hepatocellular carcinoma

	Surveillance recommendations			
	At-risk target population			
Guideline	F3 fibrosis	F4 fibrosis	Test(s)	Interval, mo
AASLD [30] (HCC guideline)	No (benefits not certain)	Yes (CP-C only if on LT waitlist)	US ± AFP	6
APASL [31] (HCC guideline)	No (benefits not certain)	Yes (CP-C only if on LT waitlist)	US + AFP	6
EASL [32] (HCC guideline)	Yes	Yes (CP-C only if on LT waitlist)	US	6
AASLD-IDSA [33] (HCV guideline)	No	Yes	US	6
APASL [34] (HCV guideline)	Yes	Yes	US + AFP	6
EASL [35] (HCV guideline)	Yes	Yes	US	6

Abbreviations: *AASLD* American association for the study of liver diseases, *AFP* alpha fetoprotein, *APASL* Asia-Pacific association for the study of the liver, *CP-C* child-pugh score, *EASL* European association for the study of the liver, *HCC* hepatocellular carcinoma, *HCV* hepatitis C virus, *IDSA* infectious diseases society of America, *LT* liver transplantation, *US* ultrasound

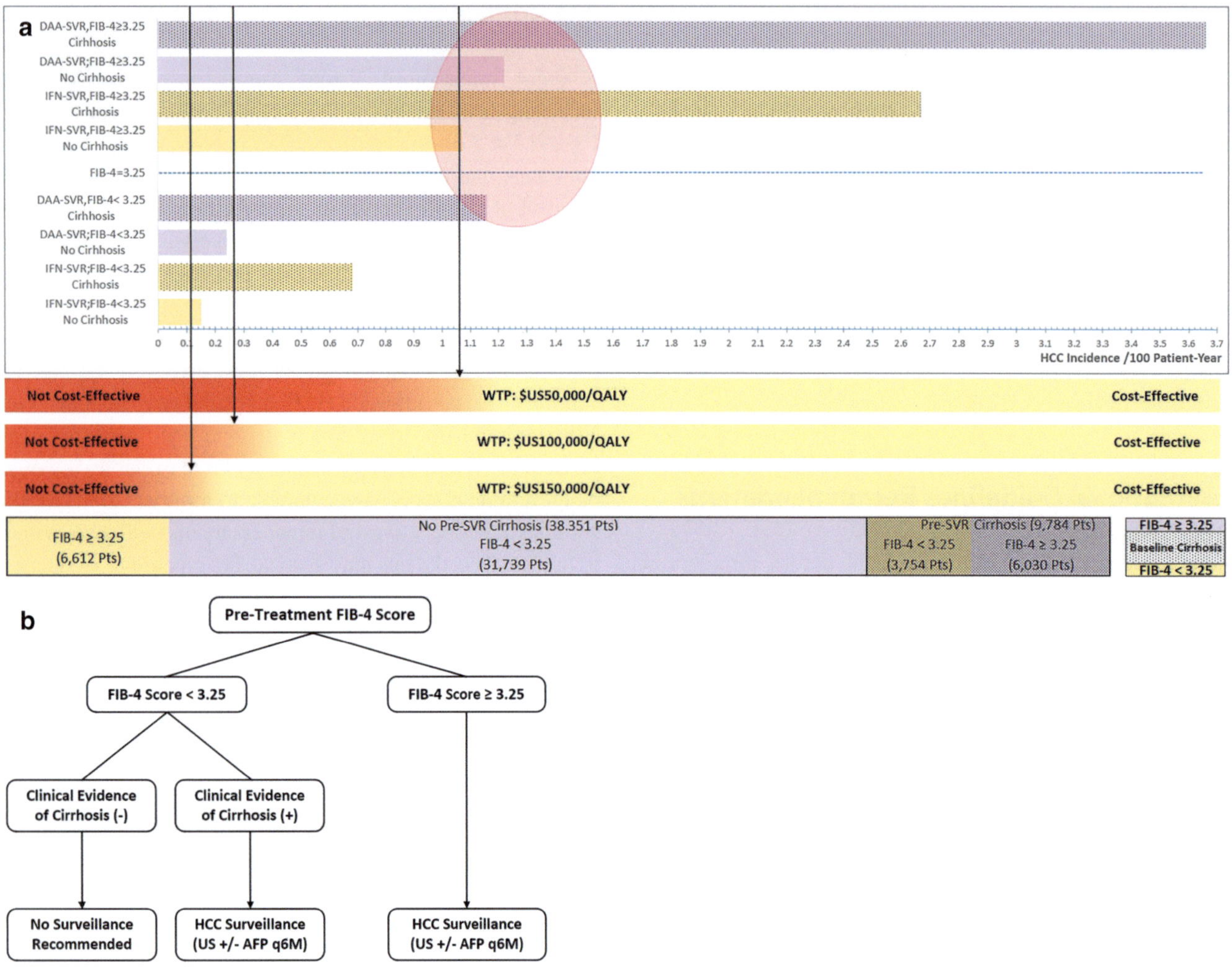

Fig. 2.1 (**a**) Annual HCC incidence post-SVR according to treatment type and FIB-4 score. Distribution of patients based on cirrhosis and treatment type combined with their annual HCC incidence according to FIB-4 score. Cost-effectiveness is schematically shown with three different willingness-to-pay thresholds for comparison. Among patients with cirrhosis (**a**), those with baseline FIB-4 ≥ 3.25 had a significantly higher incidence of HCC post-SVR (3.66/100 patient-years in DAA-treated and 2.67/100 patient-years in IFN-treated) than those with baseline FIB-4 < 3.25 (1.16/100 patient-years in DAA-treated and 0.68/100 patient-years in IFN-treated patients). *AFP* alpha-fetoprotein, *DAA* direct-acting antiviral, *FIB-4* fibrosis-4 index, *HCC* hepatocellular carcinoma, *IFN* interferon, *QALY* quality-adjusted life-years, *SVR* sustained virological response, *WTP* willingness to pay [36, 37]. (**b**) Summarized approach to HCC surveillance based on pre-treatment FIB-4 score in HCV-infected patients. *AFP* alpha fetoprotein, *DAA* direct-acting antiviral, *FIB-4* fibrosis-4 index, *HCC* hepatocellular carcinoma, *IFN* interferon, *QALY* quality-adjusted life-years, *SVR* sustained virological response, *WTP* willingness to pay

compared to 8.7% in those with F4. However, liver biopsies are rarely performed in patients with HCV and may still be unreliable to distinguish F3 from F4.

Large data sets from the Veterans Affairs (VA) health system have provided useful data to guide surveillance decision-making without biopsies. Recognizing the challenges in distinguishing F3 from F4, investigators have evaluated the ability of the aspartate transaminase to platelet ratio index (APRI) and the FIB-4 score to risk-stratify patients based on HCC risk. In the IFN-era, El-Serag and colleagues [22] showed that the incidence of HCC post SVR was markedly higher in patients with a pretreatment APRI greater than 2.0 than in patients with lower values. Similarly, Kanwal et al. [20] reported on 19,518 patients who achieved SVR with DAAs and found that the incidence of HCC post SVR was 2.16/100 person-years in those with pretreatment FIB-4 greater than or equal to 3.25 compared to 0.45 and 0.30/100 person-years in those with FIB-4 less than 3.25 and 1.45, respectively.

More recently, Ioannou and colleagues [28] evaluated the FIB-4 before and after treatment in patients treated at the VA. They showed that in patients with clear evidence of cirrhosis (leading to an International Classification of Diseases, Tenth Revision diagnosis of cirrhosis) and a FIB-4 greater than or equal to 3.25, the risk of HCC was very high (5.07%/y) and did not decrease over 10 years of follow-up. Even in patients with clinical cirrhosis who had a FIB-4 of less than 3.25 before and after treatment, the risk of HCC remained near 1%/year throughout follow-up. In those without obvious cirrhosis but a pretreatment FIB-4 greater than or equal to 3.25, the risk of HCC post SVR hovered around 1%/year throughout follow-up. However, for those without obvious cirrhosis and pretreatment FIB-4 of less than 3.25, the risk of HCC post SVR was 0.1–0.3%/year.

These data are important for many reasons. They confirm that the risk of HCC persists for at least 10 years post SVR in patients with cirrhosis, with all groups showing stable annual incidence over time. Importantly, the results confirm that in patients with obvious cirrhosis, irrespective of the FIB-4 value, surveillance is warranted. However, where they are most useful is in patients without clinically apparent cirrhosis. In this group, which includes the controversial F3 population, if the FIB-4 was above 3.25 before treatment, the risk of HCC was significant post SVR, whereas for those with a baseline FIB-4 of less than 3.25, the risk of HCC was very low and below any cost-effectiveness thresholds for surveillance. The VA is a select population that is predominantly male with overrepresentation of many other HCC risk factors. As such, tools to identify a low-risk VA population (those without obvious cirrhosis and FIB-4 < 3.25) would likely be equally or more effective in a non-VA lower-risk population. Clearly, validation of the predictive value of FIB-4 is required. Other scoring systems including albumin and/or AFP have also been proposed [39].

As an alternative to FIB-4, vibration-controlled transient elastography (VCTE) is routinely used to stage liver fibrosis. Many studies have documented that VCTE values fall with viral clearance [40]. Notably, changes in VCTE scores occur early with antiviral therapy, suggesting they more likely reflect the reduction in inflammation with viral clearance rather than true fibrosis regression. Although it is likely true that lower VCTE scores post SVR are associated with a lower risk of HCC, no clear thresholds for improvement in either absolute or relative terms have been established to identify patients at low enough risk to forego post-SVR surveillance. As such, similar to the FIB-4 approach, patients with pretreatment VCTE scores in the cirrhotic range are recommended to continue post-SVR HCC surveillance irrespective of the improvement in liver stiffness observed after treatment. Ongoing long-term follow-up studies may eventually allow for risk stratification using pretreatment and posttreatment elastography but current guidelines recommend basing surveillance decisions on the pretreatment VCTE value only.

Hepatocellular Carcinoma Surveillance Cost-Effectiveness

Cost-effectiveness analysis (CEA) uses "quality-adjusted life-years" (QALYs) gained per the cost of an intervention in monetary units. Incremental cost-effectiveness ratios, defined as the cost difference divided by the effectiveness difference (expressed in $/QALY), are calculated to compare the cost-effectiveness of given interventions. An intervention is considered "effective" if it increases the longevity for 100 days, [41] and cost-effective if it can be achieved at less than $50,000/QALY [42]. While the specific threshold for cost-effectiveness may be open to question, CEA remains very relevant for guiding decisions for cancer surveillance.

Although none of the many CEA studies of HCC surveillance is identical, [43] they have all concluded that some form of surveillance was effective and cost-effective, [5] and the cost-effectiveness was primarily dependent on HCC incidence. As such, teasing apart the incidence for different patients before and after SVR is key to determining the utility of surveillance.

Averaging out several studies suggests that for patients with chronic HCV infection, surveillance is likely effective and cost-effective if HCC incidence is above 1.5%/year but may be cost-effective at incidence rates as low as 1%/year if higher willingness-to-pay thresholds than the standard $50,000/QALY are considered. [5, 6] Recently, Farhang Zangneh et al. [44] performed the first CEA of HCC surveillance after SVR. They found that surveillance was likely to be cost-effective in patients with an HCC incidence above 1.3%/year with similar results with an incidence as low as 1%/year. Once the HCC incidence fell below approximately 0.6%/year, surveillance was unlikely to be cost-effective even with willingness-to-pay thresholds of $100,000/QALY or more. Using available incidence data, they concluded that surveillance would be cost-effective for patients with obvious cirrhosis but not for those with F3. To address the challenge of F3 diagnosis, they showed that the thresholds of baseline APRI greater than 2.0 or FIB-4 greater than or equal to 3.25 could adequately separate the populations in terms of the need for post-SVR HCC surveillance. The long-term follow-up data from Ioannou et al. [28] add further weight to this approach.

Addressing the Controversy

Determining who requires post-SVR HCC surveillance is of major importance for patients, clinicians, and health care payers. The data clearly show that patients with F3 before treatment have a low risk of post-SVR HCC. The challenge is determining who has F3 rather than F4. As more and more patients with HCV are cured and treatment moves out of specialty clinics, it is critical to develop easy-to-apply methods to determine who requires post-SVR HCC surveillance. The reflex to always err on the conservative side and favor surveillance even in those at low risk of HCC may not be prudent. Emerging data suggest that the pretreatment FIB-4 score may be adequate to guide surveillance decision-making. For those with an FIB-4 greater than or equal to 3.25, surveillance is required post SVR. For individuals with a FIB-4 less than 3.25, no post-SVR surveillance is required as long as they do not have obvious clinical evidence of cirrhosis. The biggest challenge will likely be to ensure that this caveat of "clinically obvious cirrhosis" is not missed. Ultimately, newer models, possibly including improvement in liver stiffness and other post-SVR parameters, may improve on the FIB-4 approach, but for now, this simple rule may be the best tool available to guide decision-making around HCC surveillance. The potential benefits of this risk-oriented strategy may include targeted interventions to enhance surveillance uptake, the opportunity for individualized counseling leading to improvements in adherence, and the possibility of specifying different surveillance tests for different risk groups [45].

Future Research

Measures should be sought to enhance surveillance use by health care professionals and to improve patient adherence to ensure that the potential benefits of surveillance are maximized. Although universal surveillance is unlikely to be cost-effective for patients with F3 fibrosis, there are clearly those within this category who will develop HCC before and after SVR. Developing more precise tools to identify patients with F3 fibrosis, who are at higher risk of HCC and would thus benefit from surveillance, would be very valuable. Similarly, the majority of patients with compensated cirrhosis who clear HCV will never develop HCC. Tools, whether biomarkers, risk scores, or additional assessments of fibrosis regression, that can identify patients with cirrhosis who are at low risk of HCC and can forego surveillance, would be extremely valuable for patients and for health care systems. Clearly, improved imaging modalities and better biomarkers for early detection of HCC will also be critical to optimize HCC surveillance in all populations, but particularly in the relatively low-risk post-SVR group. As these tools are developed, updated CEA models will help guide the optimal approach to HCC surveillance.

References

1. Ferlay J, Soerjomataram I, Dikshit R, et al. Cancer incidence and mortality worldwide: sources, methods and major patterns in GLOBOCAN 2012. Int J Cancer. 2015;136(5):E359–86. https://doi.org/10.1002/ijc.29210.
2. Kanwal F, Singal AG. Surveillance for HCC: current best practice and future direction. Gastroenterology. 2019;157(1):54–64. https://doi.org/10.1053/j.gastro.2019.02.049.
3. Heimbach JK, Kulik LM, Finn RS, et al. AASLD guidelines for the treatment of HCC. Hepatology. 2018;67(1):358–80. https://doi.org/10.1002/hep.29086.
4. European Association for the Study of the Liver. EASL clinical practice guidelines: management of HCC. J Hepatol. 2018;69(1):182–236. https://doi.org/10.1016/j.jhep.2018.03.019.
5. Sherman M. Surveillance for hepatocellular carcinoma. Best Pract Res Clin Gastroenterol. 2014;28(5):783–93. https://doi.org/10.1016/j.bpg.2014.08.008.
6. Bruix J, Sherman M, American Association for the Study of Liver Diseases. Management of hepatocellular carcinoma: an update. Hepatology. 2011;53(3):1020–2. https://doi.org/10.1002/hep.24199.
7. Zhang BH, Yang BH, Tang ZY. Randomized controlled trial of screening for hepatocellular carcinoma. J Cancer Res Clin Oncol. 2004;130(7):417–22. https://doi.org/10.1007/s00432-004-0552-0.
8. Kansagara D, Papak J, Pasha AS, et al. Screening for hepatocellular carcinoma in chronic liver disease: a systematic review. Ann Intern Med. 2014;161(4):261–9. https://doi.org/10.7326/M14-0558.
9. Singal AG, Volk ML, Rakoski MO, et al. Patient involvement in healthcare is associated with higher rates of surveillance for hepatocellular carcinoma. J Clin Gastroenterol. 2011;45(8):727–32. https://doi.org/10.1097/MCG.0b013e31820989d3.
10. Singal AG, Pillai A, Tiro J. Early detection, curative treatment, and survival rates for hepatocellular carcinoma surveillance in patients with cirrhosis: a meta-analysis. PLoS Med. 2014;11(4):e1001624. https://doi.org/10.1371/journal.pmed.1001624.
11. Di Bisceglie AM, Sterling RK, Chung RT, et al. Serum alpha-fetoprotein levels in patients with advanced hepatitis C: results from the HALT-C trial. J Hepatol. 2005;43(3):434–41. https://doi.org/10.1016/j.jhep.2005.03.019.
12. Tzartzeva K, Obi J, Rich NE, et al. Surveillance imaging and alpha-fetoprotein for early detection of hepatocellular carcinoma in patients with cirrhosis: a meta-analysis. Gastroenterology. 2018;154(6):1706–1718.e1. https://doi.org/10.1053/j.gastro.2018.01.064.
13. Singal AG, Yopp A, Skinner CS, Packer M, Lee WM, Tiro JA. Utilization of HCC surveillance among American patients: a systematic review. J Gen Intern Med. 2012;27(7):861–7. https://doi.org/10.1007/s11606-011-1952-x.
14. AASLD/IDSA HCV Guidance Panel. Hepatitis C guidance: AASLD-IDSA recommendations for testing, managing, and treating adults infected with hepatitis C virus. Hepatology. 2015;62(3):932–54. https://doi.org/10.1002/hep.27950.
15. van der Meer AJ, Veldt BJ, Feld JJ, et al. Association between SVR and all-cause mortality among patients with chronic hepatitis C and advanced hepatic fibrosis. JAMA. 2012;308(24):2584–93. https://doi.org/10.1001/jama.2012.144878.
16. Nahon P, Bourcier V, Layese R, et al. Eradication of hepatitis C virus infection in patients with cirrhosis reduces risk of liver and non-liver complications. Gastroenterology. 2017;152(1):142–156.e2. https://doi.org/10.1053/j.gastro.2016.09.009.
17. Bang CS, Song IH. Impact of antiviral therapy on hepatocellular carcinoma and mortality in patients with chronic hepatitis C: systematic review and meta-analysis. BMC Gastroenterol. 2017;17(1):46. https://doi.org/10.1186/s12876-017-0606-9.
18. D'Ambrosio R, Colombo M. Should surveillance for liver cancer be modified in hepatitis C patients after treatment-related cirrhosis regression? Liver Int. 2016;36(6):783–90. https://doi.org/10.1111/liv.13106.
19. Mallet V, Gilgenkrantz H, Serpaggi J, et al. Brief communication: the relationship of regression of cirrhosis to outcome in chronic hepatitis C. Ann Intern Med. 2008;149(6):399–403. https://doi.org/10.7326/0003-4819-149-6-200809160-00006.
20. Kanwal F, Kramer J, Asch SM, Chayanupatkul M, Cao Y, El-Serag HB. Risk of hepatocellular carcinoma in HCV patients treated with direct-acting antiviral agents. Gastroenterology. 2017;153(4):996–1005. https://doi.org/10.1053/j.gastro.2017.06.012.
21. Janjua NZ, Chong M, Kuo M, et al. Long-term effect of sustained virological response on hepatocellular carcinoma in patients with

hepatitis C in Canada. J Hepatol. 2017;66(3):504–13. https://doi.org/10.1016/j.jhep.2016.10.028.

22. El-Serag HB, Kanwal F, Richardson P, Kramer J. Risk of hepatocellular carcinoma after sustained virological response in veterans with hepatitis C virus infection. Hepatology. 2016;64(1):130–7. https://doi.org/10.1002/hep.28535.
23. Ioannou GN, Green PK, Berry K. HCV eradication induced by direct-acting antiviral agents reduces the risk of hepatocellular carcinoma. J Hepatol. 2018;68(1):25–32. https://doi.org/10.1016/j.jhep.2017.08.030.
24. Waziry R, Hajarizadeh B, Grebely J, et al. Hepatocellular carcinoma risk following direct-acting antiviral HCV therapy: a systematic review, meta-analyses, and meta-regression. J Hepatol. 2017;67(6):1204–12. https://doi.org/10.1016/j.jhep.2017.07.025.
25. D'Ambrosio R, Aghemo A, Rumi MG, et al. A morphometric and immunohistochemical study to assess the benefit of sustained virological response in hepatitis C virus patients with cirrhosis. Hepatology. 2012;56(2):532–43. https://doi.org/10.1002/hep.25606.
26. Bedossa P, Dargère D, Paradis V. Sampling variability of liver fibrosis in chronic hepatitis C. Hepatology. 2003;38(6):1449–57. https://doi.org/10.1016/j.hep.2003.09.022.
27. Toyoda H, Kumada T, Tada T, et al. Risk factors of hepatocellular carcinoma development in non-cirrhotic patients with sustained virological response for chronic hepatitis C virus infection. J Gastroenterol Hepatol. 2015;30(7):1183–9. https://doi.org/10.1111/jgh.12915.
28. Ioannou GN, Beste LA, Green PK, et al. Increased risk for hepatocellular carcinoma persists up to 10 years after HCV eradication in patients with baseline cirrhosis or high FIB-4 scores. Gastroenterology. 2019;157(5):1264–1278.e4. https://doi.org/10.1053/j.gastro.2019.07.033.
29. Ghany MG, Morgan TR, Hepatitis C Guidance Panel. Hepatitis C guidance 2019 Update. American Association for the study of liver diseases–infectious diseases society of America recommendations for testing, managing, and treating hepatitis C virus infection. Hepatology. 2020;71(2):686–721. https://doi.org/10.1002/hep.31060.
30. Heimbach JK, Kulik LM, Finn RS, et al. AASLD guidelines for the treatment of HCC. Hepatology. 2018;67(1):358–80.
31. Omata M, Cheng AL, Kokudo N, et al. Asia-Pacific clinical practice guidelines on the management of HCC: a 2017 update. Hepatol Int. 2017;11(4):317 70.
32. EASL Clinical Practice Guidelines. Management of HCC. J Hepatol. 2018;69(1):182–236.
33. Ghany MG, Marks KM, Morgan TR, et al. Hepatitis-C guidance 2019 update: AASLD-IDSA recommendations for testing, managing, and treating HCV infection. Hepatology. 2019;
34. Omata M, Kanda T, Wei L, et al. APASL consensus statements and recommendations for hepatitis C prevention, epidemiology, and laboratory testing. Hepatol Int. 2016;10(5):681–701.
35. Pawlotsky JM, Negro F, Aghemo A, et al. EASL recommendations on treatment of hepatitis C 2018. J Hepatol. 2018;69(2):461–511.
36. Ioannou GN, Beste LA, Green PK, et al. Increased risk for HCC persists up to 10 years after HCV eradication in patients with baseline cirrhosis or high FIB-4 scores. Gastroenterology. 2019;157:1264.
37. Farhang Zangneh H, Wong WWL, Sander B, et al. Cost-effectiveness of HCC surveillance after SVR to therapy in patients with HCV infection and advanced fibrosis. Clin Gastroenterol Hepatol. 2019;17(9):1840–9.
38. van der Meer AJ, Feld JJ, Hofer H, et al. Risk of cirrhosis-related complications in patients with advanced fibrosis following hepatitis C virus eradication. J Hepatol. 2017;66(3):485–93. https://doi.org/10.1016/j.jhep.2016.10.017.
39. Lau J, Krassenburg L, Georgie F, et al. A scoring method to identify patients with advanced fibrosis and low HCC risk after SVR in HCV infection. Paper presented at: Hepatology; 2019.
40. Trivedi HD, Lin SC, Lau DT. Noninvasive assessment of fibrosis regression in hepatitis C virus sustained virologic responders. Gastroenterol Hepatol. 2017;13(10):587–95.
41. Naimark D, Naglie G, Detsky AS. The meaning of life expectancy: what is a clinically significant gain? J Gen Intern Med. 1994;9(12):702–7. https://doi.org/10.1007/BF02599016.
42. Braithwaite RS, Meltzer DO, King JT Jr, Leslie D, Roberts MS. What does the value of modern medicine say about the $50,000 per quality-adjust life-year decision rule? Med Care. 2008;46(4):349–56. https://doi.org/10.1097/MLR.0b013e31815c31a7.
43. Sherman M, Colombo M. Hepatocellular carcinoma screening and diagnosis. Semin Liver Dis. 2014;34(4):389–97. https://doi.org/10.1055/s-0034-1394139.
44. Farhang Zangneh H, Wong WWL, Sander B, et al. Cost-effectiveness of hepatocellular carcinoma surveillance after a sustained virologic response to therapy in patients with hepatitis C virus infection and advanced fibrosis. Clin Gastroenterol Hepatol. 2019;17(9):1840–9. https://doi.org/10.1016/j.cgh.2018.12.018.
45. Singal AG, Ioannou GN. For whom is hepatocellular carcinoma surveillance after sustained virologic response cost-effective? Clin Gastroenterol Hepatol. 2019;17(9):1732–5. https://doi.org/10.1016/j.cgh.2019.02.020.

Should Hepatocellular Carcinoma Surveillance Be Performed in Patients with Noncirrhotic Metabolic Dysfunction-Associated Steatohepatitis?

3

Ruben Hernaez and Fasiha Kanwal

Given the low risk of hepatocellular carcinoma (HCC) in the absence of cirrhosis, surveillance should not be performed in patients with noncirrhotic metabolic dysfunction-associated steatotic liver disease (MASLD).

Main Points

1. Case series and cohort studies have reported that more than one-third of MASLD-related HCCs occur in the absence of cirrhosis.
2. However, the annual incidence of HCC in patients with noncirrhotic MASLD is low and HCC surveillance is not cost-effective in this subgroup.
3. Risk stratification tools are needed to identify a subgroup of patients with noncirrhotic MASLD with sufficient risk to warrant HCC surveillance.
4. Risk-benefit considerations of HCC surveillance in patients with noncirrhotic MASLD should consider screening-related harms, including risk of overdiagnosis.

MASLD and metabolic dysfunction-associated steatohepatitis (MASH) is now the most common liver disease and projected to become the leading cause of HCC in the United States. Whereas most cases of HCC develop in patients with established cirrhosis, carcinogenesis in MASLD/MASH can occur in the absence of cirrhosis. The clinical practice guidelines recommend HCC surveillance with biannual ultrasound with or without alpha fetoprotein (AFP) in individuals with MASLD/MASH cirrhosis [1]. Whether HCC surveillance should be recommended for individuals with MASLD/MASH without cirrhosis remains unclear. We address this clinical issue in this chapter. We briefly review the epidemiology of MASLD-related HCC and the data on HCC risk in the subgroup without cirrhosis. We then review the evidence (or lack thereof) for and against HCC surveillance in noncirrhotic MASLD patients. We conclude by recommending against routine HCC surveillance in patients with noncirrhotic MASLD/MASH until more evidence accrues to guide surveillance decision.

Epidemiology of Metabolic Dysfunction-Associated Steatotic Liver Disease–Related Hepatocellular Carcinoma

MASLD is now the most common liver disease. At a population prevalence of approximately 30%, MASLD affects 83.1 million individuals in the United States [2]. At least 20–30% of patients with MASLD develop MASH, characterized by liver cell injury, inflammation, and fibrosis that can lead to cirrhosis of the liver. MASLD (and MASH) is an established risk factor for HCC. In a large retrospective cohort of patients with MASLD followed in the national Veterans Administration (VA) system, we recently showed the risk of HCC was sevenfold higher in individuals with MASLD than age- and sex-matched controls (hazard ratio, HR = 7.6, 95% CI, 5.7–10.1). Given this association, and because of its high prevalence, MASLD is projected to become the greatest single contributor to the future burden of HCC in the United States. Among US patients listed for liver transplantation, MASLD is now the second most common etiology for HCC, superseding alcohol and hepatitis B virus–related HCC [3]. A recent study estimated that MASLD will result in 135,000 cases of HCC between 2015 and 2030—this could represent approximately half of HCC burden from all etiologies in this time frame [2].

R. Hernaez · F. Kanwal (✉)
Section of Gastroenterology, Michael E. DeBakey Veterans Affairs Medical Center, Houston, TX, USA

Center for Innovations in Quality, Effectiveness and Safety (IQuESt), Michael E. DeBakey Veterans Affairs Medical Center, Houston, TX, USA

Section of Gastroenterology and Hepatology, Department of Medicine, Baylor College of Medicine, Houston, TX, USA
e-mail: Ruben.Hernaez@bcm.edu; kanwal@bcm.edu

N. S. Reau, A. G. Singal (eds.), *Controversies in Hepatocellular Carcinoma*, https://doi.org/10.1007/978-3-031-90225-3_3

Hepatocellular Carcinoma Risk in Noncirrhotic Metabolic Dysfunction-Associated Steatotic Liver Disease/Metabolic Dysfunction-Associated Steatohepatitis

While it is clear that cirrhosis is an important risk factor for HCC in MASLD, HCC case series have shown that MASLD-related HCCs occurred in 49–66% patients in the absence of cirrhosis (or advanced liver fibrosis) [4, 5]. These figures are likely an overestimate due to referral and publication bias. Furthermore, it is possible that some of those patients classified as having HCC in the absence of cirrhosis (noncirrhotic HCC) indeed have cirrhosis and are misclassified (as noncirrhotic) because of sampling error. Mittal et al. [6] studied the risk factors for the development of HCC in the absence of cirrhosis in a cohort of 1500 HCC patients from the VA. The authors classified cirrhosis status as very high probability of no cirrhosis, high probability of no cirrhosis, and confirmed cirrhosis. Among patients with MASLD-related HCC ($n = 107$), 34.6% had very high or high probability evidence of no cirrhosis; 65.4% of patients had confirmed cirrhosis at the time of HCC diagnosis. More recent evidence from HCC cases that developed in longitudinal cohorts of patients with MASLD (that is incident HCCs) suggests that approximately 20–30% of MASLD-HCC develop in patients without evidence of cirrhosis [7].

These data are problematic because currently, HCC surveillance is recommended *only* in patients *with* cirrhosis (except in select patients with hepatitis B virus, who are also known to progress to HCC *without* cirrhosis), and hence may miss many at-risk MASLD patients. The lack of surveillance recommendations may continue to contribute to the underappreciation of MASLD as a risk factor for HCC, resulting in the diagnosis of HCC at advanced stages when patients are ineligible for curative treatment options. Table 3.1 displays data from selected case series that examined the distribution of risk factors, including cirrhosis, in patients with HCC and MASLD. Most HCCs were diagnosed at advanced stages, with mean tumor size ranging from 8 to 12.8 cm. Most patients exhibited high prevalence of metabolic syndrome components, which could be a target to risk-stratify these patients.

Hepatocellular Carcinoma Surveillance in Patients with Noncirrhotic Metabolic Dysfunction-Associated Steatotic Liver Disease/Metabolic Dysfunction-Associated Steatohepatitis

Can mass surveillance of all individuals with MASLD/MASH be feasible or cost-effective? We believe the answer is no. There are several reasons why mass surveillance of all individuals with MASLD/MASH cannot be recommended at this time.

First, the effectiveness of HCC surveillance depends on the individual's risk (incidence) of HCC. The precise risk of HCC in subgroups of MASLD patients without advanced fibrosis or cirrhosis is not well known but appears to be low. Table 3.2 summarizes the results of retrospective cohort studies of MASLD participants with a low prevalence of cirrhosis at baseline (0–12.3%). Studies used imaging, biopsy, or clinical criteria to define MASLD, and participants were followed for a mean of 1–13.7 years. Notably, all studies ascertained cirrhosis status at the start of follow-up. Patients without cirrhosis represented a heterogeneous group: those

Table 3.1 Selected case series of hepatocellular carcinoma in noncirrhotic metabolic dysfunction-associated steatotic liver disease

Authors	Country	Indications for specimen	No	Fibrosis	Risk factors, No. or %	Tumor size, cm
Guzman [21]	USA	Diagnostic (1), resection (2)	3	F0	BMI ≥ 25, HTN, and dyslipidemia, diabetes in all patients	6.8–10
Paradis [5]	France	Resection	20	F0-F2	Diabetes in 77%, hypertension in 84%, dyslipidemia in 65%	8.8
Hashimoto [22]	Japan	NA	34	F1-F2 (4) F3-F4 (30)	Diabetes 74%, BMI ≥ 25 (62%), dyslipidemia (29%), HTN (47%)	NA
Ertle [23]	Multinational	NA	31	NA	BMI > 25 (90%), BM 70%, HTN 80%, dyslipidemia 50%, CAD 40%	NA
Perumpail [24]	USA	Surgical resection or core biopsy	6	NA	HTN, BMI > 25, or diabetes (100%); dyslipidemia 80%	4.6–12.8
Mohamad [25]	USA	Explant or liver biopsy	36	F0 (20) F1 (6), F2 (3)	HTN (71%), BMI > 30 (53%), dyslipidemia (44%), diabetes 38%	> 5 (77.8%)
Bengtsson [26]	Sweden	Explant, resection, liver biopsy	83	Only 35 recorded F0 (1), F1 (13), F2 (16), F3 (5)	HTN (80%), diabetes (66%), mean BMI 29	> 5 (71%)
Gawrieh [27]	USA	Histology	159	NA	Overall, in cohort of 605 noncirrhotic HCC: HTN (69%), diabetes (34%), dyslipidemia (38%)	8.9 (overall cohort)

Abbreviations: *BMI* body mass index, *CAD* coronary artery disease, *HCC* hepatocellular carcinoma, *HTN* hypertension, *NA* not available, *USA* United States of America

Table 3.2 Retrospective studies of mainly noncirrhotic metabolic dysfunction-associated steatotic liver disease, chronologically ordered

Authors	Country	Source population	MASLD definition	MASLD, No.	Cirrhosis at baseline, No. (%)	Follow-up, median, y	HCC new cases, No. (%)	Incidence proportion, %
Adams [28]	USA	Clinical registry	Imaging or biopsy	420	8 (1.9)	7.6	2 (0.5)	0.5
Ekstedt [8]	Sweden	Clinical registry	Biopsy	88	4 (4.5)	13.7	3 (3.4)	3.4
Andrade [29]	Brazil	Clinical registry	Imaging or biopsy	62	8 (12.3)	1.0	1 (1.6)	1.6
Lee [30]	Korea	Insurance database	ICD codes	18,080	0 (0)	6.3	180 (1.0)	1.0
Kim [31]	Korea	Clinical registry	Imaging	8721	0 (0)	7.5	14 (0.2)	0.2
Kanwal[7]	USA	Veterans affairs	Clinical algorithm	296,707	1084 (0.4)	9.0	490 (0.2)	0.2

Abbreviations: *HCC* hepatocellular carcinoma, *ICD* international classification of diseases, *MASLD* metabolic dysfunction-associated steatotic liver disease, USA United States of America

who eventually progressed to cirrhosis (and then developed HCC) and those who never progressed to cirrhosis. With the exception of Ekstedt et al. [8], who paired biopsies in 68 participants and showed only 1 HCC, none of the studies were able to tease these 2 groups apart. Hence it is difficult to quantify the exact risk of HCC in patients with MASLD in the absence of cirrhosis. Notwithstanding this limitation, and based on available data, the risk (incidence) of HCC in noncirrhotic MASLD appears to range from 0.2% to 3.4% over 1–13.7 years of follow-up.

The largest study to date was conducted among the VA population. This study included 296,707 MASLD patients compared with 296,707 matched controls and followed for more than 9 years. The HCC incidence was statistically significantly higher among MASLD patients than controls (0.21/1000 person-years [PYs] in MASLD vs 0.02/1000 PYs in controls; adjusted hazard ratio, 7.62; 95% CI, 5.76–10.09). Among patients with MASLD, those with cirrhosis had the highest annual incidence of HCC (10.6/1000 PYs) [7]. The risk of HCC was low in all subgroups of noncirrhotic MASLD patients regardless of age, race, and metabolic traits. Specifically, the annual incidence of HCC was 0.04/1000 PYs overall. It was highest in older patients (0.41/1000 PYs) and those with diabetes (0.45/1000 PYs) but fell far below the current HCC incidence cutoff (1.5% per year), beyond which HCC surveillance may be cost-effective.

It is plausible that there are subgroup/s of individuals with noncirrhotic MASLD with HCC risk that reaches or exceeds the accepted cutoff of 1.5%/year (e.g., individuals with multiple HCC risk factors that synergistically increase HCC risk). However, currently, it is not possible to accurately risk-stratify MASLD patients at risk for HCC in the absence of cirrhosis. The only published risk prediction model for MASLD-HCC is limited to patients with MASH cirrhosis and hence not applicable to the large proportion of MASLD patients without cirrhosis [9]. However, future studies may be able to risk-stratify noncirrhotic MASLD patients and thus fill the existing evidence gap. Several demographic, metabolic, behavioral, and genetic factors are associated with HCC risk. We recently reported that among patients who progressed to HCC without cirrhosis, those with diabetes had a higher risk of HCC than patients without diabetes. These data suggest that metabolic traits may identify potential candidates for HCC surveillance. However, more than 40% of patients in our study had diabetes, underscoring the need for more precise risk stratification tools [10].

Second, successful surveillance also depends on the effectiveness of surveillance tests in detecting HCC. The American Association for the Study of Liver Diseases recommends ultrasonography with or without AFP every 6 months in patients at high risk for HCC. The performance of ultrasound can be affected by several patient-level factors, with particularly suboptimal performance in obese and MASLD patients. In a retrospective analysis of 941 cirrhotic patients undergoing HCC surveillance, more than 20% of ultrasound exams were determined to be of inadequate quality for surveillance. Factors correlated with ultrasound inadequacy included male sex, obesity and morbid obesity, alcohol or MASH etiology of cirrhosis, advanced liver disease (Child Pugh Turcotte class B cirrhosis), inpatient status, and elevated alanine transaminase [11]. Few available data show that the commonly used conventional serum AFP test loses sensitivity for early HCC detection in cirrhosis patients with varied etiologies, to the degree that necessitates using an algorithm that incorporates the underlying etiology, age, and other indicators of liver inflammation or dysfunction with AFP [12]. There is a dearth of other sensitive and specific HCC biomarkers that are well-validated in prospective studies; this gap will need to be filled before surveillance can be recommended in the growing cohort of noncirrhotic MASLD patients.

Third, surveillance can result in the overdiagnosis of HCC in patients who are not eligible for curative treatments given factors such as age and cardiovascular comorbidity (common in MASLD). Only a few studies have examined outcomes of patients with MASLD-HCC [13–15]. Collectively, these studies found that despite being diagnosed at late HCC stages, there was no statistically significant difference in the overall survival in patients with MASLD compared with those with HCV-related HCC. Thus, it is possible that if barriers in HCC surveillance are overcome, patients with MASLD-HCC may indeed have better survival compared to HCC from other risk factors. However, it is also plausible that HCC may not drive overall mortality in all patients with MASLD-HCC, many of whom have competing risks, especially cardiovascular mortality. No study has examined these competing issues. There are also no studies that balance the benefits of surveillance with its harms, such as complications from liver biopsy, liver resection, and other treatments in MASLD. There could also be other unintended consequences of surveillance, such as anxiety and distress provoked by false-positive tests [16, 17]. Ideally, a large randomized controlled trial will need to compare long-term outcomes of *routine* surveillance and *no surveillance*; this could take decades and may even be unethical for patients in the no-surveillance arm. In such a situation, decision-analytic modeling may be needed to simulate comparative effectiveness data to guide surveillance decisions in noncirrhotic MASLD. Our arguments are also supported by the International Liver Transplantation Society NASLD/alcoholic steatohepatitis (MASH) Consensus Conference [18]. Reig et al. [18] performed a systematic review to determine whether patients with MASLD/MASH should be surveyed for HCC following a specific protocol. The authors updated a prior systematic review White et al. [19] up to December 2017 including 25 papers. While they recommended HCC screening in MASLD patients with cirrhosis, they did not find any study on the true incidence of HCC considering all confounding factors in MASLD patients without cirrhosis, resulting in the conclusion "systematic HCC screening cannot be recommended based on the current data." [18]

Future Directions

Based on the aforementioned considerations, we do not recommend HCC surveillance in patients with noncirrhotic MASLD/MASH. We believe a multitiered program in which MASLD patients are first risk stratified by clinical risk prediction tools and then offered HCC surveillance based on the personalized benefits and risks may improve the eventual effectiveness and cost-effectiveness of HCC surveillance in MASLD. Future research should systematically address these gaps in the evidence surrounding HCC surveillance in MASLD.

Conclusion

HCC can develop in the absence of cirrhosis in individuals with MASLD. However, the overall risk of progression to HCC is too low in patients with noncirrhotic MASLD. We do not recommend surveillance for HCC in general MASLD/MASH patients without cirrhosis because of as-yet unclear benefits of HCC surveillance in this group. This is also consistent with the recent American Gastroenterological Association Clinical Practice update on surveillance of HCC in MASLD [20]. We also believe mass surveillance of all individuals with MASLD/MASH will never be feasible or cost-effective given the high prevalence of MASLD/MASH in the general population. Future research should focus on developing new risk-stratification tools, identifying and evaluating sensitive and specific HCC surveillance tests, and testing personalized approaches to HCC surveillance to guide HCC surveillance decisions in the growing population of individuals with MASLD.

Acknowledgments This material is based on work supported by the Cancer Prevention & Research Institute of Texas (principal investigator: Kanwal; grant No. RP150587). The work is also supported in part by the Center for Gastrointestinal Development, Infection and Injury (National Institute of Diabetes and Digestive and Kidney Diseases No. P30 DK 56338), the Veterans Administration Center for Innovations in Quality, Effectiveness and Safety (No. CIN 13-413), and the National Cancer Institute (No. NCI U01 CA230997-01).

References

1. Marrero JA, Kulik LM, Sirlin CB, et al. Diagnosis, staging, and management of hepatocellular carcinoma: 2018 practice guidance by the American Association for the study of liver diseases. Hepatology. 2018;68(2):723–50. https://doi.org/10.1002/hep.29913.
2. Estes C, Razavi H, Loomba R, Younossi Z, Sanyal AJ. Modeling the epidemic of nonalcoholic fatty liver disease demonstrates an exponential increase in burden of disease. Hepatology. 2018;67(1):123–33. https://doi.org/10.1002/hep.29466.
3. Younossi Z, Stepanova M, Ong JP, et al. Nonalcoholic steatohepatitis is the fastest growing cause of hepatocellular carcinoma in liver transplant candidates. Clin Gastroenterol Hepatol. 2019;17(4):748–755.e3. https://doi.org/10.1016/j.cgh.2018.05.057.
4. Yasui K, Hashimoto E, Komorizono Y, et al. Ministry of Health, labour, and welfare of Japan. Characteristics of patients with nonalcoholic steatohepatitis who develop hepatocellular carcinoma. Clin Gastroenterol Hepatol. 2011;9(5):428–33.
5. Paradis V, Zalinski S, Chelbi E, et al. Hepatocellular carcinomas in patients with metabolic syndrome often develop without significant liver fibrosis: a pathological analysis. Hepatology. 2009;49(3):851–9. https://doi.org/10.1002/hep.22734.
6. Mittal S, El-Serag HB, Sada YH, et al. Hepatocellular carcinoma in the absence of cirrhosis in United States veterans is associated with nonalcoholic fatty liver disease. Clin Gastroenterol Hepatol. 2016;14(1):124–131.e1. https://doi.org/10.1016/j.cgh.2015.07.019.

7. Kanwal F, Kramer JR, Mapakshi S, et al. Risk of hepatocellular cancer in patients with non-alcoholic fatty liver disease. Gastroenterology. 2018;155(6):1828–1837.e2. https://doi.org/10.1053/j.gastro.2018.08.024.
8. Ekstedt M, Franzén LE, Mathiesen UL, et al. Long-term follow-up of patients with NAFLD and elevated liver enzymes. Hepatology. 2006;44(4):865–73. https://doi.org/10.1002/hep.21327.
9. Ioannou GN, Green P, Kerr KF, Berry K. Models estimating risk of hepatocellular carcinoma in patients with alcohol or NAFLD-related cirrhosis for risk stratification. J Hepatol. 2019;71(3):523–33. https://doi.org/10.1016/j.jhep.2019.05.008.
10. Kanwal F, Kramer J, Li L, et al. Effect of metabolic traits on the risk of cirrhosis and hepatocellular cancer in nonalcoholic fatty liver disease. Hepatology. 2020;71(3):808–19. https://doi.org/10.1002/hep.31014.
11. Simmons O, Fetzer DT, Yokoo T, et al. Predictors of adequate ultrasound quality for hepatocellular carcinoma surveillance in patients with cirrhosis. Aliment Pharmacol Ther. 2017;45(1):169–77. https://doi.org/10.1111/apt.13841.
12. Tayob N, Richardson P, Kanwal F, et al. Performance of AFP-based hepatocellular carcinoma surveillance in cirrhosis patients with cured HCV. Gastroenterology. 2017;152(5):S1077–8. https://doi.org/10.1016/S0016-5085(17)33634-X.
13. Reddy SK, Steel JL, Chen HW, et al. Outcomes of curative treatment for hepatocellular cancer in nonalcoholic steatohepatitis versus hepatitis C and alcoholic liver disease. Hepatology. 2012;55(6):1809–19. https://doi.org/10.1002/hep.25536.
14. Marrero JA, Fontana RJ, Su GL, Conjeevaram HS, Emick DM, Lok AS. NAFLD may be a common underlying liver disease in patients with hepatocellular carcinoma in the United States. Hepatology. 2002;36(6):1349–54. https://doi.org/10.1053/jhep.2002.36939.
15. Mittal S, Sada YH, El-Serag HB, et al. Temporal trends of nonalcoholic fatty liver disease-related hepatocellular carcinoma in the veteran affairs population. Clin Gastroenterol Hepatol. 2015;13(3):594–601.e1. https://doi.org/10.1016/j.cgh.2014.08.013.
16. Taylor EJ, Jones RL, Guthrie JA, Rowe IA. Modeling the benefits and harms of surveillance for hepatocellular carcinoma: information to support informed choices. Hepatology. 2017;66(5):1546–55. https://doi.org/10.1002/hep.29315.
17. Kanwal F, Singal AG. Surveillance for hepatocellular carcinoma: current best practice and future direction. Gastroenterology. 2019;157(1):54–64. https://doi.org/10.1053/j.gastro.2019.02.049.
18. Reig M, Gambato M, Man NK, et al. Should patients with NAFLD/NASH be surveyed for HCC? Transplantation. 2019;103(1):39–44. https://doi.org/10.1097/TP.0000000000002361.
19. White DL, Kanwal F, El-Serag HB. Association between nonalcoholic fatty liver disease and risk for hepatocellular cancer, based on systematic review. Clin Gastroenterol Hepatol. 2012;10(12):1342–1359.e2. https://doi.org/10.1016/j.cgh.2012.10.001.
20. Loomba R, Lim JK, Patton H, El-Serag HB. AGA clinical practice update on screening and surveillance for hepatocellular carcinoma in patients with nonalcoholic fatty liver disease: expert review. Gastroenterology. 2020;158(6):1822–30. https://doi.org/10.1053/j.gastro.2019.12.053.
21. Guzman G, Brunt EM, Petrovic LM, Chejfec G, Layden TJ, Cotler SJ. Does nonalcoholic fatty liver disease predispose patients to hepatocellular carcinoma in the absence of cirrhosis? Arch Pathol Lab Med. 2008;132(11):1761–6. https://doi.org/10.5858/132.11.1761.
22. Hashimoto E, Yatsuji S, Tobari M, et al. Hepatocellular carcinoma in patients with nonalcoholic steatohepatitis. J Gastroenterol. 2009;44(Suppl 19):89–95. https://doi.org/10.1007/s00535-008-2262-x.
23. Ertle J, Dechêne A, Sowa JP, et al. Non-alcoholic fatty liver disease progresses to hepatocellular carcinoma in the absence of apparent cirrhosis. Int J Cancer. 2011;128(10):2436–43. https://doi.org/10.1002/ijc.25797.
24. Perumpail RB, Wong RJ, Ahmed A, Harrison SA. Hepatocellular carcinoma in the setting of non-cirrhotic nonalcoholic fatty liver disease and the metabolic syndrome: US experience. Dig Dis Sci. 2015;60(10):3142–8. https://doi.org/10.1007/s10620-015-3821-7.
25. Mohamad B, Shah V, Onyshchenko M, et al. Characterization of hepatocellular carcinoma (HCC) in non-alcoholic fatty liver disease (NAFLD) patients without cirrhosis. Hepatol Int. 2016;10(4):632–9. https://doi.org/10.1007/s12072-015-9679-0.
26. Bengtsson B, Stål P, Wahlin S, Björkström NK, Hagström H. Characteristics and outcome of hepatocellular carcinoma in patients with NAFLD without cirrhosis. Liver Int. 2019;39(6):1098–108. https://doi.org/10.1111/liv.14087.
27. Gawrieh S, Dakhoul L, Miller E, et al. Characteristics, aetiologies and trends of hepatocellular carcinoma in patients without cirrhosis: a United States multicentre study. Aliment Pharmacol Ther. 2019;50(7):809–21. https://doi.org/10.1111/apt.15464.
28. Adams LA, Lymp JF, St Sauver J, et al. The natural history of nonalcoholic fatty liver disease: a population-based cohort study. Gastroenterology. 2005;129(1):113–21. https://doi.org/10.1053/j.gastro.2005.04.014.
29. Costa de Andrade G, Fujise LH, de Santana JE, Oliveira F, et al. Non-alcoholic fatty liver disease (NAFLD) in different populations: a clinical and epidemiological study – sample of São José do Rio Preto. Rev Assoc Med Bras. 2016;62(3):218–26. https://doi.org/10.1590/1806-9282.62.03.218.
30. Lee TY, Wu JC, Yu SH, Lin JT, Wu MS, Wu CY. The occurrence of hepatocellular carcinoma in different risk stratifications of clinically noncirrhotic nonalcoholic fatty liver disease. Int J Cancer. 2017;141(7):1307–14. https://doi.org/10.1002/ijc.30784.
31. Kim GA, Lee HC, Choe J, et al. Association between nonalcoholic fatty liver disease and cancer incidence rate. J Hepatol. 2017;68(1):140–6. https://doi.org/10.1016/j.jhep.2017.09.012.

Can Risk Stratification Models Be Used to Identify Patients without Cirrhosis Who Might Benefit from Hepatocellular Carcinoma Surveillance?

4

George N. Ioannou

Main Points

1. There is heterogeneity in hepatocellular carcinoma (HCC) risk, both among patients with cirrhosis and those without cirrhosis. There are likely some patients with cirrhosis who have low HCC risk and surveillance is not beneficial as well as some patients without cirrhosis who have sufficient risk to benefit from surveillance.
2. There are several HCC risk calculators, ranging from simple scores such as the fibrosis-4 index (FIB-4) to more complex machine-learning models; however, these require external validation and higher discriminatory ability.
3. Sufficiently accurate and validated models would facilitate a precision screening approach instead of the current one-size-fits-all approach.

The vast majority of cases of HCC occur in patients with cirrhosis or chronic hepatitis B virus (HBV) infection. These patients have a very high risk of HCC and, consequently, are expected to benefit from effective HCC surveillance programs, unless they have Child-Pugh score class C cirrhosis or other life-limiting comorbidities. However, approximately 12% of HCC cases occur in patients without cirrhosis or HBV infection [1]. These cases arise in patients who have chronic liver disease but have not yet developed cirrhosis. HCC risk among patients with precirrhotic liver disease overall is very low and hence screening is not generally recommended. Therefore, these HCC cases often present at advanced stages because of a low index of suspicion and lack of screening. This represents a great gap in clinical care and, currently, a frustrating conundrum. On the one hand, we know that a certain proportion of HCC cases will arise in noncirrhotic livers, but on the other we do not currently have proven, reliable tools to identify those patients who are at high HCC risk despite not having cirrhosis. However, some strategies are emerging to help us identify patients with precirrhotic liver disease who have a high enough HCC risk that they may benefit from HCC screening. This chapter will outline the challenges of screening for HCC in patients with precirrhotic liver disease, strategies that might help identify patients with high HCC risk despite not having established or diagnosed cirrhosis, and future directions in HCC risk stratification.

G. N. Ioannou (✉)
Director of Hepatology, Veterans Affairs Puget Sound Health Care System, University of Washington, Seattle, WA, USA
e-mail: georgei@medicine.washington.edu

The Challenge: The Cost-Effectiveness of Hepatocellular Carcinoma Screening Depends Primarily on HCC Risk

We Cannot Screen All Patients with Precirrhotic Liver Disease

Since we know that 12% of all HCC cases will arise in patients with precirrhotic chronic liver disease, it is reasonable to ask why not screen these patients as well as those with cirrhosis, in an effort to identify these HCC cases at an early stage, when potentially curative treatments are available. The answer is that the risk of HCC is more than 100 times higher in patients with cirrhosis than in patients with precirrhotic liver disease. Among patients with established cirrhosis, the annual risk of HCC ranges from about 1% to 5% in most studies (depending on factors such as etiology of cirrhosis, age, sex, ethnicity, platelet count, and albumin level). However, among unselected patients with precirrhotic metabolic dysfunction-associated steatotic liver disease (MASLD) the annual risk of HCC was estimated to be 0.008%, whereas it was 1.1% in patients with MASLD-cirrhosis—a rate that is 130 times higher [2].

N. S. Reau, A. G. Singal (eds.), *Controversies in Hepatocellular Carcinoma*, https://doi.org/10.1007/978-3-031-90225-3_4

Cost of Screening Per Quality-Adjusted Life-Year Increases Exponentially with Decreasing Cancer Incidence

The risk (or incidence) of HCC is one of the most important determinants of the cost-effectiveness of a surveillance program. This makes sense because only patients who are destined to develop HCC can possibly benefit from surveillance, whereas all the other patients can experience only the "costs and harms" of surveillance—including financial costs as well as the potential harms of screening anxiety, incidental findings, subsequent confirmatory imaging studies, and invasive tests such as liver biopsy [3]. For example, if 1000 patients undergo HCC screening for a period of 5 years, during which approximately 110 patients are expected to develop HCC (assuming an annual incidence of 2.5% per year), then 890 out of 1000 patients will experience only the costs and harms of surveillance, while only 110 out of 1000 can potentially experience benefits, assuming that the screening test is effective at identifying HCC earlier than would have otherwise been diagnosed and that curative or life-prolonging treatment is available as a result of earlier diagnosis. These costs and harms were estimated to include a false alarm in 150, intensified ultrasound follow-up in 85, additional computed tomography/magnetic resonance imaging in 65, and liver biopsy in 39 [3].

The incremental cost-effectiveness ratio (ICER, measured as the cost per quality-adjusted life-year [QALY]) of HCC surveillance in fact increased almost exponentially with decreasing incidence of HCC, all other factors being equal. Not surprisingly, HCC incidence (or risk) is one of the most important determinants of cost-effectiveness. Among patients with HCV who have achieved sustained virological response (SVR), it has been estimated that to achieve an ICER of less than $50,000/QALY (a frequently quoted willingness-to-pay threshold), then annual incidence of HCC has to be greater than 1.32% [4]. Patients with cirrhosis have an annual HCC risk higher that is higher than that threshold (~ 1.8%) and therefore have a low estimated ICER of $43,229 per QALY for biannual surveillance. However, patients with F3 fibrosis have a much lower risk of HCC after SVR and therefore have a much higher estimated ICER of $188,157 per QALY, which is considered prohibitively high.

Hepatocellular Carcinoma Risk Above Which Surveillance Is Considered Cost-Effective

An annual HCC risk threshold greater than 1.5% is frequently quoted, including in American Association for the Study of Liver Diseases guidelines, above which HCC screening is recommended because it is considered cost-effective in patients with cirrhosis or advanced fibrosis [5–9]. This threshold is based on estimates of tumor doubling time but the supporting studies are largely outdated because treatments for HCC have improved substantially since their publication, which would be expected to improve the cost-effectiveness of screening. Therefore, it is likely that the threshold of HCC risk above which HCC surveillance with current tests (ultrasound and AFP) becomes cost-effective in patients with cirrhosis or advanced fibrosis is currently lower than 1.5% and perhaps closer to 1%. However, it is important to recognize that assumptions on the sensitivity and specificity of screening can have a great effect on such estimates as well as assumptions about the effectiveness of curative treatments.

Evidence Supporting the Effectiveness of Hepatocellular Carcinoma Screening Derives from Low-Quality/High-Bias Data

An important limitation in interpreting the field of HCC screening is that evidence supporting the effectiveness of current screening modalities (ultrasound and AFP) in reducing cancer-related mortality in patients with cirrhosis or advanced fibrosis is not based on high-quality data. Therefore, while screening for HCC in patients with cirrhosis is recommended by almost all liver societies and is the standard of care, the data supporting its effectiveness are limited. Furthermore, there are no studies of HCC surveillance in patients with precirrhotic liver disease, the subject of this chapter, other than theoretical modeling studies.

Estimating Hepatocellular Carcinoma Risk in Patients with Precirrhotic Chronic Liver Disease to Identify Patients Who May Benefit from HCC Screening

It is likely that patients with precirrhotic chronic liver disease who have an estimated annual HCC risk greater than 1% might also benefit from HCC screening. Can we accurately estimate HCC risk in patients without clear evidence of cirrhosis to identify those who may benefit from screening? A number of strategies have been considered.

Multivariable Hepatocellular Carcinoma Risk Calculators

Multivariable prediction tools have been developed in patients with hepatitis C virus (HCV) infection to estimate the risk of HCC after antiviral treatment. For example, a tool that uses 12 routinely available predictors estimated 3-year HCC risk after antiviral treatment using a multivariable Cox proportional hazards model in patients with and those without cirrhosis. These predictors are SVR (yes/no), age, sex, body mass index, race/ethnicity, HCV genotype, platelet count, aspartate transaminase (AST), alanine transaminase (ALT), albumin, international normalized ratio, and hemoglobin. This tool probably identifies patients with advanced fibrosis as well as other predictors of HCC. These models show that many patients without cirrhosis who fail to achieve SVR can have a very high annual HCC risk (well above 1%) and need to continue HCC surveillance. Given the very high SVR rates of current direct-acting antiviral regimens, this, fortunately, represents only a small absolute number of patients. These models also show that among noncirrhotic patients who achieve SVR, a small proportion can have an annual HCC risk greater than 1%, if they simultaneously have a number of HCC risk factors, such as low platelet count, low albumin, high AST/ALT ratio, male sex, advanced age and genotype 3 HCV infection before eradication. These HCC risk calculators are available as web-based tools at www.hccrisk.com and can be used to calculate HCC risk in individual patients in order to determine whether HCC surveillance is warranted. These tools were derived using large populations of patients undergoing antiviral treatment in the Veterans Affairs health care system and have not yet been validated in nonveteran populations.

Deep Learning Hepatocellular Carcinoma Prediction Models

In addition to "conventional" linear models, deep learning, machine-learning models are also being proposed for HCC risk prediction. Deep learning models are generally expected to outperform linear models in accuracy of prediction, but this usually occurs at the expense of sacrificing interpretability or risking substantial overfitting. Furthermore, deep learning models are much harder to integrate into electronic health records so that they can be implemented in real-time clinical practice. However, incorporation of deep learning prediction and diagnostic algorithms into clinical practice is currently an area of tremendous growth, promise, and innovation and it is likely that many such algorithms will become available in the next 10 years.

Simplified Hepatocellular Carcinoma Scoring Systems: Fibrosis-4 Index

The FIB-4 score may represent a simplified method to estimate HCC risk. The FIB-4 score is calculated using a simple formula $[(AST \times age)/(platelet\ count \times \sqrt{ALT})]$ and was originally developed as a noninvasive biomarker panel of fibrosis stage. However, a high FIB-4 score greater than or equal to 3.25 also appears to be a particularly strong predictor of HCC risk, both in patients with and without cirrhosis. Studies have shown that the FIB-4 score is a readily available marker of HCC risk in patients with HCV-related cirrhosis, in viremic patients with HCV, in patients with MASLD with or without cirrhosis, and in patients with chronic hepatitis B [2, 10–13].

For example, among HCV-infected patients who achieved SVR after antiviral treatment, a FIB-4 score greater than or equal to 3.25 vs less than 3.25 could accurately identify high-risk and low-risk patients, respectively [14]. Patients without cirrhosis but with a FIB-4 score greater than or equal to 3.25 had an annual HCC risk of 1.22% during long-term follow-up after SVR, which would suggest they need to continue HCC surveillance. In contrast, patients without cirrhosis with a FIB-4 score less than 3.25 had a much lower annual HCC risk of xx% after SVR, suggesting no need for ongoing surveillance.

Furthermore, the change in FIB-4 score after SVR could be used to further fine-tune HCC risk. Specifically, patients whose FIB-4 score was greater than or equal to 3.25 both before and after SVR had a very high risk (~ 2% per year) whereas those whose FIB-4 dropped from greater than or equal to 3.25 before SVR to less than 3.25 after SVR had a much lower HCC risk. Taken together, these findings suggest that the subgroup of noncirrhotic patients who continue to have a FIB-4 score greater than 3.25 both before and after SVR have an annual HCC risk that well exceeds 1% and need to continue HCC surveillance, even if they do not carry a diagnosis of cirrhosis.

Liver Elastography (Liver Stiffness)

FibroScan-derived liver stiffness (LS) is a noninvasive test that correlates well with the presence of advanced liver fibrosis and clinically significant portal hypertension in patients with cirrhosis. Multiple studies also show that LS is also independently associated with HCC risk, both in patients with known cirrhosis and in patients with precirrhotic liver disease. In patients with precirrhotic liver disease, elevated LS identifies those with advanced fibrosis who may have an elevated HCC risk. For example, in a prospective study of HCV-infected patients, annual HCC risk was 0.11% in patients with LS less than or equal to 10 KPa, 2.9% in patients with LS of 10–15 KPa, 5% in patients with LS of 15–20 KPa, 8.3% in patients with LS of 20–25 KPa, and 14.4% in patients with LS greater than 25 KPa [15]. Although FibroScan, which measures vibration-controlled transient elastography, is currently the most commonly used method of hepatic elastography, many other methods are also available, including shear-wave elastography and magnetic resonance elastography. Exactly how elastography should be combined with other even more readily available measures (e.g., FIB-4 or multivariable risk calculators) to improve HCC risk estimation remains to be determined.

Misclassification of Cirrhosis, Advanced Fibrosis, and the Fibrosis–Hepatocellular Carcinoma Risk Continuum

An important confounding factor in determining HCC risk in patients with vs without cirrhosis is that it is not always clear whether patients have cirrhosis, especially when they are in the transition zone between advanced fibrosis or cirrhosis. Although liver biopsy is considered the gold standard, it is subject to very substantial intrahepatic variability and sampling error since only 1/50000th of the liver is sampled. [16, 17] Understaging of cirrhosis is not uncommon especially in small or fragmented liver biopsies. Furthermore, the proportion of patients who undergo liver biopsy has diminished dramatically in the last few years because of the development of noninvasive methods of fibrosis assessment. A more fundamental issue is that the extent of fibrosis is in reality a continuous measure that we dichotomize into cirrhosis and noncirrhosis for convenience and terminology. It is inevitable that patients who are "just below" the threshold for cirrhosis will have very similar HCC risk to those who are just above. Therefore, patients with advanced, F3 fibrosis may attain a sufficiently high HCC risk, especially if they have an underlying liver disease that continues to progress (e.g., untreated HCV infection).

Considerations in Special Patient Populations

The following special populations are addressed in more detail in other chapters and summarized briefly here.

Hepatocellular Carcinoma Risk After Hepatitis C Virus Eradication

Patients with HCV-related cirrhosis have a very high HCC risk, almost two to three times higher than the risk in patients with other common etiologies of cirrhosis even after adjusting for potential confounders [11]. Eradication of HCV has been shown to reduce the risk of HCC and may also cause regression of fibrosis and even early cirrhosis [10, 18–21]. This raises the question of whether HCC risk may decline sufficiently in some patients with cirrhosis after SVR to such a degree that HCC surveillance is no longer warranted. Unfortunately, a recent study showed that HCC risk remained high for many years after SVR in patients with pre-SVR cirrhosis, and even in patients without pre-SVR cirrhosis who had a high FIB-4 score greater than or equal to 3.25. It is possible that HCV-related preneoplastic changes that occur before SVR may persist "indefinitely" after SVR, [22] predisposing a patient to develop HCC many years after SVR. Therefore, fibrosis regression may not correlate perfectly with HCC risk reduction after SVR. Unfortunately, this means that patients with cirrhosis and even patients without cirrhosis who have a FIB-4 score greater than or equal to 3.25 have to continue HCC surveillance "indefinitely" after SVR, until better markers of dynamic HCC risk are discovered.

Hepatocellular Carcinoma Risk in Patients with Precirrhotic Metabolic Dysfunction-Associated Steatohepatitis

Gawrieh et al. reported that metabolic dysfunction-associated steatohepatitis (MASH) was the most common underlying liver disease in noncirrhotic HCCs, accounting for 26.3% of such cases, while MASH accounted for only 13.4% of cirrhotic HCCs [1]. This suggests there is something unique about the pathogenesis of HCC in precirrhotic MASH. It will be imperative to determine in the future what specifically drives HCC in patients with noncirrhotic MASH. MASH is associated with obesity, diabetes, and hyperinsulinemia, all of which are independent risk factors for HCC. In addition, MASH-related dysbiosis of gut microbiota, changes in bile acids, genetic polymorphisms, and lipotoxicity might be implicated. However, it is critical to point out that the absolute risk of HCC in precirrhotic MASH still remains well below the threshold for justifying HCC surveillance.

Future Directions: Risk Stratification and Risk-Based Surveillance Strategies

Currently we have a "one-size-fits-all" surveillance strategy, whereby all patients with cirrhosis (or HBV) are recommended to undergo surveillance by the same relatively ineffective strategy (ultrasonography ± serum AFP) irrespective of their underlying risk of HCC. This is problematic because there is great variability in HCC risk in patients with cirrhosis or HBV ranging from as little as 0.1% per year to more than 5% per year [23–25]. Conversely, screening is not recommended in any patients with pr-cirrhotic liver disease (other than HBV). However, some patients with precirrhotic liver disease may actually have high enough HCC risk to merit surveillance.

Instead of this one-size-fits-all strategy, a more effective HCC surveillance strategy in the future might be risk-based surveillance. This implies that the individual HCC risk of a patient may be calculated and an appropriate surveillance strategy recommended according to HCC risk. The principle is that the comparative effectiveness or cost-effectiveness of a surveillance strategy depends on the patient's HCC risk.

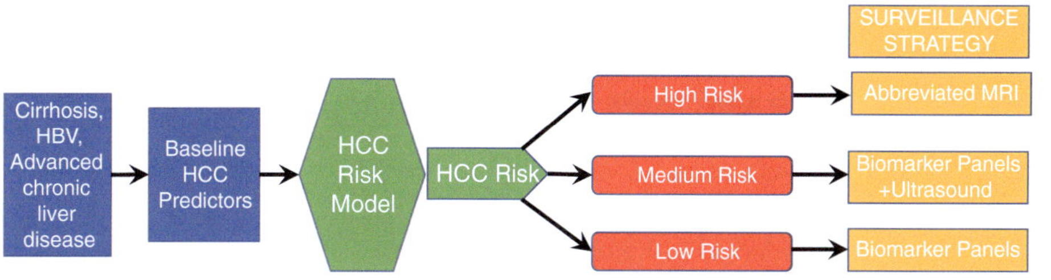

Fig. 4.1 Future directions in hepatocellular carcinoma risk stratification and risk-based surveillance. *HBV* hepatitis B virus, *HCC* hepatocellular carcinoma, *MRI* magnetic resonance imaging

More effective strategies that are also more expensive (such as screening by abbreviated magnetic resonance imaging) or more invasive/harmful, could be more cost-effective if they focus on the high-risk groups [8]. Conversely, patients at lower risk of HCC may undergo screening using only serum/plasma biomarker panels without any ultrasonography, while patients at intermediate risk may undergo screening with both ultrasound and biomarkers.

For risk-based surveillance to become a reality, HCC risk calculators and prediction models need to improve and be validated in diverse populations so we can reliably estimate HCC risk. Furthermore, the performance characteristics of different screening modalities (including novel imaging tests and biomarker panels) need to be described so they can be applied according to HCC risk, as shown in a hypothetical example in Fig. 4.1.

References

1. Gawrieh S, Dakhoul L, Miller E, et al. Characteristics, aetiologies and trends of hepatocellular carcinoma in patients without cirrhosis: a United States multicentre study. Aliment Pharmacol Ther. 2019;50(7):809–21. https://doi.org/10.1111/apt.15464.
2. Kanwal F, Kramer JR, Mapakshi S, et al. Risk of hepatocellular cancer in patients with non-alcoholic fatty liver disease. Gastroenterology. 2018;155(6):1828–1837.e2. https://doi.org/10.1053/j.gastro.2018.08.024.
3. Taylor EJ, Jones RL, Guthrie JA, Rowe IA. Modeling the benefits and harms of surveillance for hepatocellular carcinoma: information to support informed choices. Hepatology. 2017;66(5):1546–55. https://doi.org/10.1002/hep.29315.
4. Farhang Zangneh H, Wong WWL, Sander B, et al. Cost effectiveness of hepatocellular carcinoma surveillance after a sustained virologic response to therapy in patients with hepatitis C virus infection and advanced fibrosis. Clin Gastroenterol Hepatol. 2019;17(9):1840–1849.e16. https://doi.org/10.1016/j.cgh.2018.12.018.
5. Bruix J, Sherman M. American Association for the study of liver diseases: management of hepatocellular carcinoma: an update. Hepatology. 2011;53(3):1020–2. https://doi.org/10.1002/hep.24199.
6. Sarasin FP, Giostra E, Hadengue A. Cost-effectiveness of screening for detection of small hepatocellular carcinoma in Western patients with child-Pugh class a cirrhosis. Am J Med. 1996;101(4):422–34. https://doi.org/10.1016/S0002-9343(96)00197-0.
7. Arguedas MR, Chen VK, Eloubeidi MA, Fallon MB. Screening for hepatocellular carcinoma in patients with hepatitis C cirrhosis: a cost-utility analysis. Am J Gastroenterol. 2003;98(3):679–90. https://doi.org/10.1111/j.1572-0241.2003.07327.x.
8. Goossens N, Singal AG, King LY, et al. Cost-effectiveness of risk score-stratified hepatocellular carcinoma screening in patients with cirrhosis. Clin Transl Gastroenterol. 2017;8(6):e101. https://doi.org/10.1038/ctg.2017.26.
9. Heimbach JK, Kulik LM, Finn RS, et al. AASLD guidelines for the treatment of hepatocellular carcinoma. Hepatology. 2018;67(1):358–80. https://doi.org/10.1002/hep.29086.
10. Kanwal F, Kramer J, Asch SM, Chayanupatkul M, Cao Y, El-Serag HB. Risk of hepatocellular cancer in HCV patients treated with direct-acting antiviral agents. Gastroenterology. 2017;153(4):996–1005.e1. https://doi.org/10.1053/j.gastro.2017.06.012.
11. Ioannou GN, Green P, Lowy E, Mun EJ, Berry K. Differences in hepatocellular carcinoma risk, predictors and trends over time according to etiology of cirrhosis. PLoS One. 2018;13(9):e0204412. https://doi.org/10.1371/journal.pone.0204412.
12. Zou H, Yang X, Li QL, Zhou QX, Xiong L, Wen Y. A comparative study of albumin-bilirubin score with child-Pugh score, model for end-stage liver disease score and indocyanine green R15 in predicting posthepatectomy liver failure for hepatocellular carcinoma patients. Dig Dis. 2018;36(3):236–43. https://doi.org/10.1159/000486590.
13. Vergniol J, Foucher J, Terrebonne E, et al. Noninvasive tests for fibrosis and liver stiffness predict 5-year outcomes of patients with chronic hepatitis C. Gastroenterology. 2011;140(7):1970–9. https://doi.org/10.1053/j.gastro.2011.02.058.
14. Ioannou GN, Beste LA, Green PK, et al. Increased risk for hepatocellular carcinoma persists up to 10 years after HCV eradication in patients with baseline cirrhosis or high FIB-4 xcores. Gastroenterology. 2019;157(5):1264–1278.e4. https://doi.org/10.1053/j.gastro.2019.07.033.
15. Masuzaki R, Tateishi R, Yoshida H, et al. Prospective risk assessment for hepatocellular carcinoma development in patients with chronic hepatitis C by transient elastography. Hepatology. 2009;49(6):1954–61. https://doi.org/10.1002/hep.22870.
16. Goldstein NS, Hastah F, Galan MV, Gordon SC. Fibrosis heterogeneity in nonalcoholic steatohepatitis and hepatitis C virus needle core biopsy specimens. Am J Clin Pathol. 2005;123(3):382–7. https://doi.org/10.1309/EY72-F1EN-9XCB-1KXX.
17. Ratziu V, Charlotte F, Heurtier A, et al. Sampling variability of liver biopsy in nonalcoholic fatty liver disease. Gastroenterology. 2005;128(7):1898–906. https://doi.org/10.1053/j.gastro.2005.03.084.
18. Morgan RL, Baack B, Smith BD, Yartel A, Pitasi M, Falck-Ytter Y. Eradication of hepatitis C virus infection and the development of hepatocellular carcinoma: a meta-analysis of observational studies. Ann Intern Med. 2013;158(5 Pt 1):329–37. https://doi.org/10.7326/0003-4819-158-5-201303050-00005.
19. Singal AK, Singh A, Jaganmohan S, et al. Antiviral therapy reduces risk of hepatocellular carcinoma in patients with hepatitis C virus-related cirrhosis. Clin Gastroenterol Hepatol. 2010;8(2):192–9. https://doi.org/10.1016/j.cgh.2009.10.026.
20. Singer AW, Reddy KR, Telep LE, et al. Direct-acting antiviral treatment for hepatitis C virus infection and risk of incident liver

cancer: a retrospective cohort study. Aliment Pharmacol Ther. 2018;47(9):1278–87. https://doi.org/10.1111/apt.14593.
21. Ioannou GN, Green PK, Berry K. HCV eradication induced by direct-acting antiviral agents reduces the risk of hepatocellular carcinoma. J Hepatol. 2018;68:25–32. https://doi.org/10.1016/j.jhep.2017.08.030.
22. Paradis V, Dargere D, Bonvoust F, et al. Clonal analysis of micronodules in virus C-induced liver cirrhosis using laser capture microdissection (LCM) and HUMARA assay. Lab Investig. 2000;80(10):1553–9. https://doi.org/10.1038/labinvest.3780165.
23. Su F, Green P, Berry K, Ioannou GN. Estimating hepatocellular carcinoma risk in patients with chronic hepatitis B infection on antiviral therapy: a prediction model derived from a Veterans Affairs cohort. Abstract 2081. Presented at: The Liver Meeting. American Association for the Study of Liver Diseases; November 12, 2018. San Francisco, CA.
24. Ioannou GN, Green PK, Beste LA, Mun EJ, Kerr KF, Berry K. Development of models estimating the risk of hepatocellular carcinoma after antiviral treatment for hepatitis C. J Hepatol. 2018;69(5):1088–98. https://doi.org/10.1016/j.jhep.2018.07.024.
25. Ioannou GN, Green P, Kerr KF, Berry K. Models estimating risk of hepatocellular carcinoma in patients with alcohol or NAFLD-related cirrhosis for risk stratification. J Hepatol. 2019;71(3):523–33. https://doi.org/10.1016/j.jhep.2019.05.008.

Should Hepatocellular Carcinoma Surveillance Be Performed Only for Patients with Decompensated Cirrhosis If They Are Listed for Liver Transplantation?

5

Dharani Guttikonda and Sheila Eswaran

Hepatocellular carcinoma (HCC) surveillance can be used in select patients with decompensated cirrhosis who may be eligible for treatment if found to have early-stage HCC.

Main Points

1. Many therapies, such as surgical resection and systemic therapy, have been studied only in patients with compensated cirrhosis and have limited efficacy and poor safety in patients with decompensated cirrhosis.
2. There is heterogeneity in prognosis among patients with decompensated cirrhosis, with select patients with Child-Pugh score (CPS) class B cirrhosis having fair survival.
3. Increasing data suggest selective delivery of locoregional therapies may be tolerated and offer benefit to select patients with CPS class B cirrhosis. Real-world data similarly suggest the safety of systemic therapies in select patients with CPS class B cirrhosis.

Liver cancer is the fifth most incident cancer and the fourth leading cause of cancer-related deaths worldwide [1, 2]. HCC represents about 90% of primary liver cancers and a substantial global health challenge. In the United States, HCC is a rising cause of cancer death, owing in part to the obesity epidemic. As many as one-third of cirrhotic individuals will develop HCC, with 1–8% of patients diagnosed with HCC per year [3, 4]. There is a higher incidence of cancer that parallels the severity of portal pressure and liver stiffness [3, 5–10].

A primary driver of HCC prognosis is tumor stage at diagnosis, with a 44% 5-year survival for localized liver cancer compared to an 18% 5-year survival for cases diagnosed with distant-stage disease [11]. Curative options for HCC, such as liver transplant, can provide 5-year survival rates greater than 75% with liver transplantation when within T2 criteria [12]. It would stand to reason that patients at risk for HCC should undergo surveillance to detect cancer at an early stage to provide the longest survival [13].

However, there is a controversy regarding surveillance for decompensated patients with cirrhosis who are not transplant candidates [14]. European Association for the Study of the Liver 2018, American Association for the Study of Liver Diseases 2018, and Asian Pacific Association for the Study of the Liver 2017 guidelines all recommend surveillance for compensated cirrhotic patients at CPS classes A and B, but recommend screening in decompensated cirrhotic patients only if they are listed for orthotopic liver transplantation [3, 15, 16] (Table 5.1). Transition from compensated to decompensated cirrhosis can occur at a rate of 5–7%, making the implications of restrictive surveillance in these patients quite serious [17].

The changing landscape for HCC treatment, including palliative therapies that have a durable effect on survival, is compelling the medical community to rethink this restriction.

D. Guttikonda
Endeavor Health Glenview, Chicago, IL, USA
e-mail: RGuttikonda@northshore.org

S. Eswaran (✉)
Department of Medicine, Rush University Medical Center, Chicago, IL, USA
e-mail: Sheila_Eswaran@rush.edu

N. S. Reau, A. G. Singal (eds.), *Controversies in Hepatocellular Carcinoma*, https://doi.org/10.1007/978-3-031-90225-3_5

Table 5.1 Summary of recommended surveillance practice guidelines

Society	Publication y	Surveillance recommendation	Cirrhosis population
AASLD	2018	Ultrasound with or without AFP every 6 mo	CPS class A and B
EASL	2018	Ultrasound every 6 mo	CPS class C if listed for liver transplant
APASL	2017	Ultrasound and AFP every 6 mo	

Abbreviations: *AASLD* American Association for the study of liver diseases *AFP* alpha fetoprotein, *APASL* Asian Pacific Association for the study of the liver, *CPS* child-pugh score, *EASL* European Association for the study of the liver

Pro: Hepatocellular Carcinoma Surveillance Should Not Be Performed in Patients with Decompensated Cirrhosis Unless They Are Liver Transplant Candidates

Surveillance goals are aimed at decreasing HCC-related and overall liver-related deaths. Historically, surveillance is ineffective in decompensated cirrhotic patients who are not transplant candidates because poor liver function limits HCC treatment and adversely affects overall mortality. Accordingly, surveillance for HCC is not cost-effective in these patients.

The ITA.LI.CA group demonstrated this by identifying 140 CPS class C patients with HCC detected during surveillance vs those that did not undergo surveillance. Although cancer stage (P = 0.001) and treatment distribution (P = 0.021) were better in the group that underwent surveillance, median survival did not differ: 7.1 (2.1–12.1) vs 6.0 (4.1–7.9) months (P = 0.740) [14]. The authors concluded that early detection does not correlate with mortality benefit in decompensated cirrhotic patients with HCC.

Patients with hepatic decompensation and poor performance status have a median survival of 3 months or 11% at 1 year [18]. Although nontransplant therapies, such as resection, ablation, stereotactic body radiation therapy (SBRT), chemotherapy, and transarterial therapy, are options for HCC, individuals with decompensated cirrhosis often do not tolerate the treatments because of side effects and progressive liver failure, limiting the efficacy of therapy and subsequently, overall survival (OS).

For example, although HCC resection outcomes are favorable, with a 75% 5-year OS, patients with decompensated cirrhosis cannot tolerate resection [19]. Clinically, significant portal hypertension and an elevated serum bilirubin are associated with a much lower 25% 5-year survival.

Radiofrequency ablation has also been evaluated as a first-line therapy in early HCCs. In a series of 162 patients with cirrhosis, the OS and recurrence-free survival were 67.9% and 25.9%, respectively. A meta-analysis showed OS of 76% at 3 years for a single HCC smaller than 3 cm, with recurrence-free rates of 46%. However, 2 major predictors of poor OS in this group were CPS class B (compared to CPS class A) and the presence of portosystemic collaterals, which limit the efficacy for patients diagnosed with HCC detected in a surveillance program [20].

Although transarterial therapies like radioembolization and chemoembolization (TARE and TACE) offer palliative survival benefit in intermediate-stage cancer, these treatments are often detrimental in patients with poor performance status or severe hepatic decompensation (CPS class C or CPS class B decompensated cirrhosis). Inadequate hepatic function, such as serum bilirubin greater than 2 mg/dL, and a tumor burden of more than 50% of total liver volume, increase the risk of hepatic decompensation after TACE [21].

Current palliative systemic chemotherapy, although evolving, still has limited OS rates and may not be tolerated by CPS class C cirrhotic patients. Most chemotherapy clinical trials for unresectable HCC exclude enrollment of CPS class C patients, and therefore are inadequate to determine these outcomes.

Surveillance is considered cost-effective if it can be performed at a cost of less than $50,000 for quality-adjusted life-year gained. Surveillance with alpha fetoprotein (AFP) and ultrasound in patients with decompensated cirrhosis does not have a survival benefit, and is therefore deemed not to be cost-effective [14].

Con: Patients with Decompensated Cirrhosis Should Undergo Surveillance for Hepatocellular Carcinoma Whether or Not They Are Transplant Candidates

Surveillance for HCC, even in patients with decompensated cirrhosis who are not transplant candidates, still offers improvement in survival. Thanks to advances in early detection, chemotherapeutic options, interventional radiology techniques, and the surgical landscape, there are now broader options both for curative and palliative treatment beyond liver transplant alone. In fact, changes in liver transplant organ allocation nationwide are leading to a decrease in the transplant rate for HCC and should encourage the medical community to consider durable, innovative, nontransplant therapeutic options for patients with HCC in every stage.

The Barcelona Clinic Liver Cancer staging (Fig. 5.1) system is based on the size and number of tumors in addition to liver function, functional status, and cancer symptoms

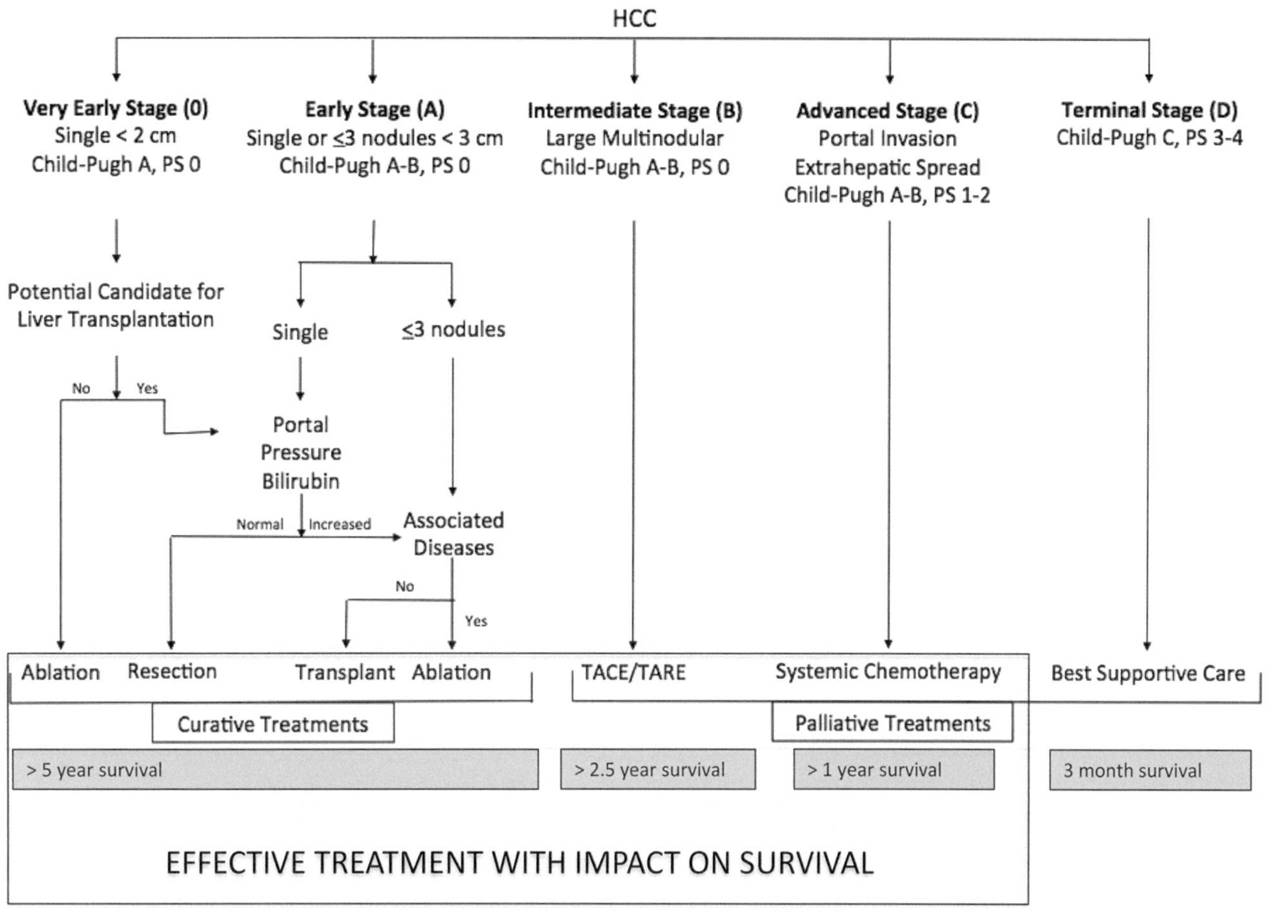

Fig. 5.1 Barcelona Clinic Liver Cancer Staging System. HCC, hepatocellular carcinoma; TACE, transarterial chemoembolization; TARE, transarterial radioembolization. (Adapted from Forner et al. [24])

[22–24]. Expected survival outcomes in each stage are affected by liver function, which is not static, and not all patients fulfill all the criteria for each stage definition. Owing to improved management of cirrhosis complications, such as hepatitis C treatment, liver function may improve over time. As intervention studies are designed, there has been increasing consideration of migration of stage and stratification of patients within each stage according to liver function (e.g., CPS class A vs B, presence of ascites, albumin-bilirubin score) and other prognostic molecular biomarkers, or evolutionary events (pattern of progression, development of side effects), all of which are nuances that may affect the quality of liver function and therefore affect expected survival.

Recent developments in the approach to HCC that have led to more aggressive treatment in patients with more advanced liver dysfunction include (1) more sensitive diagnostic tumor markers (des-γ-carboxyprothrombin [DCP] and lectin-bound AFP [AFP-L3]) that can identify "very small" HCCs (≤ 2 cm) that are curable with ablation even for patients with relatively advanced cirrhosis; (2) development in surgical approaches for patients with previously considered prohibitive features such as hepatic dysfunction and portal hypertension; (3) segmental or subsegmental TACE and TARE procedures that are safer and more effective; (4) biomarker-based SBRT; and (5) multiple systemic multikinase inhibitors and immunotherapy chemotherapy options [25].

Patients with historically prohibitive features such as hepatic dysfunction and portal hypertension may now be candidates for resection at experienced centers (i.e., postoperative mortality and severe postsurgical morbidity of < 3% and < 30%, respectively [26]) (Fig. 5.2). This has been demonstrated across multiple studies in patients with one or more factors that are traditionally associated with liver decompensation [27–30]. With improved liver functional assessment using liver stiffness, indocyanine green clearance testing, more precise definitions of tumor extent using advanced imaging and minimally invasive options, there have been great advances in the field of liver resection [31]. In some cases after surgical resection, in the absence of recurrence, a patient may become a transplant candidate, further demonstrating that HCC surveillance, even in a decompensated cirrhotic individual, can be of value.

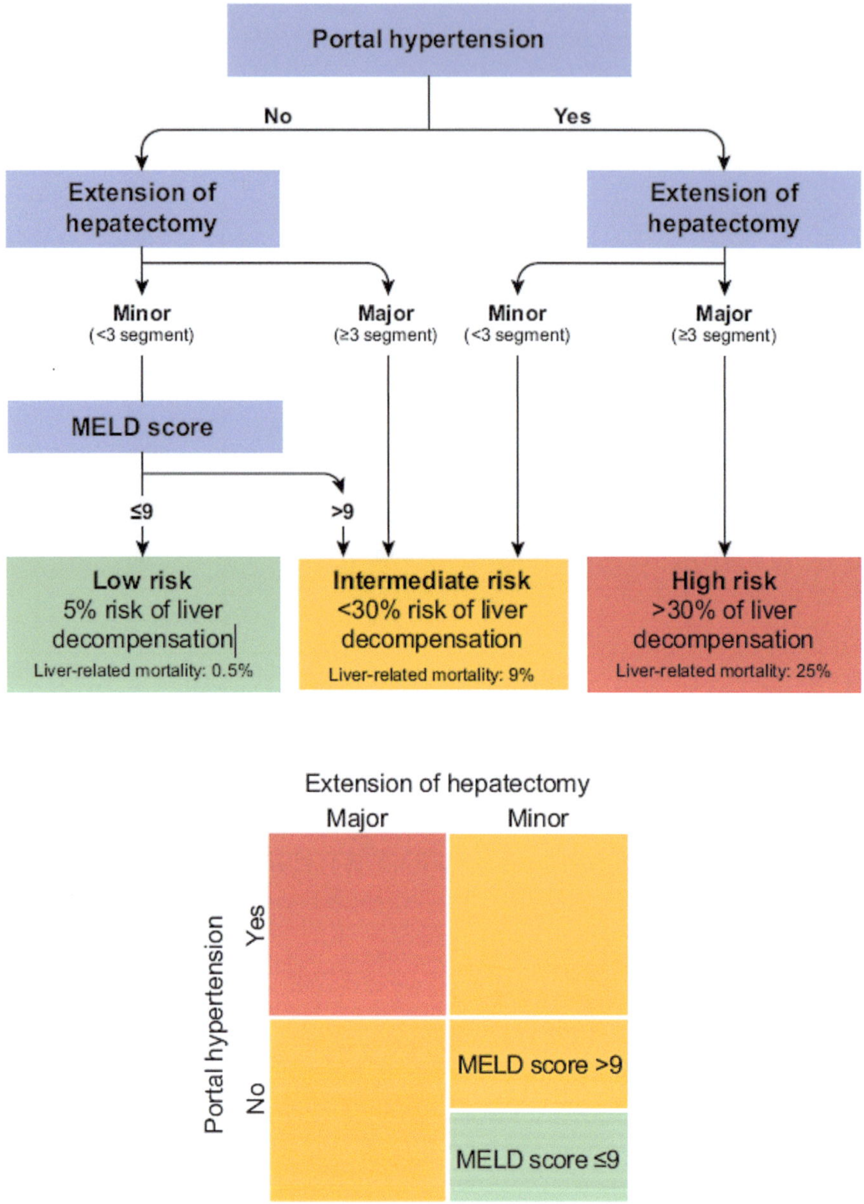

Fig. 5.2 Multiparametric assessment of the risk of decompensation based on 3 determinants of liver insufficiency. MELD, model for end-stage liver disease. (Used with permission [30])

The Liver Cancer Study Group of Japan aimed to answer the question of whether locoregional therapy was an option for HCC within a Milan Criteria in CPS class C cirrhotic patient (score < 11). They found that TACE and ablation both offered a survival benefit over best supportive care, and can be used as salvage therapy when transplantation is not an option [32]. In addition, cone-beam computed tomography imaging and superselective chemoembolization and radioembolization can be used to increase treatment efficacy and minimize the ischemic insult to nontumoral tissue for some patients with decompensated cirrhosis [33–35]. The outcomes of TACE in CPS class C patients and TARE in patients with elevated bilirubin greater than 3 have been evaluated with promising results [36, 37].

Biomarker-based, individualized SBRT as a treatment strategy for HCC has recently been evaluated. In a study published in *JAMA Oncology*, the authors evaluated the efficacy of SBRT in patients at higher risk for liver damage, and included 62 cirrhotic patients with a median CPS class B, average score 6. The group used change in indocyanine green clearance as a biomarker of liver function. The results were remarkable: a 99% rate of local control at 1 year and 95% at 2 years. Though further study is required, the tolerability and relative safety in CPS class B cirrhotic patients is promising enough to support surveillance of HCC in decompensated cirrhotic individuals [38]. Novel approaches like using biomarkers as a surrogate measure of liver function may allow for consideration of therapy, including SBRT, for patients previously not considered able to undergo HCC treatment.

In addition, multiple recent studies have demonstrated that chemotherapeutic agents may have a role in the treatment of HCCs with more advanced liver dysfunction. For example, CPS class B cirrhotic patients taking sorafenib had no difference in response rate, safety, and tolerability, although they did have a shorter OS compared to that in CPS class A cirrhotic patients. Although there was a difference in survival in the 2 groups, this appeared to be tied to functional status, hepatitis B positivity, and advanced stage of disease. For this reason, the authors of a meta-analysis advocate for HCC surveillance, arguing that OS may improve for HCC regardless of CPS status [39, 40]. The GIDEON study also determined that sorafenib had a similar safety profile across CPS class A and B cirrhosis, and could therefore be safely used in both populations. As expected, the group also found that CPS class A did have a longer median OS [41]. One small study of nivolumab demonstrated similar findings of comparable rates of adverse effects in CPS class A and B cirrhotic individuals with HCC, and with a median OS of 5.9 months (higher than that of sorafenib), although further studies of larger populations are necessary [42]. Because the original data looking into the use of chemotherapeutic agents such as sorafenib and nivolumab excluded CPS class C cirrhotic patients, further study into the use of both tyrosine kinase inhibitors and immunotherapy agents is needed in this population.

In addition, limiting HCC surveillance to transplant-eligible patients does not consider the fluidity of transplant candidacy. Some patients may not be a candidate for varying reasons such as insurance, social, and medical, which are inherently modifiable, and could therefore change with time. The median survival time of a decompensated cirrhotic patient without liver cancer is 2 years [43]. A provider can offer medical therapy for the complications of cirrhosis and HCC treatment options that can help to prolong survival long enough for the patient to modify factors to become a transplant candidate, acting as a bridge to curative therapy. The ability to creatively use these different tools and options in the care of all patients, including patients with decompensated cirrhosis, should encourage providers to consider HCC surveillance in all patients with cirrhosis.

Current guidelines advise screening only for compensated CPS class A and B cirrhotic patients, or decompensated cirrhotic individuals who are transplant candidates/listed for transplant. However, these guidelines do not account for the heterogeneity of patients within these classifications. For example, a decompensated CPS class B cirrhotic patient with ascites controlled by diuretics who otherwise has a good functional status and could tolerate HCC treatment is quite different from a decompensated CPS class B cirrhotic patient with ascites who requires weekly paracentesis and has a poorer functional status, making them less likely to tolerate treatment. Furthermore, the classification of decompensated cirrhosis can be modifiable as therapies are available to treat underlying liver disease (metabolic dysfunction-associated steatohepatitis, alcohol-related cirrhosis, hepatitis C virus [HCV]), and stage migration may occur as palliative treatments are becoming more effective and better tolerated. For example, a patient with HCV-decompensated cirrhosis may be treated for HCV with direct-acting antivirals, improve liver function, and tolerate a PD1 inhibitor to achieve progression-free survival with an adequate quality of life. Another scenario is to treat a patient with a chemotherapeutic agent, allowing them enough survival to complete an alcohol-cessation program that would thereby open up transplantation as a definitive therapeutic option. Such strategies would ultimately achieve the goal of surveillance: reducing HCC-related mortality and improving quality-adjusted life-years gained.

Conclusion

Historically the case for restricting HCC surveillance to patients with decompensated cirrhosis only if they were listed for transplant was based on the lack of benefit in HCC-related and overall liver-related deaths. However, with improvements in clinical care for patients with decompensated cirrhosis, early detection of HCC, advances in palliative and curative therapies for HCC, and the dedication of providers to help mitigate modifiable factors for transplant candidacy, the HCC treatment paradigm has shifted. Now, with effective treatment options for patients with decompensated cirrhosis that affect survival, we recommend screening compensated and select decompensated cirrhosis patients for HCC.

References

1. Akinyemiju T, Abera S, Ahmed M, Alam N, Alemayohu MA, Allen C, Global Burden of Disease Liver Cancer Collaboration. The burden of primary liver cancer and underlying etiologies from 1990 to 2015 at the global, regional, and national level: results from the Global Burden of Disease Study 2015. JAMA Oncol. 2017;3(12):1683–91. https://doi.org/10.1001/jamaoncol.2017.3055.
2. Bray F, Ferlay J, Soerjomataram I, Siegel RL, Torre LA, Jemal A. Global cancer statistics 2018: GLOBOCAN estimates of incidence and mortality worldwide for 36 cancers in 185 countries. CA Cancer J Clin. 2018;68(6):394–424. https://doi.org/10.3322/caac.21492.
3. European Association for the Study of the Liver. EASL clinical practice guidelines: management of hepatocellular carcinoma. J Hepatol. 2018;69(1):182–236. https://doi.org/10.1016/j.jhep.2018.03.019.
4. Ioannou GN, Splan MF, Weiss NS, McDonald GB, Beretta L, Lee SP. Incidence and predictors of hepatocellular carcinoma in patients with cirrhosis. Clin Gastroenterol Hepatol. 2007;5(8):938–45. https://doi.org/10.1016/j.cgh.2007.02.039.

5. Masuzaki R, Tateishi R, Yoshida H, et al. Risk assessment of hepatocellular carcinoma in chronic hepatitis C patients by transient elastography. J Clin Gastroenterol. 2008;42(7):839–43. https://doi.org/10.1097/mcg.0b013e318050074f.
6. Ripoll C, Groszmann RJ, Garcia-Tsao G, et al. Hepatic venous pressure gradient predicts development of hepatocellular carcinoma independently of severity of cirrhosis. J Hepatol. 2009;50(5):923–8. https://doi.org/10.1016/j.jhep.2009.01.014.
7. Jung KS, Kim SU, Ahn SH, et al. Risk assessment of hepatitis B virus-related hepatocellular carcinoma development using liver stiffness measurement (FibroScan). Hepatology. 2011;53(3):885–94. https://doi.org/10.1002/hep.24121.
8. Singh S, Fujii LL, Murad MH, et al. Liver stiffness is associated with risk of decompensation, liver cancer, and death in patients with chronic liver diseases: a systematic review and meta-analysis. Clin Gastroenterol Hepatol. 2013;11(12):1573–84. https://doi.org/10.1016/j.cgh.2013.07.034.
9. Wong GLH, Chan HLY, Wong CKY, et al. Liver stiffness-based optimization of hepatocellular carcinoma risk score in patients with chronic hepatitis B. J Hepatol. 2014;60(2):339–45. https://doi.org/10.1016/j.jhep.2013.09.029.
10. Kim MN, Kim SU, Kim BK, et al. Increased risk of hepatocellular carcinoma in chronic hepatitis B patients with transient elastography-defined subclinical cirrhosis. Hepatology. 2015;61(6):1851–9. https://doi.org/10.1002/hep.27735.
11. Cancer Stat Facts: Liver and Intrahepatic Bile Duct Cancer. In: NIH National Cancer Institute.
12. Kwong A, Kim WR, Lake JR, et al. OPTN/SRTR 2018 annual data report: liver. Am J Transplant. 2020;20(Suppl 1):193–299. https://doi.org/10.1111/ajt.15674.
13. Hanouneh IA, Alkhouri N, Singal AG. Hepatocellular carcinoma surveillance in the 21st century: saving lives or causing harm? Clin Mol Hepatol. 2019;25(3):264–9. https://doi.org/10.3350/cmh.2019.1001.
14. Trevisani F, Santi V, Gramenzi A, et al. Surveillance for early diagnosis of hepatocellular carcinoma: is it effective in intermediate/advanced cirrhosis? Am J Gastroenterol. 2007;102(11):2448–57. https://doi.org/10.1111/j.1572-0241.2007.01395.x.
15. Marrero JA, Kulik LM, Sirlin CB, et al. Diagnosis, staging, and management of hepatocellular carcinoma: 2018 practice guidance by the American Association for the study of liver diseases. Hepatology. 2018;68(2):723–50. https://doi.org/10.1002/hep.29913.
16. Omata M, Cheng AL, Kokudo N, et al. Asia-Pacific clinical practice guidelines on the management of hepatocellular carcinoma: a 2017 update. Hepatol Int. 2017;11(4):317–70. https://doi.org/10.1007/s12072-017-9799-9.
17. D'Amico G, Morabito A, Pagliaro L, Marubini E. Survival and prognostic indicators in compensated and decompensated cirrhosis. Dig Dis Sci. 1986;31(5):468–75. https://doi.org/10.1007/BF01320309.
18. Cabibbo G, Enea M, Attanasio M, Bruix J, Craxì A, Cammà C. A meta-analysis of survival rates of untreated patients in randomized clinical trials of hepatocellular carcinoma. Hepatology. 2010;51(4):1274–83. https://doi.org/10.1002/hep.23485.
19. Llovet JM, Fuster J, Bruix J. Intention-to-treat analysis of surgical treatment for early hepatocellular carcinoma: resection versus transplantation. Hepatology. 1999;30(6):1434–40. https://doi.org/10.1002/hep.510300629.
20. Lee DH, Lee JM, Lee JY, et al. Radiofrequency ablation of hepatocellular carcinoma as first-line treatment: long-term results and prognostic factors in 162 patients with cirrhosis. Radiology. 2014;270(3):900–9. https://doi.org/10.1148/radiol.13130940.
21. Raoul JL, Sangro B, Forner A, et al. Evolving strategies for the management of intermediate-stage hepatocellular carcinoma: available evidence and expert opinion on the use of transarterial chemoembolization. Cancer Treat Rev. 2011;37(3):212–20. https://doi.org/10.1016/j.ctrv.2010.07.006.
22. Befeler AS, Di Bisceglie AM. Hepatocellular carcinoma: diagnosis and treatment. Gastroenterology. 2002;122(6):1609–19. https://doi.org/10.1053/gast.2002.33411.
23. Pons F, Varela M, Llovet JM. Staging systems in hepatocellular carcinoma. HPB (Oxford). 2005;7(1):35–41. https://doi.org/10.1080/13651820410024058.
24. Forner A, Reig M, Bruix J. Hepatocellular carcinoma. Lancet. 2018;391(10127):1301–14. https://doi.org/10.1016/S0140-6736(18)30010-2.
25. Choi J, Kim GA, Han S, Lee W, Chun S, Lim YS. Longitudinal assessment of three serum biomarkers to detect very early-stage hepatocellular carcinoma. Hepatology. 2019;69(5):1983–94. https://doi.org/10.1002/hep.30233.
26. Citterio D, Facciorusso A, Sposito C, Rota R, Bhoori S, Mazzaferro V. Hierarchic interaction of factors associated with liver decompensation after resection for hepatocellular carcinoma. JAMA Surg. 2016;151(9):846–53. https://doi.org/10.1001/jamasurg.2016.1121.
27. Cucchetti A, Ercolani G, Vivarelli M, et al. Is portal hypertension a contraindication to hepatic resection? Ann Surg. 2009;250(6):922–8. https://doi.org/10.1097/SLA.0b013e3181b977a5.
28. Capussotti L, Ferrero A, Viganò L, Muratore A, Polastri R, Bouzari H. Portal hypertension: contraindication to liver surgery? World J Surg. 2006;30(6):992–9. https://doi.org/10.1007/s00268-005-0524-9.
29. Romagnoli R, Mazzaferro V, Bruix J. Surgical resection for hepatocellular carcinoma: moving from what can be done to what is worth doing. Hepatology. 2015;62(2):340–2. https://doi.org/10.1002/hep.27831.
30. Mazzaferro V, Lencioni R, Majno P. Early hepatocellular carcinoma on the procrustean bed of ablation, resection, and transplantation. Semin Liver Dis. 2014;34(4):415–26. https://doi.org/10.1055/s-0034-1394365.
31. Kudo M, Izumi N, Kokudo N, et al. HCC Expert Panel of Japan Society of Hepatology. Management of hepatocellular carcinoma in Japan: consensus-based clinical practice guidelines proposed by the Japan Society of Hepatology (JSH) 2010 updated version. Dig Dis. 2011;29(3):339–64. https://doi.org/10.1159/000327577.
32. Kitai S, Kudo M, Nishida N, et al. Survival benefit of locoregional treatment for hepatocellular carcinoma with advanced liver cirrhosis. Liver Cancer. 2016;5(3):175–89. https://doi.org/10.1159/000367765.
33. Vouche M, Habib A, Ward TJ, et al. Unresectable solitary hepatocellular carcinoma not amenable to radiofrequency ablation: multicenter radiology-pathology correlation and survival of radiation segmentectomy. Hepatology. 2014;60(1):192–201. https://doi.org/10.1002/hep.27057.
34. Lewandowski RJ, Gabr A, Abouchaleh N, et al. Radiation segmentectomy: potential curative therapy for early hepatocellular carcinoma. Radiology. 2018;287(3):1050–8. https://doi.org/10.1148/radiol.2018171768.
35. Miyayama S. Ultraselective conventional transarterial chemoembolization: when and how? Clin Mol Hepatol. 2019;25(4):344–53. https://doi.org/10.3350/cmh.2019.0016.
36. Choi TW, Kim HC, Lee JH, et al. The safety and clinical outcomes of chemoembolization in Child-Pugh class C patients with hepatocellular carcinomas. Korean J Radiol. 2015;16(6):1283–93. https://doi.org/10.3348/kjr.2015.16.6.1283.
37. Tasse J, Chauhan N, Ahmed O, Madassery S, Tabriz D, Arslan B. Abstract No. 439 Safety of Yttrium-90 radioembolization for the treatment of hepatocellular carcinoma in patients with hyperbilirubinemia. J Vasc Interv Radiol. 2019;30(Suppl):S192. https://doi.org/10.1016/j.jvir.2018.12.519.
38. Feng M, Suresh K, Schipper MJ, et al. Individualized adaptive stereotactic body radiotherapy for liver tumors in patients at

high risk for liver damage: a phase 2 clinical trial. JAMA Oncol. 2018;4(1):40–7. https://doi.org/10.1001/jamaoncol.2017.2303.
39. McNamara MG, Slagter AE, Nuttall C, et al. Sorafenib as first-line therapy in patients with advanced Child-Pugh B hepatocellular carcinoma – a meta-analysis. Eur J Cancer. 2018;105:1–9. https://doi.org/10.1016/j.ejca.2018.09.031.
40. Gellert L, Jalaludin B, Levy M. Hepatocellular carcinoma in Sydney South West: late symptomatic presentation and poor outcome for most. Intern Med J. 2007;37(8):516–22. https://doi.org/10.1111/j.1445-5994.2007.01392.x.
41. Marrero JA, Kudo M, Venook AP, et al. Observational registry of sorafenib use in clinical practice across Child-Pugh subgroups: the GIDEON study. J Hepatol. 2016;65(6):1140–7. https://doi.org/10.1016/j.jhep.2016.07.020.
42. Kambhampati S, Bauer KE, Bracci PM, et al. Nivolumab in patients with advanced hepatocellular carcinoma and Child-Pugh class B cirrhosis: safety and clinical outcomes in a retrospective case series. Cancer. 2019;125(18):3234–41. https://doi.org/10.1002/cncr.32206.
43. D'Amico G, Garcia-Tsao G, Pagliaro L. Natural history and prognostic indicators of survival in cirrhosis: a systematic review of 118 studies. J Hepatol. 2006;44(1):217–31. https://doi.org/10.1016/j.jhep.2005.10.013.

Should Statins and Aspirin Be Used for Hepatocellular Carcinoma Chemoprevention in Patients with Cirrhosis?

6

Ashwini Arvind and Tracey G. Simon

Aspirin and statins have been identified as potential agents for chemoprevention of hepatocellular carcinoma (HCC) in patients with chronic liver disease. Preclinical and observational studies are encouraging. Yet, protective associations have yet to be confirmed using randomized controlled trials (RCTs).

Main Points

1. Aspirin and statins are among the most commonly prescribed medications worldwide and have been identified as candidate chemopreventive agents against HCC.
2. Statins exert pleiotropic anti-inflammatory, antineoplastic, and immunomodulatory effects, and may inhibit HCC development through 3-hydroxy-3-methylglutaryl coenzyme A (HMG-CoA) reductase-dependent and independent pathways.
3. It has been posited that lipophilic statins may offer greater anti-HCC effects compared to hydrophilic statins.
4. Preclinical data suggest that aspirin may prevent HCC development via blockade of proinflammatory COX2.
5. Findings from large-scale population studies indicate a protective effect of aspirin against the development of gastrointestinal cancers, but its effect on HCC incidence is less clear.

Hepatocellular carcinoma (HCC) is the third leading contributor to cancer-related mortality worldwide [1] and is a major cause of death among patients with cirrhotic liver disease. While overall cancer mortality has downtrended in recent years, the incidence and mortality of HCC have progressively risen [2]. Mathematical modeling based on the SEER (Surveillance, Epidemiology, and End Results) registry has projected continued increases in HCC incidence until 2030 [3]. Screening strategies for patients at high risk remain suboptimal, and HCC is typically diagnosed at a late stage. Current treatment approaches yield poor survival benefit and are not cost-effective [4]: At present, the 5-year survival rate for incident HCC stands at less than 15%, [5] with 70% of patients experiencing tumor recurrence within 5 years [6]. Recurrent HCC is frequently refractory to treatment. Thus, given the growing burden and poor prognosis of HCC, there is an urgent need to introduce effective preventive strategies to avert HCC development.

Aspirin and statins are among the most commonly prescribed medications worldwide [7] and have been identified as candidate chemopreventive agents against HCC [8]. Findings from in vitro studies and animal models strongly indicate protective effects of aspirin and statins against the development and progression of HCC. Results from some observational studies have supported this association in humans. However, logistical challenges have thus far precluded any large-scale RCTs from being conducted. Further research is warranted before either aspirin or statins can be recommended as a chemopreventive standard in patients with chronic liver disease.

Chemopreventive Effects of Statins: Mechanistic Underpinnings

Statins are a class of cholesterol-lowering medication that competitively inhibit the 3-hydroxy-3-methylglutaryl coenzyme A (HMG-CoA) reductase enzyme, which regulates the rate-limiting step in the generation of mevalonate [9]. Statins are traditionally prescribed for primary and secondary prevention of cardiovascular disease. However, beyond their

A. Arvind
Division of Gastroenterology, University of Texas at Southwestern (UTSW) Medical Center, Dallas, TX, USA
e-mail: Damal.Arvind@UTSouthwestern.edu

T. G. Simon (✉)
Division of Gastroenterology, Massachusetts General Hospital, Harvard Medical School, Boston, MA, USA
e-mail: TGSIMON@MGH.HARVARD.EDU

N. S. Reau, A. G. Singal (eds.), *Controversies in Hepatocellular Carcinoma*, https://doi.org/10.1007/978-3-031-90225-3_6

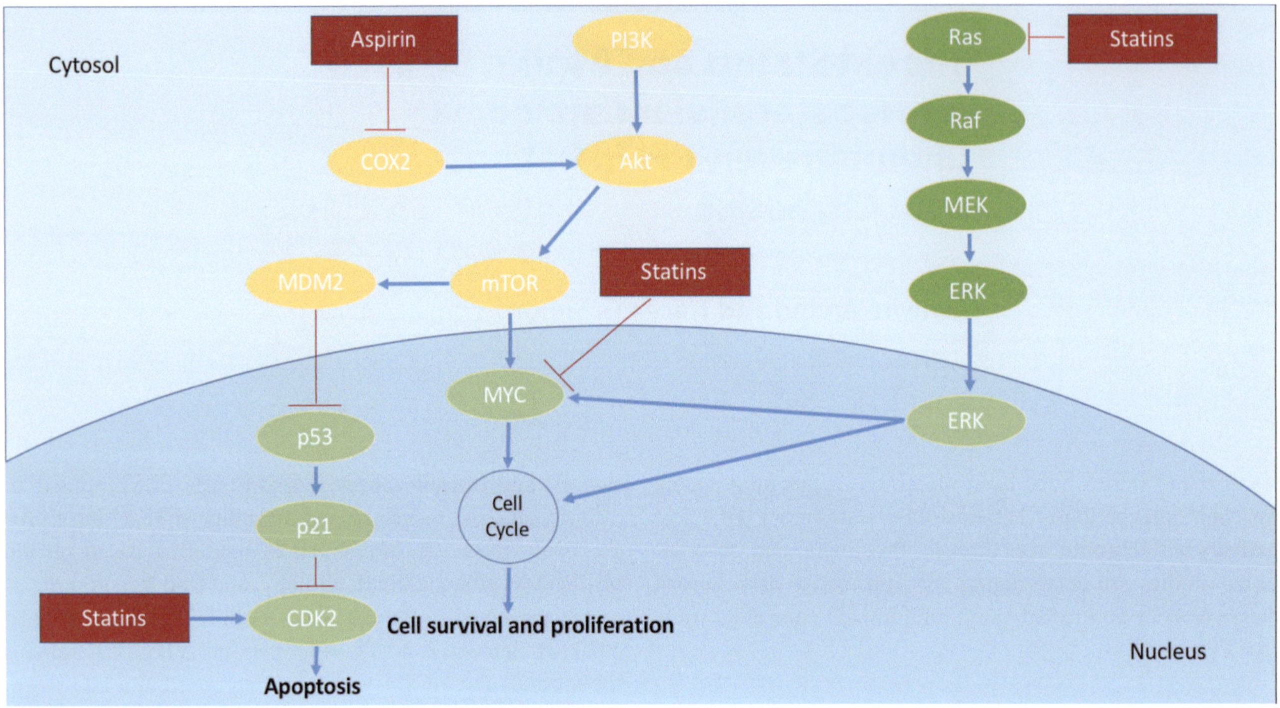

Fig. 6.1 Molecular targets for putative chemopreventive agents

effects on cholesterol biosynthesis, statins exert pleiotropic anti-inflammatory, antineoplastic, and immunomodulatory effects, and may inhibit HCC development through HMG-CoA reductase-dependent and independent pathways (Fig. 6.1). By curtailing the production of mevalonate, statins block downstream pathways, including posttranslational modification of Ras/Rho signaling proteins, which regulate cellular survival and growth. Statins promote apoptosis by regulating the Ras-Raf-MEK-ERK pathway, and they also inhibit proteasomal breakdown of cyclin-dependent kinase inhibitors p21 and p27, thus enabling these molecules to exert growth-inhibitory effects [10]. Further, atorvastatin-induced blockade of MYC phosphorylation and activation was found to inhibit tumor growth in a human HCC cell line, and this effect was reversed by mevalonate administration [11–13].

The potential chemopreventive benefits of statins may derive from their direct lipid-lowering actions. Cancer cells demonstrate upregulation of enzymes within the mevalonate pathway and augmented lipid biosynthesis, [14] suggesting these lipid compounds play important roles in cancer cell growth and proliferation. Hence, statins may impede tumor development by curbing the generation of mevalonate and its downstream products.

Statins also appear to exert robust antifibrotic effects that may further potentiate anti-HCC effects. Liver fibrosis is driven by the activation of hepatic stellate cells (HSCs), which undergo a phenotypic change from a quiescent state to become proliferative myofibroblasts. It has been hypothesized that statins exert antifibrotic effects in the liver in part through direct inhibition of proliferation and collagen production by HSCs [15–17]. Moreover, statins upregulate expression of transcription factors, such as Kruppel-like factor 2, that promote HSC quiescence [18]. Finally, the administration of statins has been shown to reduce pressures in the portal circulation, which may further inhibit liver fibrosis [19].

Chemopreventive Effects of Statins: Evidence in Humans

Although findings from preclinical studies suggest that statins inhibit HCC development, evidence in humans is conflicting, and support for the chemopreventive efficacy of statins mainly derives from observational studies. Results from recent meta-analyses [20–22] indicated that statins exert a protective effect against HCC development. This association was primarily seen in observational studies; findings from post hoc analysis of RCTs showed no statistically significant chemoprotective effect. However, these RCTs were designed to evaluate the effects of statin use on cardiovascular outcomes, and studied a population at low risk for HCC [20, 21]. Hence, they may have been underpowered to evaluate HCC incidence. Furthermore, participants were not screened to exclude HCC, giving rise to an appreciable risk of undetected cases of HCC with resultant misclassification bias and type 1 error.

More recent observational data support a role for statins in the prevention of incident HCC. Simon et al. [23] identified a dose-dependent association between statin use and lower incidences of cirrhosis and HCC in the setting of chronic HCV infection, within a prospective cohort of US veterans. Atorvastatin and fluvastatin showed the greatest statistical significance in impeding the development of hepatic fibrosis, an important precursor for HCC development. In a different cohort, the authors also found statin use to be associated with a statistically significantly reduced risk of further fibrosis progression in patients with advanced underlying fibrosis [24].

Despite those promising findings, observational data nevertheless remain mixed. In a dichotomized analysis involving a cohort of veterans with newly diagnosed cirrhosis, Kaplan et al. [25] found that statin exposure (defined as minimum 90-d usage) was associated with a statistically significant decrease in HCC incidence compared to nonexposure. When statin exposure was analyzed as a time-dependent covariate, cumulative statin use was associated with a trend toward reduced HCC. This suggests the effect of statin therapy on HCC risk to be dose-dependent, albeit more modest than reported in previous retrospective studies.

Collectively, these observational data are encouraging but must be interpreted with caution. The limitations inherent to these observational studies may lead to an overestimation of chemopreventive efficacy. Participants are not randomly assigned to either a statin use or nonuse group, leading to potential selection or indication bias. Indeed, the prevalence of key risk factors for HCC (such as hepatitis B virus [HBV] and hepatitis C virus [HCV] infection, alcohol-related liver disease, diabetes, and smoking status) was balanced between treatment groups in only a few prior studies, [25, 26] but not others [27–30]. Addressing confounding by indication is particularly important because providers may still be deterred from prescribing statins to patients with chronic liver disease, the subgroup at greatest risk for HCC development, because of a perceived risk of statin-induced hepatotoxicity. Furthermore, it is highly plausible that an appreciable number of statin users were prescribed other medications with putative anti-HCC effects, such as metformin, given the frequent co-occurrence of diabetes with hyperlipidemia and other cardiovascular risk factors. Metformin is known to possess anti-inflammatory and chemopreventive properties [31]; however, very few prior observational studies accounted carefully for potential confounding by antidiabetic therapy. Encouragingly, those that have addressed concomitant metformin use [24, 26, 28] have found that the observed protective effects of statins nevertheless persist. Finally, an added limitation of many prior studies is the failure to account for potential immortal time bias, which in turn can lead to a spurious protective effect of treatment.

Statins can be broadly divided into 2 subclasses: lipophilic (including atorvastatin, lovastatin, and simvastatin) and hydrophilic (including pravastatin, rosuvastatin, and fluvastatin). It has been posited that lipophilic statins may offer greater anti-HCC effects compared to hydrophilic statins. This hypothesis is supported by 4 key observations: (1) In preclinical models, lipophilic statins suppress viral replication, potentiate antiviral therapy, and stimulate antitumor immunity to a greater degree than hydrophilic statins [32–36]; (2) in the setting of chronic liver disease, reduced hepatocyte surface expression of organic ion transporters may preclude hydrophilic statins from accessing hepatocytes (in contrast, lipophilic statins enter hepatocytes via passive diffusion) [37, 38]; (3) lipophilic statins exert potent antitumor effects, including induction of G0/G1 cell cycle arrest and regulation of Ras/Raf/MEK/ERK signaling, [39] thereby driving apoptosis; and (4) treatment of HepG2 cells with simvastatin upregulates the proapoptotic *BAX* gene and suppresses the antiapoptotic *BCL-2* gene, suggesting that lipophilic statins operate at the pretranslational level to induce apoptosis [40]. Despite such supportive preclinical data, earlier population-based studies did not report differences in HCC risk reduction with lipophilic compared hydrophilic statins [21, 28]. However, more recently, Simon et al. [26] identified a statistically significantly lower risk of HCC among lipophilic statin users compared to statin nonusers in a propensity-matched cohort of adults in Sweden with viral hepatitis. In contrast, no association was found between hydrophilic statin users and HCC incidence. Hence, the lipophilic subclass of statins may have primarily contributed to previously reported inverse associations between statin use and decreased HCC incidence.

Current clinical evidence is not sufficiently robust to serve as a basis for recommending statins for HCC chemoprevention, and there is a crucial lack of data from RCTs designed to evaluate the effect of statin use on HCC risk. However, conducting an RCT would pose substantial logistical challenges, as it would require a large sample size and a long follow-up period, given the relatively low incidence of HCC and the established latency of hepatocarcinogenesis. Leveraging a cohort of patients at very high risk of HCC (e.g., Asian patients with HBV-related cirrhosis or patients with a history of resected HCC) would enrich the studied population and may represent a feasible strategy to overcome the risk of type 2 error. Importantly, it may also be prudent to reserve treatment for patients deemed at highest risk of HCC development. For example, the number of patients requiring regular statin treatment to prevent 1 case of HCC per year was estimated to be 5209 in the general East Asian male population, compared to 57 in East Asian men with chronic HBV-associated cirrhosis [20]—the latter may represent an appropriate subgroup to target. In addition, this mode of chemoprevention may be especially suitable in patients with metabolic dysfunction-associated steatotic liver disease fibrosis or cirrhosis, as this population frequently exhibits other features of metabolic syndrome, including dyslipidemia, for which statins may be indicated.

Chemopreventive Effects of Aspirin: Mechanistic Underpinnings

Preclinical data suggest that aspirin may prevent HCC development via blockade of proinflammatory COX2 [41–43]. Aspirin irreversibly inhibits both COX1 (which is constitutively expressed) and COX2, which is induced by proinflammatory stimuli, [44] resulting in decreased production of prostaglandins (see Fig. 6.1). This may in turn increase apoptosis, decrease tumor cell proliferation, and limit angiogenesis [45–47]. Activity of the PI3K kinase signaling pathway may be induced by COX2 expression and results in disruption of the mTOR (mammalian target of rapamycin) pathway, which promotes hepatocarcinogenesis. COX2 overexpression also upregulates protein kinase 3 and nuclear factor κ B signaling, which are associated with hepatocarcinogenesis, and drive inflammatory and cellular proliferation processes, respectively [41]. Aspirin has been shown to dampen protein kinase 3 and nuclear factor κ B signaling in a dose-dependent manner [48, 49]. COX2 overexpression is seen in inflammatory cancers, including HCC [41–43]. It predicts reduced disease-free survival in patients with HCC, while COX2 inhibition has been shown to limit proliferation and promote apoptosis of HCC tumor cells in vitro [42].

The inflammatory microenvironment is believed to nonspecifically promote neoplastic processes [50], and inflammatory mediators (such as COX2) may advance tumor formation. The chronic inflammatory state perpetuates a cycle of injury and regeneration, frequently resulting in fibrotic and cirrhotic changes [51]. The development of activating mutations in oncogenes and inactivating mutations in tumor-suppressor genes may culminate in HCC. Aspirin and statins both possess anti-inflammatory properties, which may enable them to suppress this process. In a study using the HBV transgenic mouse model, antiplatelet therapy attenuated inflammatory reactions, fibrotic changes in the liver, and HCC development [52]. Aspirin was also shown to prevent activation of HSCs by blocking the COX2 enzyme in rodent models [53]. Furthermore, inhibiting the activities of COX2 or prostaglandin E2 was found to promote fibrosis resolution [54, 55].

There is evidence that the effects of aspirin in limiting inflammation, fibrosis, and injury to the liver may be partly mediated by inhibition of platelet activity [52]. Platelets play a central role in promoting accumulation of $CD8^+$ T cells in the liver during chronic viral infection. They also generate platelet-derived growth factor, which activates HSCs and promotes fibrosis progression in rodent models [56]. Hence, aspirin-induced platelet inhibition is likely to impede fibrosis progression and the development of HCC.

Chemopreventive Effects of Aspirin: Evidence in Humans

Findings from large-scale population studies indicate a protective effect of aspirin against the development of gastrointestinal cancers, but its effect on HCC incidence is less clear. Preclinical findings indicate a protective effect of aspirin against HCC development, [43, 57] hence it seems biologically plausible that this association is observable in humans. Findings from population-based studies are broadly supportive (Table 6.1) [58–74]. Lee et al. [66] conducted a hospital-based retrospective analysis to characterize the relationship between aspirin use and HCC incidence. In propensity score-matched pairs, aspirin was found to statistically significantly reduce the risk of HCC in chronic HBV patients with suppressed viral loads. These findings were supported by the results of a Taiwanese nationwide cohort study that showed a reduced risk of HBV-related HCC in patients on daily aspirin therapy [59]. A national cohort study of Korean adults [61] found that aspirin's protective effect was amplified in patients who were also using nonaspirin nonsteroidal anti-inflammatory drugs (NSAIDs). Sahasrabuddhe et al. [72] conducted a large population-based study in the National Institutes of Health's Association of American Retired Persons Diet and Health Study cohort. They concluded that aspirin use was associated with a statistically significant reduction in HCC risk when compared to nonuse. A further population-based study by Petrick et al. [71] reported consistent findings. A recent meta-analysis of 5 studies reported a statistically significant association between aspirin use and decreased risk of HCC development [75]. Notably, the studies included in the meta-analysis did not consistently report rates of other HCC risk factors (including HBV and HCV infection status) among treated and untreated patients.

In a more recent prospective cohort study of patients with biopsy-confirmed metabolic dysfunction-associated steatotic liver disease, Simon et al. [76] identified a duration-dependent association between daily aspirin use and reduced incidence of hepatic fibrosis, an important precursor for HCC development. Patients with baseline F0 to F2 fibrosis also experienced lower rates of progression to advanced fibrosis. Aspirin use was also found to predict reduced incidence of HCC in 2 prospective cohorts of US patients with and without underlying cirrhosis [63]. This association became statistically significant after 5 or more years of use, with 1.5 or more standard-dose tablets per week. The benefits of aspirin use progressively diminished on stopping treatment and ceased to be evident 8 years after discontinuation.

Table 6.1 Association between aspirin use and development of hepatocellular carcinoma: evidence from observational studies

Study, author, y	Country	Study design	Mean age, y	No. of HCC cases	Aspirin group	Total No.	Exposure assessment	Definition of HCC	Etiology of liver disease	Effect estimate for HCC
Du [58]	China	Retrospective	45	41	59	264	Prescription	AFP and imaging	Viral hepatitis	OR (95% CI), 0.16 (0.04–0.71)
Lee [59]	Korea	Retrospective	58.8	63	343	10,615	Prescription	Billing code	Mixed etiology	HR 95% CI, 0.70 (0.58–0.86)
Simon [60]	Sweden	Prospective	54	1612	14,205	50,275	Prescription	Billing code; chart review	Viral hepatitis	HR 95% CI, 0.69 (0.62–0.76)
Hwang, [61]	Korea	Prospective	50	2336	64,782	460,755	Prescription	Billing code	Mixed etiology	HR 95% CI, 0.87 (0.77–0.98)
Tsoi [62]	Hong Kong	Retrospective	67.5	9370	204,170	612,509	Prescription	Billing code	Unknown	RR 95% CI, 0.49 (0.45–0.53)
Simon [63]	USA	Prospective	62	108	133,371	133,371	Self-reported	Biopsy or imaging	Mixed etiology	HR 95% CI, 0.51 (0.34–0.77)
Lin [64]	Taiwan	Retrospective	–	110	3576	18,234	Prescription	Billing code	Unknown	HR 95% CI, 0.75 (0.42–1.08)
Oh [65]	Korea	Retrospective	–	133	222	973	Prescription	–	Mixed etiology	HR 95% CI, 0.27 (0.09–0.76)
Lee [66]	Korea	Retrospective	55.2	63	343	1674	Prescription	AFP and imaging	Viral hepatitis	HR 95% CI, 0.34 (0.15–0.77)
Lee [67]	Taiwan	Retrospective	52.69	19	5602	5602	Prescription	Billing code	Unknown	HR 95% CI, 0.70 (0.37–1.36)
Kim [68]	Korea	Case control	–	54	321	1371	Prescription	Billing code	Unknown	OR 95% CI, 0.34 (0.15–0.78)
Tseng [69]	Taiwan	Retrospective	–	236	9574	9574	Prescription	Billing code	Mixed etiology	HR 95% CI, 0.83 (0.69–0.99)
Yang, [70]	UK	Case control	67	1195	376	7030	Prescription	Billing code	Mixed etiology	OR 95% CI, 1.11 (0.86–1.44)
Petrick [71]	USA	Prospective	–	679	477,470	1,084,133	Self-reported	Biopsy confirmed	Viral hepatitis	HR 95% CI, 0.68 (0.57–0.81
Sahasrabuddhe [72]	USA	Prospective	62.8	250	89,585	300,504	Self-reported	Billing code	Unknown	RR 95% CI, 0.51 (0.35–0.75)
Chiu [29]	Taiwan	Retrospective	66.1	81	81	1166	Prescription	Billing code	Mixed etiology	OR 95% CI, 1.0 (0.73–1.38)
Friis [73]	Denmark	Prospective	70	21	32,794	32,794	Prescription	Billing code	Unknown	SIR 95% CI, 1.0 (0.6–1.5)
Coogan [74]	USA	Retrospective	–	49	34	1149	Self-reported	Biopsy confirmed	–	NA

Adapted with permission from Zoe Memel

Abbreviations: *AFP* alpha fetoprotein, *HCC* hepatocellular carcinoma, *HR* hazard ratio, *NA* not available, *OR* odds ratio, *RR* relative risk, *SIR* standardized incidence ratio, *USA* United States of America

[a]The authors reported OR for nonaspirin users (compared to aspirin users) of 6.21 (95% CI, 1.41–27.32)

Gastrointestinal bleeding is a frequently encountered adverse effect of aspirin use. It is a particular concern in patients with chronic liver disease, who have a high likelihood of developing complications that predispose them to serious bleeding events, such as esophageal varices or portal hypertensive gastropathy [77]. It is reassuring to note that aspirin monotherapy was not associated with an increased risk of bleeding in a cohort of Korean patients with chronic HBV [66]. However, the risks vs potential benefits conferred by aspirin use must be evaluated on an individual patient basis. Additional research is necessary to further characterize the risks associated with aspirin use in cirrhotic patients and inform clinical decision making.

There have been conflicting reports regarding an association between nonaspirin NSAIDs and HCC development [61, 71, 72, 78]. Aspirin irreversibly blocks COX enzymes, whereas nonaspirin NSAIDs do so in a reversible manner, producing transient effects that may not substantially impede HCC development. Furthermore, nonaspirin NSAIDs disrupt the intestinal barrier, increasing the flux of proinflammatory cytokines to the liver and inducing a hepatic inflammatory response. A similar effect has not been observed with aspirin use [79]. Finally, aspirin may exert additional hepatoprotective effects via COX-independent pathways, for example, through enhancing biosynthesis of proresolving lipid mediators and suppressing the production of proinflammatory lipids [80].

Conclusion

Aspirin and statins have been identified as potential agents for chemoprevention of HCC in patients with chronic liver disease. Findings from preclinical and observational studies are generally encouraging. However, protective associations have yet to be confirmed using RCTs. Comprehensive risk-benefit analyses are necessary to evaluate whether potential chemoprotective effects conferred by aspirin and/or statins outweigh the adverse effects of medication use. The ultimate goal is to implement safe and effective preventive strategies that improve outcomes among patients with end-stage liver disease.

References

1. Bertuccio P, Turati F, Carioli G, et al. Global trends and predictions in hepatocellular carcinoma mortality. J Hepatol. 2017;67(2):302–9. https://doi.org/10.1016/j.jhep.2017.03.011.
2. Mokdad AH, Dwyer-Lindgren L, Fitzmaurice C, et al. Trends and patterns of disparities in cancer mortality among US counties, 1980-2014. JAMA. 2017;317(4):388–406. https://doi.org/10.1001/jama.2016.20324.
3. Petrick JL, Kelly SP, Altekruse SF, McGlynn KA, Rosenberg PS. Future of hepatocellular carcinoma incidence in the United States forecast through 2030. J Clin Oncol. 2016;34(15):1787–94. https://doi.org/10.1200/JCO.2015.64.7412.
4. Parikh ND, Singal AG, Hutton DW. Cost effectiveness of regorafenib as second-line therapy for patients with advanced hepatocellular carcinoma. Cancer. 2017;123(19):3725–31. https://doi.org/10.1002/cncr.30863.
5. Goossens N, Sun X, Hoshida Y. Molecular classification of hepatocellular carcinoma: potential therapeutic implications. Hepat Oncol. 2015;2(4):371–9. https://doi.org/10.2217/hep.15.26.
6. Nakagawa S, Wei L, Song WM, et al. Precision liver cancer prevention consortium. Molecular liver cancer prevention in cirrhosis by organ transcriptome analysis and lysophosphatidic acid pathway inhibition. Cancer Cell. 2016;30(6):879–90. https://doi.org/10.1016/j.ccell.2016.11.004.
7. Schuppan D, Afdhal NH. Liver cirrhosis. Lancet. 2008;371(9615):838–51. https://doi.org/10.1016/S0140-6736(08)60383-9.
8. Singh S, Singh PP, Roberts LR, Sanchez W. Chemopreventive strategies in hepatocellular carcinoma. Nat Rev Gastroenterol Hepatol. 2014;11(1):45–54. https://doi.org/10.1038/nrgastro.2013.143.
9. Demierre MF, Higgins PDR, Gruber SB, Hawk E, Lippman SM. Statins and cancer prevention. Nat Rev Cancer. 2005;5(12):930–42. https://doi.org/10.1038/nrc1751.
10. Rao S, Porter DC, Chen X, Herliczek T, Lowe M, Keyomarsi K. Lovastatin-mediated G1 arrest is through inhibition of the proteasome, independent of hydroxymethyl glutaryl-CoA reductase. Proc Natl Acad Sci USA. 1999;96(14):7797–802. https://doi.org/10.1073/pnas.96.14.7797.
11. Cao Z, Fan-Minogue H, Bellovin DI, et al. MYC phosphorylation, activation, and tumorigenic potential in hepatocellular carcinoma are regulated by HMG-CoA reductase. Cancer Res. 2011;71(6):2286–97. https://doi.org/10.1158/0008-5472.CAN-10-3367.
12. Farazi PA, DePinho RA. Hepatocellular carcinoma pathogenesis: from genes to environment. Nat Rev Cancer. 2006;6(9):674–87. https://doi.org/10.1038/nrc1934.
13. Shachaf CM, Kopelman AM, Arvanitis C, et al. MYC inactivation uncovers pluripotent differentiation and tumour dormancy in hepatocellular cancer. Nature. 2004;431(7012):1112–7. https://doi.org/10.1038/nature03043.
14. Freed-Pastor WA, Prives C. Mutant p53: one name, many proteins. Genes Dev. 2012;26(12):1268–86. https://doi.org/10.1101/gad.190678.112.
15. Trebicka J, Hennenberg M, Odenthal M, et al. Atorvastatin attenuates hepatic fibrosis in rats after bile duct ligation via decreased turnover of hepatic stellate cells. J Hepatol. 2010;53(4):702–12. https://doi.org/10.1016/j.jhep.2010.04.025.
16. Miyaki T, Nojiri S, Shinkai N, et al. Pitavastatin inhibits hepatic steatosis and fibrosis in non-alcoholic steatohepatitis model rats. Hepatol Res. 2011;41(4):375–85. https://doi.org/10.1111/j.1872-034X.2010.00769.x.
17. Wang W, Zhao C, Zhou J, Zhen Z, Wang Y, Shen C. Simvastatin ameliorates liver fibrosis via mediating nitric oxide synthase in rats with non-alcoholic steatohepatitis-related liver fibrosis. PLoS One. 2013;8(10):e76538. https://doi.org/10.1371/journal.pone.0076538.
18. Marrone G, Maeso-Díaz R, García-Cardena G, et al. KLF2 exerts antifibrotic and vasoprotective effects in cirrhotic rat livers: behind the molecular mechanisms of statins. Gut. 2015;64(9):1434–43. https://doi.org/10.1136/gutjnl-2014-308338.
19. Abraldes JG, Albillos A, Bañares R, et al. Simvastatin lowers portal pressure in patients with cirrhosis and portal hypertension: a randomized controlled trial. Gastroenterology. 2009;136(5):1651–8. https://doi.org/10.1053/j.gastro.2009.01.043.

20. Singh S, Singh PP, Singh AG, Murad MH, Sanchez W. Statins are associated with a reduced risk of hepatocellular cancer: a systematic review and meta-analysis. Gastroenterology. 2013;144(2):323–32. https://doi.org/10.1053/j.gastro.2012.10.005.
21. Shi M, Zheng H, Nie B, Gong W, Cui X. Statin use and risk of liver cancer: an update meta-analysis. BMJ Open. 2014;4(9):e005399. https://doi.org/10.1136/bmjopen-2014-005399.
22. Pradelli D, Soranna D, Scotti L, et al. Statins and primary liver cancer: a meta-analysis of observational studies. Eur J Cancer Prev. 2013;22(3):229–34. https://doi.org/10.1097/CEJ.0b013e328358761a.
23. Simon TG, Bonilla H, Yan P, Chung RT, Butt AA. Atorvastatin and fluvastatin are associated with dose-dependent reductions in cirrhosis and hepatocellular carcinoma, among patients with hepatitis C virus: results from ERCHIVES. Hepatology. 2016;64(1):47–57. https://doi.org/10.1002/hep.28506.
24. Simon TG, King LY, Zheng H, Chung RT. Statin use is associated with a reduced risk of fibrosis progression in chronic hepatitis C. J Hepatol. 2015;62(1):18–23. https://doi.org/10.1016/j.jhep.2014.08.013.
25. Kaplan DE, Serper MA, Mehta R, et al. Effects of hypercholesterolemia and statin exposure on survival in a large national cohort of patients with cirrhosis. Gastroenterology. 2019;156(6):1693–706. https://doi.org/10.1053/j.gastro.2019.01.026.
26. Simon TG, Duberg AS, Aleman S, et al. Lipophilic statins and risk for hepatocellular carcinoma and death in patients with chronic viral hepatitis: results from a nationwide Swedish population. Ann Intern Med. 2019;171(5):318–27. https://doi.org/10.7326/M18-2753.
27. El-Serag HB, Johnson ML, Hachem C, Morgana RO. Statins are associated with a reduced risk of hepatocellular carcinoma in a large cohort of patients with diabetes. Gastroenterology. 2009;136(5):1601–8. https://doi.org/10.1053/j.gastro.2009.01.053.
28. Tsan YT, Lee CH, Wang JD, Chen PC. Statins and the risk of hepatocellular carcinoma in patients with hepatitis B virus infection. J Clin Oncol. 2012;30(6):623–30. https://doi.org/10.1200/JCO.2011.36.0917.
29. Chiu HF, Ho SC, Chen CC, Yang CY. Statin use and the risk of liver cancer: a population-based case-control study. Am J Gastroenterol. 2011;106(5):894–8. https://doi.org/10.1038/ajg.2010.475.
30. Lonardo A, Loria P. Potential for statins in the chemoprevention and management of hepatocellular carcinoma. J Gastroenterol Hepatol. 2012;27(11):1654–64. https://doi.org/10.1111/j.1440-1746.2012.07232.x.
31. Kisseleva T, Brenner DA. Anti-fibrogenic strategies and the regression of fibrosis. Best Pract Res Clin Gastroenterol. 2011;25(2):305–17. https://doi.org/10.1016/j.bpg.2011.02.011.
32. Syed GH, Amako Y, Siddiqui A. Hepatitis C virus hijacks host lipid metabolism. Trends Endocrinol Metab. 2010;21(1):33–40. https://doi.org/10.1016/j.tem.2009.07.005.
33. Kapadia SB, Chisari FV. Hepatitis C virus RNA replication is regulated by host geranylgeranylation and fatty acids. Proc Natl Acad Sci USA. 2005;102(7):2561–6. https://doi.org/10.1073/pnas.0409834102.
34. Dorobantu C, Macovei A, Lazar C, Dwek RA, Zitzmann N, Branza-Nichita N. Cholesterol depletion of hepatoma cells impairs hepatitis B virus envelopment by altering the topology of the large envelope protein. J Virol. 2011;85(24):13373–83. https://doi.org/10.1128/JVI.05423-11.
35. Xia Y, Xie Y, Yu Z, et al. The mevalonate pathway is a druggable target for vaccine adjuvant discovery. Cell. 2018;175(4):1059–1073.e21. https://doi.org/10.1016/j.cell.2018.08.070.
36. Bader T, Korba B. Simvastatin potentiates the anti-hepatitis B virus activity of FDA-approved nucleoside analogue inhibitors in vitro. Antivir Res. 2010;86(3):241–5. https://doi.org/10.1016/j.antiviral.2010.02.325.
37. Kato S, Smalley S, Sadarangani A, et al. Lipophilic but not hydrophilic statins selectively induce cell death in gynaecological cancers expressing high levels of HMGCoA reductase. J Cell Mol Med. 2010;14(5):1180–93. https://doi.org/10.1111/j.1582-4934.2009.00771.x.
38. Thakkar N, Slizgi JR, Brouwer KLR. Effect of liver disease on hepatic transporter expression and function. J Pharm Sci. 2017;106(9):2282–94. https://doi.org/10.1016/j.xphs.2017.04.053.
39. Sutter AP, Maaser K, Höpfner M, Huether A, Schuppan D, Scherübl H. Cell cycle arrest and apoptosis induction in hepatocellular carcinoma cells by HMG-CoA reductase inhibitors. Synergistic antiproliferative action with ligands of the peripheral benzodiazepine receptor. J Hepatol. 2005;43(5):808–16. https://doi.org/10.1016/j.jhep.2005.04.010.
40. Spampanato C, De Maria S, Sarnataro M, et al. Simvastatin inhibits cancer cell growth by inducing apoptosis correlated to activation of Bax and down-regulation of BCL-2 gene expression. Int J Oncol. 2012;40(4):935–41. https://doi.org/10.3892/ijo.2011.1273.
41. Chen H, Cai W, Chu ESH, et al. Hepatic cyclooxygenase-2 overexpression induced spontaneous hepatocellular carcinoma formation in mice. Oncogene. 2017;36(31):4415–26. https://doi.org/10.1038/onc.2017.73.
42. Kern MA, Schubert D, Sahi D, et al. Proapoptotic and antiproliferative potential of selective cyclooxygenase-2 inhibitors in human liver tumor cells. Hepatology. 2002;36(4 Pt 1):885–94. https://doi.org/10.1053/jhep.2002.36125.
43. Foderà D, D'Alessandro N, Cusimano A, et al. Induction of apoptosis and inhibition of cell growth in human hepatocellular carcinoma cells by COX-2 inhibitors. Ann N Y Acad Sci. 2004;1028:440–9. https://doi.org/10.1196/annals.1322.052.
44. Cervello M, Montalto G. Cyclooxygenases in hepatocellular carcinoma. World J Gastroenterol. 2006;12(32):5113–21. https://doi.org/10.3748/wjg.v12.i32.5113.
45. Kim AK, Dziura J, Strazzabosco M. Nonsteroidal anti-inflammatory drug use, chronic liver disease, and hepatocellular carcinoma: the egg of Columbus or another illusion? Hepatology. 2013;58(2):819–21. https://doi.org/10.1002/hep.26498.
46. Hossain MA, Kim DH, Jang JY, et al. Aspirin enhances doxorubicin-induced apoptosis and reduces tumor growth in human hepatocellular carcinoma cells in vitro and in vivo. Int J Oncol. 2012;40(5):1636–42. https://doi.org/10.3892/ijo.2012.1359.
47. Fajardo AM, Piazza GA. Chemoprevention in gastrointestinal physiology and disease. Anti-inflammatory approaches for colorectal cancer chemoprevention. Am J Physiol Gastrointest Liver Physiol. 2015;309(2):G59–70. https://doi.org/10.1152/ajpgi.00101.2014.
48. Chan TA, Morin PJ, Vogelstein B, Kinzler KW. Mechanisms underlying nonsteroidal antiinflammatory drug-mediated apoptosis. Proc Natl Acad Sci USA. 1998;95(2):681–6. https://doi.org/10.1073/pnas.95.2.681.
49. Leng J, Han C, Demetris AJ, Michalopoulos GK, Wu T. Cyclooxygenase-2 promotes hepatocellular carcinoma cell growth through Akt activation: evidence for Akt inhibition in celecoxib-induced apoptosis. Hepatology. 2003;38(3):756–68. https://doi.org/10.1053/jhep.2003.50380.
50. Weber A, Boege Y, Reisinger F, Heikenwälder M. Chronic liver inflammation and hepatocellular carcinoma: persistence matters. Swiss Med Wkly. 2011;141:w13197. https://doi.org/10.4414/smw.2011.13197.
51. Berasain C, Castillo J, Perugorria MJ, Latasa MU, Prieto J, Avila MA. Inflammation and liver cancer: new molecular links. Ann N Y Acad Sci. 2009;1155:206–21. https://doi.org/10.1111/j.1749-6632.2009.03704.x.
52. Sitia G, Aiolfi R, Di Lucia P, et al. Antiplatelet therapy prevents hepatocellular carcinoma and improves survival in a mouse model of chronic hepatitis B. Proc Natl Acad Sci

USA. 2012;109(32):E2165–72. https://doi.org/10.1073/pnas.1209182109.
53. Paik YH, Kim JK, Lee JI, et al. Celecoxib induces hepatic stellate cell apoptosis through inhibition of Akt activation and suppresses hepatic fibrosis in rats. Gut. 2009;58(11):1517–27. https://doi.org/10.1136/gut.2008.157420.
54. Kanai S, Ishihara K, Kawashita E, et al. ASB14780, an orally active inhibitor of group IVA phospholipase A2, is a pharmacotherapeutic candidate for nonalcoholic fatty liver disease. J Pharmacol Exp Ther. 2016;356(3):604–14. https://doi.org/10.1124/jpet.115.229906.
55. Hu X, Cifarelli V, Sun S, Kuda O, Abumrad NA, Su X. Major role of adipocyte prostaglandin E2 in lipolysis-induced macrophage recruitment. J Lipid Res. 2016;57(4):663–73. https://doi.org/10.1194/jlr.M066530.
56. Yoshida S, Ikenaga N, Liu SB, et al. Extrahepatic platelet-derived growth factor-β, delivered by platelets, promotes activation of hepatic stellate cells and biliary fibrosis in mice. Gastroenterology. 2014;147(6):1378–92. https://doi.org/10.1053/j.gastro.2014.08.038.
57. Cervello M, Foderàa D, Florena AM, et al. Correlation between expression of cyclooxygenase-2 and the presence of inflammatory cells in human primary hepatocellular carcinoma: possible role in tumor promotion and angiogenesis. World J Gastroenterol. 2005;11(30):4638–43. https://doi.org/10.3748/wjg.v11.i30.4638.
58. Du ZQ, Zhao JZ, Dong J, et al. Effect of low-dose aspirin administration on long-term survival of cirrhotic patients after splenectomy: a retrospective single-center study. World J Gastroenterol. 2019;25(28):3798–807. https://doi.org/10.3748/wjg.v25.i28.3798.
59. Lee TY, Hsu YC, Tseng HC, et al. Association of daily aspirin therapy with risk of hepatocellular carcinoma in patients with chronic hepatitis B. JAMA Intern Med. 2019;179(5):633–40. https://doi.org/10.1001/jamainternmed.2018.8342.
60. Simon TG, Duberg AS, Aleman S, Chung RT, Ludvigsson J. Aspirin use is associated with reduced risk for incident hepatocellular carcinoma in patients with chronic viral hepatitis: results from a nationwide population. Gastroenterology. 2019;156(6 Suppl 1):S-1200. https://doi.org/10.1016/S0016-5085(19)39981-0.
61. Hwang IC, Chang J, Kim K, Park SM. Aspirin use and risk of hepatocellular carcinoma in a national cohort study of Korean adults. Sci Rep. 2018;8(1):4968. https://doi.org/10.1038/s41598-018-23343-0.
62. Tsoi KKF, Ho JMW, Chan FCH, Sung JJ. Long-term use of low-dose aspirin for cancer prevention: a 10-year population cohort study in Hong Kong. Int J Cancer. 2019;145(1):267–73. https://doi.org/10.1002/ijc.32083.
63. Simon TG, Ma Y, Ludvigsson JF, et al. Association between aspirin use and risk of hepatocellular carcinoma. JAMA Oncol. 2018;4(12):1683–90. https://doi.org/10.1001/jamaoncol.2018.4154.
64. Lin YS, Yeh CC, Huang SF, et al. Aspirin associated with risk reduction of secondary primary cancer for patients with head and neck cancer: a population-based analysis. PLoS One. 2018;13(8):e0199014. https://doi.org/10.1371/journal.pone.0199014.
65. Oh S, Shin S, Lee H, et al. Aspirin and the risk of hepatocellular carcinoma development in patients with compensated alcoholic cirrhosis. J Hepatol. 2017;66(1):S629–30. https://doi.org/10.1016/S0168-8278(17)31706-3.
66. Lee M, Chung GE, Lee JH, et al. Antiplatelet therapy and the risk of hepatocellular carcinoma in chronic hepatitis B patients on antiviral treatment. Hepatology. 2017;66(5):1556–69. https://doi.org/10.1002/hep.29318.
67. Lee TY, Wu JC, Yu SH, Lin JT, Wu MS, Wu CY. The occurrence of hepatocellular carcinoma in different risk stratifications of clinically noncirrhotic nonalcoholic fatty liver disease. Int J Cancer. 2017;141(7):1307–14. https://doi.org/10.1002/ijc.30784.
68. Kim G, Jang SY, Han E, et al. Effect of statin on hepatocellular carcinoma in patients with type 2 diabetes: a nationwide nested case-control study. Int J Cancer. 2017;140(4):798–806. https://doi.org/10.1002/ijc.30506.
69. Tseng CH. Metformin and risk of hepatocellular carcinoma in patients with type 2 diabetes. Liver Int. 2018;38(11):2018–27. https://doi.org/10.1111/liv.13872.
70. Yang B, Petrick JL, Chen J, et al. Associations of NSAID and paracetamol use with risk of primary liver cancer in the clinical practice research datalink. Cancer Epidemiol. 2016;43:105–11. https://doi.org/10.1016/j.canep.2016.06.009.
71. Petrick JL, Sahasrabuddhe VV, Chan AT, et al. NSAID use and risk of hepatocellular carcinoma and intrahepatic cholangiocarcinoma: the liver cancer pooling project. Cancer Prev Res (Phila). 2015;8(12):1156–62. https://doi.org/10.1158/1940-6207.CAPR-15-0126.
72. Sahasrabuddhe VV, Gunja MZ, Graubard BI, et al. Nonsteroidal anti-inflammatory drug use, chronic liver disease, and hepatocellular carcinoma. J Natl Cancer Inst. 2012;104(23):1808–14. https://doi.org/10.1093/jnci/djs452.
73. Friis S, Sørensen HT, McLaughlin JK, Johnsen SP, Blot WJ, Olsen JH. A population-based cohort study of the risk of colorectal and other cancers among users of low-dose aspirin. Br J Cancer. 2003;88(5):684–8. https://doi.org/10.1038/sj.bjc.6600760.
74. Coogan PF, Rosenberg L, Palmer JR, et al. Nonsteroidal anti-inflammatory drugs and risk of digestive cancers at sites other than the large bowel. Cancer Epidemiol Biomarkers Prev. 2000;9(1):119–23.
75. Tao Y, Li Y, Liu X, Deng Q, Yu Y, Yang Z. Nonsteroidal anti-inflammatory drugs, especially aspirin, are linked to lower risk and better survival of hepatocellular carcinoma: a meta-analysis. Cancer Manag Res. 2018;10:2695–709. https://doi.org/10.2147/CMAR.S167560.
76. Simon TG, Henson J, Osganian S, et al. Daily aspirin use associated with reduced risk for fibrosis progression in patients with nonalcoholic fatty liver disease. Clin Gastroenterol Hepatol. 2019;17(13):2776–84. https://doi.org/10.1016/j.cgh.2019.04.061.
77. Valkhoff VE, Sturkenboom MCJM, Kuipers EJ. Risk factors for gastrointestinal bleeding associated with low-dose aspirin. Best Pract Res Clin Gastroenterol. 2012;26(2):125–40. https://doi.org/10.1016/j.bpg.2012.01.011.
78. Pang Q, Jin H, Qu K, et al. The effects of nonsteroidal anti-inflammatory drugs in the incident and recurrent risk of hepatocellular carcinoma: a meta-analysis. Onco Targets Ther. 2017;10:4645–56. https://doi.org/10.2147/OTT.S143154.
79. Tugendreich S, Pearson CI, Sagartz J, Jarnagin K, Kolaja K. NSAID-induced acute phase response is due to increased intestinal permeability and characterized by early and consistent alterations in hepatic gene expression. Toxicol Pathol. 2006;34(2):168–79. https://doi.org/10.1080/01926230600611752.
80. Börgeson E, Johnson AMF, Lee YS, et al. Lipoxin A4 attenuates obesity-induced adipose inflammation and associated liver and kidney disease. Cell Metab. 2015;22(1):125–37. https://doi.org/10.1016/j.cmet.2015.05.003.

Part II

Liver Cancer Surveillance

7 Should a Randomized Controlled Trial of Hepatocellular Carcinoma Surveillance in Patients with Cirrhosis Be Performed?

Ian A. Rowe

Main Points

1. Cohort studies suggest HCC surveillance is associated with increased early HCC detection, curative treatment receipt, and improved survival among patients with cirrhosis; however, these studies have notable limitations including lead time and length time biases.
2. HCC surveillance is associated with risks of physical, psychological, and financial harms as well as risk of overdiagnosis.
3. Given uncertain benefits in patients with cirrhosis and risk of potential harms, there is sufficient equipoise to warrant a randomized clinical trial evaluating HCC surveillance.
4. High-quality data from a large randomized clinical trial evaluating HCC surveillance benefits and harms may increase surveillance use in clinical practice.
5. However, there are notable barriers to conducting such a trial including the required time scale to demonstrate a mortality benefit as well as concerns about acceptance by patients and providers.

The prevalence of liver disease is increasing in many Western countries largely because of an increase both in alcohol-associated liver disease and metabolic dysfunction-associated steatotic liver disease. Patients with liver disease are at risk of progressive liver fibrosis culminating in cirrhosis. Once cirrhosis has developed, individuals are at risk of complications both from progressive liver disease and liver failure and also from the development of HCC. The risk of each of these complications increases with the duration and severity of cirrhosis, and this dual mortality risk has important implications for the survival. Surveillance for HCC for individuals with cirrhosis is proposed to reduce the mortality in this patient group through early diagnosis of HCC and subsequent treatment with curative intent.

I. A. Rowe (✉)
Leeds Institute for Medical Research, University of Leeds, Leeds, England, UK

Leeds Liver Unit, Leeds Teaching Hospitals NHS Trust, Leeds, England, UK
e-mail: I.A.C.Rowe@leeds.ac.uk

Rationale for Surveillance

Individuals with cirrhosis are at risk of developing HCC at a rate of approximately 1% to 3% per year. The incidence of HCC varies between different liver disease groups, with patients with alcohol-associated liver disease having relatively low incidence and patients with cirrhosis due to viral hepatitis having the highest incidence. Historically, patients diagnosed with HCC by symptoms had a very poor prognosis, often measured in a short number of weeks. For patients in whom the HCC was diagnosed before symptoms, there was often treatment available and that treatment was considered to improve survival.

Surveillance is conducted according to recommended guidelines at 6-month intervals by liver ultrasonography with or without serum alpha-fetoprotein monitoring. The acceptance of surveillance for the development of HCC in patients with cirrhosis followed the publication of observational, modeling, and randomized studies supporting its use since the identification of early HCC allows the use of effective treatments that aim to cure the cancer and improve survival. The optimal path to improved survival with surveillance in cirrhosis would be the accurate identification of all HCCs smaller than 2 cm for which a minimally invasive technique such as ablation could be used to facilitate cure of cancer in patients with otherwise early cirrhosis and a low risk of death from liver failure. Unfortunately, that is all too infrequently the scenario in which surveillance diagnoses HCC.

This clear rationale, while supported by some evidence, is open to question. The studies that have been conducted in the observational setting have often only included patients with cirrhosis and HCC, not the population that is considered for

N. S. Reau, A. G. Singal (eds.), *Controversies in Hepatocellular Carcinoma*, https://doi.org/10.1007/978-3-031-90225-3_7

surveillance, that being patients with cirrhosis as a whole. While HCC is an important complication, cirrhosis is not universal and the majority of patients who have surveillance will not develop HCC in the long run. In addition, the randomized control trials that have been performed were developed in the East where the nature of the underlying liver disease is different, and where participants did not always have cirrhosis. The applicability of these trials therefore to Western practice where surveillance is conducted almost exclusively in patients with cirrhosis is questionable. Furthermore, these trials have been questioned based on inadequate adjustment for important aspects of trial design and inconsistencies of reporting [1, 2].

Current Provision of Surveillance

It is widely recognized that the provision of surveillance to patients with cirrhosis is often poor. While provision is improving in some areas, available pooled data suggest only a minority of eligible patients receive surveillance of any degree and even when it is done, it is often suboptimal. The barriers to providing high-quality surveillance are many. From the provider perspective these vary from poor provider knowledge to, perhaps most important in the context of this piece, doubts over its overall effectiveness [3]. Just as important, there are similar barriers from a patient perspective that might be informed by intervention [4].

There are additional issues around the tests that are used for surveillance. Ultrasound (US) has a limited sensitivity for small lesions with a sensitivity of approximately only 50% for early-stage HCC [5]. The target of surveillance is to identify cancers that are amenable to curative treatment, meaning they should be less than 2 cm in diameter at the time of diagnosis. The concurrent measurement of alpha fetoprotein increases this sensitivity somewhat but at the same time decreases the overall specificity of surveillance, meaning that the number of patients experiencing false-positive tests or "false alarms" is increased.

Benefits of Surveillance

Patients with cirrhosis who enter surveillance are at risk of both liver failure and liver cancer. These risks increase together and it can be challenging to determine the ultimate cause of death in patients who develop HCC. In that context, the principal benefit of surveillance in cirrhosis is best defined by an increase in overall survival (OS) of individuals who receive that intervention. This is not an end point that has been measured in observational studies in surveillance for HCC because these have most often been of a case-control design in patients with HCC, nor was this defined in the randomized trials of surveillance that have been conducted. Rather, these studies have looked at the relative risk of death of patients who have already developed HCC and, since the majority of patients enrolled in surveillance will not develop HCC, the risk reduction associated with surveillance defined in these studies will necessarily be an overestimate [6].

In a synthesis of observational studies reporting survival with surveillance in patients with HCC, the overall estimated odds ratio for increased survival with surveillance was 1.90 [7]. More important from the patients' perspective, in absolute terms this represented a difference in survival of 10%, increasing from 30% at 3 years to 40% at 3 years in those studies that considered the role of lead time bias in the patients who had received surveillance. This bias is an important consideration when evaluating nonrandomized studies in this area. Indeed, the lead time considered was often short and when extended in many studies the benefits of surveillance are lost, suggesting that this may be an important confounder of the observed apparent survival benefit.

Moon and colleagues [8] recently made an important contribution in this area. Using methods previously developed to identify the potential effects of population screening for breast, colorectal, and prostate cancer, they sought to understand the benefits of surveillance for HCC. These methods have face validity in that they accurately recapitulate the findings of RCTs of population screening for each of those disease sites. Important in the context of patients diagnosed with HCC, however, Moon et al. [8] found that surveillance as it was performed in the Veterans Affairs population was not effective at reducing mortality from HCC and by implication was therefore not effective at reducing mortality in the general population with cirrhosis.

In addition, two recent modeling studies have questioned the magnitude of the benefit that might be seen with surveillance for HCC in a population with cirrhosis. These studies sought to define the benefits of surveillance in patients with cirrhosis, and each found rather modest benefits in terms of improved OS. The first of these from Taylor and colleagues [9] evaluated a short (5-y) horizon. The proportion of patients in whom early diagnosis of HCC was associated with an OS benefit was approximately 1%. The second, from Yang and colleagues, [10] indicated a similar survival benefit over an extended time horizon; 15 years after the start of surveillance, 3.5% of patients had a survival benefit and mean OS increased from 10.4 to 11.2 years.

In summary, the evidence that supports surveillance for HCC in patients with cirrhosis is weaker than is often acknowledged in guidelines. Recognizing this, and understanding that a survival benefit has not been clearly demonstrated in prospective studies of patients with cirrhosis who have not yet developed HCC, is an important first step in the journey toward a randomized trial of surveillance.

Risks of Surveillance

It is axiomatic that all screening or surveillance tests carry harms in addition to potential benefits. In the absence of an RCT of surveillance that is applicable to Western patients with cirrhosis, the potential harms related to surveillance in cirrhosis are poorly studied and understood. The idea of harms relating to US-based surveillance has generated increasing interest over the last few years. The harms that have been described share similarities with those seen in other screening programs. These relate to the performance characteristics of the surveillance test conducted, in this case liver US with or without alpha fetoprotein. It is clear from the data described earlier that US-based surveillance is not fully sensitive for early HCC, meaning that some patients present with HCC beyond curative criteria despite surveillance. In perhaps the most relevant study, a randomized trial of surveillance intervals, [11] the proportion of patients diagnosed with HCC outside curative criteria was approximately 30% further, highlighting the relative insensitivity of US for the diagnosis of early HCC. At the same time a small but important number of individuals undergoing surveillance will experience a false-positive test or a false alarm. The consequence of this false alarm may range from intensification of US-based surveillance or more commonly will result in additional imaging with radiation and contrast exposure to liver biopsy, for which the imaging findings remain uncertain [12]. In previous diagnostic algorithms it was estimated that the number of individuals experiencing harm from surveillance, including the number of individuals having a liver biopsy that did not show HCC, was never greater than those benefiting from surveillance in terms of improved survival [9]. In addition, there are emerging data to indicate that phenomena such as overdiagnosis that are important in other population screening programs may also be important in surveillance for HCC in patients with cirrhosis [13, 14]. On balance this indicates there is potential for substantial harm to the population under surveillance where the benefits of that surveillance may be limited and applicable to only a relatively few individuals.

Arguments in Favor of a Randomized Controlled Trial of Surveillance in Cirrhosis

RCTs should be considered when there is uncertainty as to the overall clinical and cost-effectiveness of health care intervention. There is clear uncertainty as to the magnitude of the benefit (if any) of surveillance in individuals with cirrhosis, and there is the potential to cause significant harm. This suggests there is equipoise regarding surveillance in cirrhosis and that it is an ideal topic for randomized studies.

Moreover, the surveillance frequency for HCC in patients with cirrhosis is an important aspect of the care of these individuals as a major driver of hospital attendance with an associated caregiver and financial burden both to the individual and to the payer. In this context, and where there is uncertainty as to the benefits of surveillance, an RCT would provide clarity as to how surveillance might best be undertaken to the benefit of this increasing patient group in a clinically and cost-effective way.

Perhaps the most compelling argument in favor of an RCT is the limited availability of surveillance to many patients. It has been shown repeatedly that the uptake of surveillance is low, [15] which, if surveillance were shown to be effective, would mean that many patients with cirrhosis are being denied access to an intervention that is of benefit. The low rates of surveillance in pooled analyses have been shown to be below 50% of the potentially eligible population, and even where surveillance is conducted it is often not done well. There are deficiencies in the regular scheduling of surveillance investigations, recall policies, and urgent access to treatment, which further disadvantages patients who are enrolled in surveillance but not in formalized programs [3]. As the estimated benefits of surveillance may be small, a randomized trial would need to be large, including many patients who would need to be systematically followed in surveillance for many years. The surveillance intervention in such a trial would necessarily be protocolized and established within a formal program with downstream triggers for recall investigations and involvement of local multidisciplinary teams or tumor boards to determine treatment selection. A trial run in this way would provide access to high-quality surveillance for a large number of patients allocated to the intervention arm. Arguably this would provide effective surveillance to a larger number of patients than is currently the case with *ad hoc* provision-negating arguments that such a trial would be "unethical." Furthermore, a trial showing clinical and cost-effectiveness of surveillance would then support the widespread implementation of robust and formalized surveillance programs at a pace that, despite the long duration of any trial, exceeds the current slow progress toward complete coverage of the population with cirrhosis.

Finally, the natural history of cirrhosis is such that even after the detection of a small or early HCC, long-term survival is not predicted for many patients because of the competing risk of death from liver failure. It is therefore not clear that early diagnosis is sufficient to provide a lasting survival advantage to large proportions of the at-risk population. There are few large-scale studies to inform this natural history and the collateral information gained through a population-based, RCT of surveillance in cirrhosis will provide invaluable data on the effect both of HCC and liver failure in terms of the OS of patients with advanced liver disease.

Arguments Against a Randomized Controlled Trial of Surveillance in Cirrhosis

There are a large number of clinical guidelines from specialist societies concerning liver disease written by topic experts that endorse the role of surveillance for individuals who have cirrhosis. This expert consensus, which grades the evidence presented previously as high quality, provides a major barrier to a randomized trial of surveillance. This consensus is carried through to specialty care, where the rates of surveillance and cirrhosis are higher than in less-specialized settings [15]. A trial that would need to be large to recruit sufficient numbers of patients would require the involvement of specialty and topic experts, and it would need those individuals to be prepared for their patients to be randomly assigned to no surveillance. In the current environment where specialist societies recommend surveillance, it seems particularly challenging to overcome this barrier. Indeed, it has been said by guideline writers that, "failing to provide screening will result in greater harm than providing screening." [16] Against that backdrop, it would require a full-scale reassessment of the surveillance literature and an acceptance that there is considerable uncertainty regarding the benefits of surveillance among guideline writers. It should be emphasized that this evidence has been challenged previously without a change in the interpretation of the existing evidence [2].

The scale of a randomized trial of surveillance for HCC in cirrhosis has also been described as a barrier to the development of such a study, though this should be caused to reflect on the reasons that such a trial would need to be so large. Using modeling data from recent attempts to understand the overall benefits of surveillance, the number needed to screen to prevent 1 death for each 6-month US is estimated at 100 per year. Taking that figure forward to power an RCT suggests that such a trial would need approximately 20,000 randomly assigned participants with cirrhosis followed for 10 years to confirm that degree of benefit in a statistically rigorous way. Cirrhosis has and will continue to increase in incidence and prevalence over the coming years and can no longer be considered in the realm of a rare disease for which small and bespoke trials are the standard. Indeed, major steps forward in preventive cardiology have come through the use of so-called megatrials of agents that have limited effectiveness at the individual level. The benefits to the population, however, are sufficient to justify the investment in those trials and the information that is gained and this approach should be embraced for patients with liver disease.

The complexity of the patient population under study is also substantial. The competing risk of death from liver failure would require that patients enrolled in the study be balanced between trial arms with regard to liver disease etiology, severity of underlying cirrhosis as measured by the model for end-stage liver disease score or similar, and additionally by risk of development of HCC. Stratifying patients in this way would be challenging within centers to maintain adherence to trial interventions and to prevent crossover between the nonsurveillance and surveillance arms. From a trial delivery perspective, it would be easier to consider a cluster randomized design but this risks differences in the incidence of HCC and also the incidence of liver failure between arms if the clusters are not well matched.

An additional concern regarding such a trial taking a decade or longer to complete is that the intervention arm, in this case US-based surveillance, would no longer be considered the standard of care as new and improved diagnostics come to market and are implemented in the context of surveillance for HCC. We have seen over the course of the last few years an acceleration in the development of protein biomarkers, [17] often in combination, with the aim to improve the early diagnosis of HCC. It is foreseeable that these will come to be used in surveillance within the lifetime of a randomized study of US-based surveillance.

Last, it has often been said that patients would not participate in a randomized trial of surveillance when presented with all the information regarding its effectiveness. A frequently cited study indicated that less than 10% of patients approached to discuss a potential trial would not have consented to participate when they stood a 50% chance of being allocated to a no-surveillance intervention [18]. This is surprising where there is uncertainty regarding the benefits of surveillance and the harms of surveillance are real. It is important to note that in that study approximately one-third of patients reporting the decision said that accompanied the consent was biased toward surveillance, suggesting that the information provided was, at the very least, not presented in a way that recognized the considerable uncertainty in the evidence that supports surveillance.

Scenarios in Which a Trial of Surveillance Might Be Accepted

The development of novel diagnostics for early HCC suggests an opportunity to better understand the benefits of surveillance in patients who have cirrhosis. Accepting that the best way to provide care to our patients is through a rigorous evidence-based medicine approach would mean that a randomized trial of these diagnostics should be the first step in their path to use in the clinic. Developing large-scale trials now in which patients are offered balanced information as to the potential risks and harms of surveillance offers the chance for registration of patients into 1 of 3 arms of a study understanding that in the current environment a trial that includes a formal no-surveillance arm is unlikely to be acceptable. These arms are, in turn, current "standard of care" US-based surveillance, standard of care plus novel diagnostic, and an observational, no-intervention arm in which patients who choose not to have surveillance based on balanced information are enrolled. Such a design would permit some comparison of intervention vs no intervention. It seems unlikely that sufficient patients would enter the no-intervention arm to provide a group as large as those that would be allocated to the surveillance interventions, and in addition, there would be concerns that this group may not reflect those individuals who were included in the intervention arms. Despite this, there are important data that would accumulate from this patient group and collection of that information would inform future studies in HCC surveillance and cirrhosis more generally.

Conclusion

An RCT of surveillance is desirable since it would provide a clear estimate of the benefits of surveillance to individuals with cirrhosis and also an estimate of the harms that accrue. While a minority of patients with cirrhosis receive effective surveillance in current clinical practice, a necessarily large randomized trial would likely increase the number of patients who receive effective surveillance in the population and result in robust data to support the widespread implementation of formalized surveillance interventions. Despite this, in the current climate, it seems unlikely that a trial would receive sufficient support among topic experts without a major shift in the interpretation of the current evidence base.

References

1. Kansagara D, Papak J, Pasha AS, et al. Screening for hepatocellular carcinoma in chronic liver disease: a systematic review. Ann Intern Med. 2014;161(4):261–9. https://doi.org/10.7326/M14-0558.
2. Lederle FA, Pocha C. Screening for liver cancer: the rush to judgment. Ann Intern Med. 2012;156(5):387–9. https://doi.org/10.7326/0003-4819-156-5 201203060-00012.
3. Cross TJS, Villanueva A, Shetty S, et al. A national survey of the provision of ultrasound surveillance for the detection of hepatocellular carcinoma. Frontline Gastroenterol. 2016;7(2):82–9. https://doi.org/10.1136/flgastro-2015-100617.
4. Farvardin S, Patel J, Khambaty M, et al. Patient-reported barriers are associated with lower hepatocellular carcinoma surveillance rates in patients with cirrhosis. Hepatology. 2017;65(3):875–84. https://doi.org/10.1002/hep.28770.
5. Tzartzeva K, Obi J, Rich NE, et al. Surveillance imaging and alpha fetoprotein for early detection of hepatocellular carcinoma in patients with cirrhosis: a meta-analysis. Gastroenterology. 2018;154(6):1706–1718.e1. https://doi.org/10.1053/j.gastro.2018.01.064.
6. Roskilly A, Rowe IA. Surveillance for hepatocellular cancer. Clin Med (Lond). 2018;18(Suppl 2):s66–9. https://doi.org/10.7861/clinmedicine.18-2-s66.
7. Singal AG, Pillai A, Tiro J. Early detection, curative treatment, and survival rates for hepatocellular carcinoma surveillance in patients with cirrhosis: a meta-analysis. PLoS Med. 2014;11(4):e1001624. https://doi.org/10.1371/journal.pmed.1001624.
8. Moon AM, Weiss NS, Beste LA, et al. No association between screening for hepatocellular carcinoma and reduced cancer-related mortality in patients with cirrhosis. Gastroenterology. 2018;155(4):1128–1139.e6. https://doi.org/10.1053/j.gastro.2018.06.079.
9. Taylor EJ, Jones RL, Guthrie JA, Rowe IA. Modeling the benefits and harms of surveillance for hepatocellular carcinoma: information to support informed choices. Hepatology. 2017;66(5):1546–55. https://doi.org/10.1002/hep.29315.
10. Yang JD, Mannalithara A, Piscitello AJ, et al. Impact of surveillance for hepatocellular carcinoma on survival in patients with compensated cirrhosis. Hepatology. 2018;68(1):78–88. https://doi.org/10.1002/hep.29594.
11. Trinchet JC, Chaffaut C, Bourcier V, et al. Groupe d'Etude et de Traitement du Carcinome Hépatocellulaire (GRETCH). Ultrasonographic surveillance of hepatocellular carcinoma in cirrhosis: a randomized trial comparing 3- and 6-month periodicities. Hepatology. 2011;54(6):1987–97. https://doi.org/10.1002/hep.24545.
12. Atiq O, Tiro J, Yopp AC, et al. An assessment of benefits and harms of hepatocellular carcinoma surveillance in patients with cirrhosis. Hepatology. 2017;65(4):1196–205. https://doi.org/10.1002/hep.28895.
13. Rich NE, Parikh ND, Singal AG. Overdiagnosis: an understudied issue in hepatocellular carcinoma surveillance. Semin Liver Dis. 2017;37(4):296–304. https://doi.org/10.1055/s-0037-1608775.
14. Rich NE, John BV, Parikh ND, et al. Hepatocellular carcinoma demonstrates heterogeneous growth patterns in a multi-center cohort of patients with cirrhosis. Hepatology. 2020;72(5):1654–65. https://doi.org/10.1002/hep.31159.
15. Wolf E, Rich NE, Marrero JA, Parikh ND, Singal AG. Utilization of hepatocellular carcinoma surveillance in patients with cirrhosis: a systematic review and meta-analysis. Hepatology. 2021;73(2):713–25. https://doi.org/10.1002/hep.31309.
16. Sherman M, Bruix J, Porayko M, Tran T. AASLD practice guidelines committee. Screening for hepatocellular carcinoma: the rationale for the American Association for the Study of Liver Diseases recommendations. Hepatology. 2012;56(3):793–6. https://doi.org/10.1002/hep.25869.
17. Berhane S, Toyoda H, Tada T, et al. Role of the GALAD and BALAD-2 serologic models in diagnosis of hepatocellular carcinoma and prediction of survival in patients. Clin Gastroenterol Hepatol. 2016;14(6):875–886.e6. https://doi.org/10.1016/j.cgh.2015.12.042.
18. Poustchi H, Farrell GC, Strasser SI, Lee AU, McCaughan GW, George J. Feasibility of conducting a randomized control trial for liver cancer screening: is a randomized controlled trial for liver cancer screening feasible or still needed? Hepatology. 2011;54(6):1998–2004. https://doi.org/10.1002/hep.24581.

8 Should Contrast-Enhanced Computed Tomography and/or Magnetic Resonance Imaging Play a Routine Role in Hepatocellular Carcinoma Surveillance in Patients with Cirrhosis?

Jorge A. Marrero

Computed tomography (CT) or magnetic resonance imaging (MRI) should not be routinely used for surveillance because of costs, capacity issues, and unknown performance characteristics for hepatocellular carcinoma (HCC) surveillance.

Main Points

1. Surveillance using ultrasound (US) and alpha fetoprotein (AFP) has the highest level of supporting evidence, so this should be the preferred strategy.
2. CT imaging should not be used for surveillance for HCC given insufficient data demonstrating benefits compared to US and AFP as well as risks of radiation exposure and toxicity from the intravenous contrast.
3. Although studies with short-term follow-up have shown that MRI is more sensitive than US for early detection of HCC, these data require validation in Western patients with cirrhosis.
4. MRI can be considered in patients with inadequate US and patients on a transplant waiting list given the higher priority for liver transplantation if HCC is detected.
5. If paired with an accurate risk-stratification biomarker, MRI will likely be cost-effective in those at the highest risk of developing HCC.

Patients with cirrhosis of the liver are at the highest risk for developing HCC. It has been estimated that more than 90% of patients with HCC have underlying cirrhosis, and the annual incidence among cirrhosis has been reported to be 2% to 4% per 100,000 [1]. Since such a high-risk group exists, surveillance of patients with cirrhosis represents an important strategy to improve outcomes, mainly overall survival. An ideal surveillance test would be highly sensitive and specific for early HCC, and undetectable in premalignant liver disease. We will discuss surveillance for HCC and specifically, if CT or MRI can be used for this aspect.

Surveillance for Hepatocellular Carcinoma

Cancer surveillance is the continued application of a test over time with the goal of improving overall survival. One of the important aspects of whether a cancer surveillance program can be instituted is if a high-risk group can be readily identified, and cirrhosis is such a group for HCC. We are currently living in an era with an active change in the epidemiology of HCC. The rise in HCC incidence was fueled by chronic hepatitis B infection worldwide, while in the United States it was driven by chronic hepatitis C virus (HCV) infection [2]. However, the development of novel direct-acting antiviral therapies with sustained virological responses greater than 90% and to some extent the aging of the cohort of people infected with HCV 40 to 50 years ago has led to a statistically significant reduction in HCC incidence from 3.6% per year to 1.8% per year [3, 4]. Over the last 5 years, metabolic dysfunction-associated steatotic liver disease (MASLD) and alcoholic liver disease have replaced HCV infection as the most common causes of cirrhosis and of needing liver transplantation. Importantly, the annual HCC incidence rate among patients with cirrhosis from MASLD is between 0.5% and 1.8% per 100,000, which is similar to treated HCV infection [5]. The annual HCC incidence rate among patients with alcohol cirrhosis has been shown to be 1.8% [6]. Thus, there is a large patient population with cirrhosis from these common etiologies who require long-term follow-up and surveillance for HCC and, owing to the lower risk of developing this tumor, when compared to the late 1990s at the height of the HCV epidemic, will place a substantial burden on practicing

J. A. Marrero (✉)
Division of Gastroenterology and Hepatology, University of Pennsylvania, Philadelphia, PA, USA
e-mail: Jorge.Marrero@Pennmedicine.upenn.edu

N. S. Reau, A. G. Singal (eds.), *Controversies in Hepatocellular Carcinoma*, https://doi.org/10.1007/978-3-031-90225-3_8

physicians and health systems. Surveillance tests that are easily measurable, reproducible, and minimally invasive are going to be critically important at this time.

A source of controversy is whether patients without cirrhosis should undergo surveillance for HCC. It is well known that patients with chronic HCV and treated HCV infection can develop HCC, as well as those with MASLD-related cirrhosis. However, given that their incidence rates are lower than 1% per year, it is highly unlikely that it will be cost-effective or improve outcomes for these patients. For those with advanced fibrosis with these etiologies, if the provider believes the patient may actually have cirrhosis then providing surveillance may be beneficial but would be an individual decision and not suitable for all patients.

Surveillance has been recommended for HCC, and US with or without AFP has been recommended because it can improve survival [7]. The modalities recommended for surveillance are liver US with or without AFP every 6 months. Their performance has been estimated to have a sensitivity of 65% and a specificity of 90% [1]. A systematic review showed that the combination of US and AFP leads to improved survival among patients who underwent surveillance. Unfortunately, the use of HCC surveillance continues to be quite low, with only 20% to 30% of those eligible getting surveillance [1]. Innovative methods, such as mailing outreach letters to at-risk patients to schedule their surveillance testing, have increased use by 30% [8]. There is a need to have surveillance tests with better performance characteristics and increased surveillance use.

It is important to pay attention to the harms of surveillance (mostly related to false positives and indeterminate tests), and a recent study showed that harms were more often associated with US when compared to AFP [9]. It has been estimated that about 20% of USs are classified as inadequate for surveillance, and alternative surveillance modalities may be needed in patients with inadequate surveillance US such as in individuals with obesity, alcohol, and MASLD-related cirrhosis [10]. This highlights the need for surveillance tests that limit the harms, especially in those with inadequate surveillance US.

Imaging for Surveillance of Hepatocellular Carcinoma

The relatively poor performance, low use, and harms of the current surveillance strategy of AFP and US lead to a need to seek alternative surveillance strategies. One of these strategies is the development of new serum markers, and ongoing studies are addressing this important aspect as discussed in controversy 2. The other strategy is using MRI or CT scans, the most important diagnostic tests for HCC, for the surveillance of HCC among patients with cirrhosis.

When evaluating CT or MRI scans for the surveillance of HCC, one has to review the performance characteristics of these as diagnostic tests. A recent systematic review showed that the sensitivity of MRI was 82% and higher than that of CT, 66%, while there was no difference in specificity (91% for MRI and 92% for CT) [11]. However, when evaluating HCC between 1 and 2 cm, the sensitivity of CT and MRI decreases to 54% and 78%, respectively, and specificities of 91% and 83%, respectively. Therefore, when CT and MRI are used as diagnostic tests there is a decrease in their performance characteristics in patients with small tumors, which is the goal of surveillance. The performance characteristics of CT and MRI may be insufficient for surveillance use in patients with cirrhosis.

A couple of studies have compared the use of CT and MRI for HCC surveillance. A study of 163 patients with compensated cirrhosis were randomly assigned to a biannual US vs a yearly CT scan [12]. The overall incidence rate was 6.6% per year, and 9 HCCs were detected by US and 8 by CT scan. Although there was no difference between US and CT for surveillance of HCC, CT would not be recommended for surveillance. The overall performance of CT may not differ from US as indicated in this study and may be worse than MRI, so it is unlikely that CT will provide a benefit over other modalities. An important limitation for the widespread use of CT scans is the radiation exposure for patients. Using a CT scan for the surveillance for HCC would require 1 to 2 scans annually for years, leading to substantial amounts of radiation exposure. Another limitation of CT is the need for iodine contrast and the potential for developing renal insufficiency, especially in patients with cirrhosis who may be at a higher risk for this toxicity. At this time, the lack of benefits and the potential for significant harm does not support the use of CT scan for the surveillance of HCC.

An important prospective study compared the performance characteristics of US and MRI for the surveillance of patients with cirrhosis [13]. A total of 407 patients with cirrhosis, the majority with chronic hepatitis B infection, underwent 1 to 3 biannual paired examinations with US and live-specific contrast MRI, and at the end of the examination period all patients underwent CT scan to assess for confirmation bias. A total of 43 patients with HCC were detected, and as shown in Table 8.1, the overall detection with MRI was 86%, which was statistically significantly higher than that of US at 27.9% ($P < 0.001$), and the detection rate for early-stage HCC was 59.5% for MRI and 26.2% for US. There was no difference in survival between those with HCC and those without HCC. Surveillance imaging testing would be evaluated as a categorical variable, whether there is a liver mass, and whether it is suspicious. The criteria used to classify liver masses on US and MRI have not been validated, a weakness of these data. However, this study

Table 8.1 Hepatocellular detection with magnetic resonance imaging

Lesion characterization	Total no. of tests	Sensitivity any HCC, %	Specificity, %	False positive, %	False negative, %
Ultrasound					
Suspicious lesion	71	27.9	94.4	5.6	72.1
Equivocal lesion	76	27.9	93.9	6.1	72.1
Probably benign	108	32.6	91.1	8.9	67.4
MRI					
Highly suggestive	33	60.5	99.3	0.7	39.5
Suspicious	69	86	97	3.0	14
Equivocal	84	88.4	95.6	4.4	11.6

Abbreviations: *HCC* hepatocellular carcinoma, *MRI* magnetic resonance imaging
Adapted from Kim et al. [13]

shows the potential of MRI in a surveillance setting compared to US. There are some issues to consider. First, the study was biased for MRI as the patients underwent 1 to 3 paired biannual surveillance examinations with US and MRIs, thereby getting more frequent MRIs than would normally be required. Second, this study is not generalizable to cirrhosis patients seen outside Asia, who tend to have MASLD, alcoholic liver disease, and treated HCV infection. Third, there is no mention of the effect of AFP and its complementary nature on the imaging tests. Fourth, the authors did not assess MRI capacity in the study, which will be a major issue due to the competition for MRI slots for other indications. Fifth, while it was not powered for outcomes, there was no difference in survival among those with and without HCC. Sixth, US performance was much lower than what has been reported in other studies [7] and was likely underestimated since surveillance was terminated when a lesion was detected by MRI (often < 2 cm), and perhaps could have been detected at the next US surveillance testing. This study [13] highlights the benefit of using MRI for surveillance of HCC, but major hurdles exist before it can be applied to all patients with cirrhosis.

The high operational costs of a full MRI will likely restrict its use for surveillance of HCC among patients with cirrhosis. Abbreviated MRI (AMRI) has been proposed as an alternative using only the sequences needed, such as coronal T2-weighted, axial dynamic contrast-enhanced T1-weighted, fat-suppressed sequences [14]. When compared to complete-protocol MRI, AMRI showed comparable performance characteristics on a per-lesion basis. When AMRI is compared to US, it has a higher per-lesion sensitivity and negative predictive value similar to complete-contrast MRI [15]. A retrospective study evaluated noncontrast AMRI for the diagnosis of HCC and compared it to US and contrast-enhanced MRI [16]. Twenty-eight out of 188 patients with cirrhosis had malignant liver lesions. Noncontrast AMRI had a sensitivity of 85% and specificity of 93% for HCC, comparable to US and complete-contrast MRI. Prospective studies assessing noncontrast MRI and AMRI are ongoing, and we will learn of their efficacy in the future.

Kim et al. recently evaluated the cost-effectiveness of biannual US vs MRI in patients with cirrhosis [17]. Their study showed that MRI led to an increase in life-years and quality-adjusted life-years despite an increase in costs of more than $5000. However, the importance of this study is that the annual HCC incidence was the most influential factor in the cost-effectiveness of this strategy. When the HCC incidence rate was greater than 1.81%, the incremental cost-effectiveness ratio was below $50,000/quality-adjusted life-years but it was not cost-effective at lower incidence rates. As reviewed earlier, the HCC incidence rates in MASLD, alcohol-related and treated HCV cirrhosis, the most common liver diseases at this time, have lower incidence rates than what this study shows to be cost-effective and therefore this strategy of obtaining MRI every 6 months may not be feasible. This study confirms the need for and importance of stratifying the risk of HCC among patients with cirrhosis to use MRI for those at the highest risk.

An important consideration for surveillance for HCC in cirrhosis is whether patients can be stratified reliably according to their risk of developing this tumor. Calculators to evaluate the risk of HCC are available but their performance is currently poor and unable to account for all etiologies of liver disease. A 186-gene signature has been developed that can reliably stratify the risk of HCC among patients with HCV cirrhosis [18]. This signature requires large-scale validation not only for patients with HCV but also for other etiologies of cirrhosis. Using these data, a recent Markov model was performed with an initial strategy of stratifying HCC risk into a high-, intermediate-, or low-risk group using the molecular signature [19]. Once stratified into a risk group, a surveillance modality was designed for each group. Compared with biannual US and AFP, the risk-stratified approach was more cost-effective. The surveillance of the high-risk group with MRI or AMRI was cost-effective with no surveillance of the low-risk group. While this study needs prospective validation, it shows that surveillance strategies tailored to HCC risk are cost-effective and may improve overall use of surveillance.

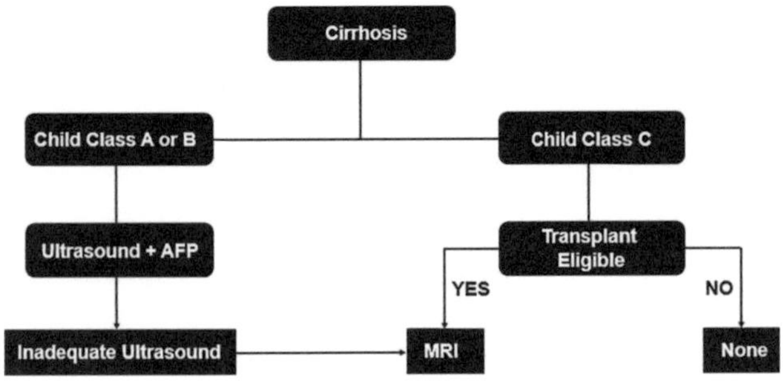

Fig. 8.1 The strategy for using magnetic resonance imaging (MRI) for the surveillance of hepatocellular carcinoma (HCC)

Conclusion

Surveillance for HCC should be performed in patients with cirrhosis with US and AFP as it has been shown to improve survival. Despite the issues with this strategy, such as relatively poor performance characteristics and the harms that they can cause, these tests are readily available, cheap, reproducible, and accepted by patients and health care providers. So, the question is who among patients with cirrhosis should undergo surveillance using CT or MRI? First, CT scan does not appear to be better than US for surveillance, and because of its risks of radiation exposure and toxicity from the intravenous contrast, CT scan should not be used for surveillance for HCC. Second, MRI should be performed in those with inadequate US, and those patients already on the transplant waiting list for decompensation, where they can receive a higher priority for liver transplantation if an HCC is detected (Fig. 8.1). Third, once a reliable way to stratify the risk of HCC in patients with cirrhosis is established, MRI will likely be the best option to use in those at highest risk.

References

1. Marrero JA, Kulik L, Sirlin CB, et al. Diagnosis, staging and management of hepatocellular carcinoma: 2018 practice guidance by the American association for the Study of liver disease. Hepatology. 2018;68(2):723–50. https://doi.org/10.1002/hep.29913.
2. El-Serag HB, Mason AC, Key C. Trends in survival of patients with hepatocellular carcinoma between 1977 and 1996 in the United States. Hepatology. 2001;33(1):62–5. https://doi.org/10.1053/jhep.2001.21041.
3. Kanwal F, Kramer JR, Asch SM, Chayanupatkul M, Cao Y, El-Serag HB. Risk of hepatocellular cancer in HCV patients treated with direct-acting antiviral agents. Gastroenterology. 2017;153(4):996–1005. https://doi.org/10.1053/j.gastro.2017.06.012.
4. Ioannou GN, Beste LA, Green PK, et al. Increased risk for hepatocellular carcinoma persists up to 10 years after HCV eradication in patients with baseline cirrhosis or high FIB-4 scores. Gastroenterology. 2019;157(5):1264–78. https://doi.org/10.1053/j.gastro.2019.07.033.
5. Bertot LC, Adams LA. Trends in hepatocellular carcinoma due to non-alcoholic fatty liver disease. Expert Rev Gastroenterol Hepatol. 2019;13(2):179–87. https://doi.org/10.1080/17474124.2019.1549989.
6. Ganne-Carrié N, Chaffaut C, Bourcier V, CIRRAL Group, et al. Estimate of hepatocellular carcinoma incidence in patients with alcoholic cirrhosis. J Hepatol. 2018;69(6):1274–83. https://doi.org/10.1016/j.jhep.2018.07.022.
7. Heimbach JK, Kulik LM, Finn RS, et al. AASLD guidelines for the treatment of hepatocellular carcinoma. Hepatology. 2018;67(1):358–80. https://doi.org/10.1002/hep.29086.
8. Singal AG, Tiro JA, Murphy CC, et al. Mailed outreach invitations significantly improve HCC surveillance rates in patients with cirrhosis: a randomized clinical trial. Hepatology. 2019;69(1):121–30. https://doi.org/10.1002/hep.30129.
9. Atiq O, Tiro J, Yopp AC, et al. An assessment of benefits and harms of hepatocellular carcinoma surveillance in patients with cirrhosis. Hepatology. 2017;65(4):1196–205. https://doi.org/10.1002/hep.28895.
10. Simmons O, Fetzer DT, Yokoo T, et al. Predictors of adequate ultrasound quality for hepatocellular carcinoma surveillance in patients with cirrhosis. Aliment Pharmacol Ther. 2017;45(1):169–77. https://doi.org/10.1111/apt.13841.
11. Roberts LR, Sirlin CB, Zaiem F, et al. Imaging for the diagnosis of hepatocellular carcinoma: a systematic review and meta-analysis. Hepatology. 2018;67(1):401–21. https://doi.org/10.1002/hep.29487.
12. Pocha C, Dieperink E, McMaken KA, Knott A, Thuras P, Ho SB. Surveillance for hepatocellular cancer with ultrasonography vs. computed tomography—a randomised study. Aliment Pharmacol Ther. 2013;38(3):303–12. https://doi.org/10.1111/apt.12370.
13. Kim SY, An J, Lim YS, et al. MRI with liver-specific contrast for surveillance of patients with cirrhosis at high risk of hepatocellular carcinoma. JAMA Oncol. 2017;3(4):456–63. https://doi.org/10.1001/jamaoncol.2016.3147.
14. Khatri G, Pedrosa I, Anathakrishnan L, et al. Abbreviated MRI protocol screening MRI vs. complete protocol diagnostic MRI for detection of hepatocellular carcinoma in patients with cirrhosis: an equivalence study using LI-RADS v2018. J Magn Reson Imaging. 2020;51(2):415–25. https://doi.org/10.1002/jmri.26835.
15. Tillman BG, Gorman JD, Hru JM, et al. Diagnostic per-lesion performance of a simulated gadoxetate disodium-enhanced abbreviated MRI protocol for hepatocellular carcinoma screening. Clin Radiol. 2018;73(5):485–93. https://doi.org/10.1016/j.crad.2017.11.013.
16. Chan MV, McDonald SJ, Ong YY, et al. HCC screening: assessment of an abbreviated non-contrast MRI protocol. Eur Radiol Exp. 2019;3(1):49. https://doi.org/10.1186/s41747-019-0126-1.
17. Kim HL, An J, Park JA, Park SH, Lim YS, Lee EK. Magnetic resonance imaging is cost-effective for hepatocellular carcinoma surveillance in high-risk patients with cirrhosis. Hepatology. 2019;69(4):1599–613. https://doi.org/10.1002/hep.30330.
18. Hoshida Y, Villanueva A, Sangiovanni A, et al. Prognostic gene expression signature for patients with hepatitis C-related early-stage cirrhosis. Gastroenterology. 2013;144(5):1024–30. https://doi.org/10.1053/j.gastro.2013.01.021.
19. Goossens N, Singal AG, King LY, et al. Cost-effectiveness of risk score-stratified hepatocellular carcinoma screening in patients with cirrhosis. Clin Transl Gastroenterol. 2017;8(6):e101. https://doi.org/10.1038/ctg.2017.26.

Should Biomarkers, Such as Alpha Fetoprotein, Play a Routine Role in Hepatocellular Carcinoma Surveillance in Patients with Cirrhosis?

9

Ashwin Rao and Amit G. Singal

Ultrasound (US) alone has insufficient sensitivity for detection of early-stage hepatocellular carcinoma (HCC) and the preferred strategy is to use ultrasound with biomarkers, such as alpha fetoprotein (AFP).

Main Points

1. Ultrasound alone has a sensitivity below 50% for the detection of early-stage HCC.
2. Ultrasound visualization is impaired in patients with alcohol and metabolic dysfunction-associated steatohepatitis (MASH)-related cirrhosis, suggesting inadequate ultrasound image quality and related failure will likely become more problematic in the future.
3. Adding AFP to ultrasound statistically significantly increases sensitivity for early-stage HCC detection compared to ultrasound alone.
4. Several other promising blood-based biomarkers have been proposed and are undergoing phase 3 and 4 biomarker validation.

Prognosis and therapeutic options for HCC are largely driven by tumor stage at diagnosis. Patients with early-stage HCC are eligible for curative therapies, such as surgical resection or liver transplantation, with 5-year survival rates approaching 70% [1]. In contrast, patients with advanced HCC are typically managed with palliative therapies and have a poor prognosis, with a median survival of 1 to 2 years. Given that symptomatic presentation is indicative of advanced tumor burden, implementing surveillance to detect HCC at a subclinical stage is critical. A large randomized controlled trial (RCT) demonstrated HCC surveillance increases early HCC detection and reduces HCC-related mortality in patients with chronic hepatitis B virus infection [2]. Several cohort studies have similarly shown that HCC surveillance is associated with statistically significant improvements in early tumor detection, curative treatment receipt, and overall survival in patients with cirrhosis [3].

Although professional societies, such as American Association for the Study of Liver Disease (AASLD), European Association for the Study of the Liver (EASL), and Asian Pacific Association for the Study of the Liver (APASL), agree on the need for HCC surveillance in at-risk patients—including those with cirrhosis—they differ as to recommended surveillance tests. While all societies recommend abdominal ultrasound every 6 months, they disagree on the use of serum biomarkers as adjunct surveillance tests. The updated AASLD guidelines recommend ultrasound with AFP given data reporting higher sensitivity for early-stage tumor detection using the two tests together [4]. Similarly, the APASL guidelines acknowledge conflicting data but recommend ultrasound with AFP given data suggesting increased sensitivity for early HCC detection and results of cost-effectiveness analyses [5]. In contrast, EASL continues to discourage the use of AFP and recommends ultrasound alone for surveillance because of reports of minimal improvement in sensitivity, concerns about decreased specificity, and potential increase in cost [6].

Ultrasound has been the cornerstone of HCC surveillance for many years given several advantages including being readily available, cost-effective, and noninvasive. However, recent data highlight several shortcomings. Studies demonstrate nearly 1 in 5 ultrasounds in patients with cirrhosis are inadequate to exclude HCC, with poor visualization and lower sensitivity significantly associated with obesity, increasing liver nodularity, and parenchymal heterogeneity, and alcohol-associated or MASH etiologies of liver disease [7–9]. In addition to patient characteristics, ultrasound image

A. Rao (✉)
Section of Gastroenterology and Hepatology, Baylor College of Medicine, Houston, TX, USA
e-mail: Ashwin.Rao@bcm.edu

A. G. Singal
Department of Medicine, University of Texas Southwestern Medical Center, Dallas, TX, USA

N. S. Reau, A. G. Singal (eds.), *Controversies in Hepatocellular Carcinoma*, https://doi.org/10.1007/978-3-031-90225-3_9

quality is dependent on the skill and expertise of the operator, leading to marked variability in ultrasound performance between patients and centers [10]. In the United States, ultrasound examinations are performed by technicians, with static ultrasound images interpreted later by radiologists, whereas in other parts of the world, physicians perform surveillance examinations themselves. Ultrasound examination quality is likely in part dictated by operator knowledge of liver anatomy as well as vigilance for excluding any HCC nodules. The role of expertise and skill has been well described in screening mammography, where experienced radiologists displayed greater accuracy, more efficient scanning patterns, and overall faster lesion detection times when compared with less-experienced readers [11]. Owing to the aforementioned limitations, poor quality or indeterminate exams can lead to surveillance-related harms by necessitating further use of cross-sectional imaging, such as computed tomography (CT) or magnetic resonance imaging (MRI). In a retrospective cohort study among 680 patients with cirrhosis undergoing surveillance, nearly one-third of cases requiring additional cross-sectional imaging after ultrasound were due to the presence of heterogeneous liver echotexture precluding adequate visualization [12]. Overall, these factors contribute to concerns about using ultrasound alone as a surveillance strategy for detecting early-stage HCC. Given the increasing prevalence of metabolic syndrome and MASH cirrhosis, inadequate ultrasound image quality and related failure will likely become more problematic [13]. Interventions such as using expert ultrasound operators, developing specific protocols for HCC surveillance, and standardizing reporting methods may improve ultrasound quality but would not address patient-related issues and are therefore likely to be insufficient. Surveillance strategies other than, or in addition to, ultrasound should be considered.

There has been increasing interest in CT- or MRI-based imaging, as these modalities address some of the limitations of ultrasound highlighted earlier [14]. A full discussion of alternative imaging modalities is beyond the scope of this chapter (see Chap. 1); however, it is clear there are notable constraints that likely preclude their widespread adoption for HCC surveillance. Specifically, CT imaging is associated with radiation exposure and potential contrast injury, particularly worrisome if performed semi-annually. While MRI avoids radiation exposure, there remains limited MRI radiologic capacity for all at-risk patients requiring HCC surveillance, marginal patient tolerance in some cases, unclear long-term clinical significance of gadolinium accumulation in the dentate nucleus and globus pallidus, and concerns about cost-effectiveness if applied to all patients with cirrhosis. There is increasing mention of the possibility of using cross-sectional imaging for surveillance in cirrhotic patients at high risk for HCC and prone to ultrasound failure, but such a strategy is yet to be developed or evaluated. Therefore, use of serum biomarkers, which are thought to be more objective, provides a potential solution to this growing issue.

The best-studied serum biomarker has been AFP, which has completed all 5 phases of biomarker development [15]. AFP, at the traditional surveillance cutoff of 20 ng/mL, has sensitivities of 41% to 65% and specificities of 80% to 94% for HCC in patients with cirrhosis [16]. It is clear AFP has insufficient sensitivity for HCC surveillance when used in isolation, as up to 80% of small HCC (tumors < 3 cm) fail to express AFP. However, several studies have evaluated the potential role of AFP as an adjunct biomarker when used in combination with ultrasound and have suggested a benefit of using the 2 tests in combination.

An RCT comparing ultrasound alone and ultrasound with AFP was attempted as part of a French multicenter study; however, high rates of AFP contamination in the ultrasound-alone group precluded meaningful comparisons [17]. This high rate of contamination reflects providers' hesitance to abandon AFP and suggests that an RCT will not be possible. Therefore, we are forced to rely on comparisons between ultrasound with and without AFP from cohort studies. A meta-analysis of cohort studies evaluating the benefit of using AFP with ultrasound for HCC surveillance in patients with cirrhosis found a statistically significant increase in sensitivity for early-stage HCC when the two tests were used in combination (63%; 95% CI, 48%–75%) compared to ultrasound alone (45%; 95% CI, 30%–62%) [18]. This increased sensitivity was consistently noted across studies despite variation in reported sensitivities for ultrasound. Specifically, the benefit of AFP was observed both in retrospective and prospective studies, studies in the United States and beyond the United States, and studies conducted before and after the year 2000. Although studies evaluated only early HCC detection and not downstream benefits, such as improved survival, it is reasonable to infer that increased early tumor detection would translate into a survival benefit.

As expected, the increase in sensitivity for early HCC detection was accompanied by a small but statistically significant decrease in specificity. Several studies have highlighted that AFP can be elevated in other clinical scenarios, including other malignancies, or undergo benign variability, contributing to false-positive and indeterminate results. AFP-related false-positive results are particularly common in patients with active chronic hepatitis C virus infection or elevated alanine transaminase (ALT) levels [19, 20]. However, it is possible concerns about AFP's specificity may become less noteworthy as the epidemiology of cirrhosis and HCC changes to reflect both the increasing treatment and cure of hepatitis C virus infection and rising prevalence of MASH [21]. Further, the diagnostic odds ratio, which incorporates both sensitivity and specificity, remained higher for ultrasound and AFP than ultrasound alone, suggesting that the gain in sensitivity for early HCC detection outweighed the loss in specificity.

Further evaluation is needed to discern whether the observed decrease in specificity translates clinically into a statistically significant increase in physical harms (e.g., further diagnostic imaging or biopsy), financial harms (e.g., direct costs of testing or indirect costs of missed work), or psychological harms (e.g., anxiety or depression related to false-positive results) [22–24]. A single-center study characterizing screening-related harms suggested AFP-related physical harms due to false positives are lower than those related to US, likely because of providers using clinical judgment when interpreting positive AFP levels but not being able to do so with ultrasound results [12]; however, these results still require validation in other practice settings. Several methods can be used to help mitigate AFP-related false-positive results, improving its value when implemented in clinical practice. First, studies have suggested that the AFP cutoff for upper limit of normal (usually set at a single threshold of 20 ng/mL) can be tailored to liver disease etiology, with higher cutoffs in viral than nonviral cirrhosis [19]. Alternatively, interpreting AFP trends over time rather than as individual values can better guide clinical interpretation of this biomarker. Longitudinal interpretation of AFP values more closely mirrors how it is interpreted in clinical practice, with decreasing values often providing reassurance even if above 20 ng/mL and increasing values even if below 20 ng/mL potentially raising concerns for development of an early-stage HCC. In addition, Tayob and colleagues showed that a parametric empirical Bayes screening algorithm, incorporating prior AFP results, statistically significantly increased sensitivity for HCC detection from 60.4% to 77.1% compared to single-threshold AFP measurements at a traditional cutoff of 20 ng/mL [25]. Further refinement and validation of longitudinal models, such as the parametric empirical Bayes algorithm, may facilitate their incorporation into electronic medical records to make this process more formalized and consistent between providers. There has also been increased interest in AFP-adjusted algorithms, incorporating patient demographics or routinely available clinical laboratory values, to increase performance characteristics. For example, the Hepatocellular Carcinoma Early Detection Screening (HES) algorithm includes age, AFP, AFP change over time, ALT, and platelet count. In a cohort of 4804 patients with cirrhosis, the HES algorithm had a sensitivity of 53% at a 10% false-positive rate, compared to 48% for AFP alone [26].

Although combining AFP with ultrasound appears to provide clinical benefit, the two tests in combination still miss up to one-third of HCCs at an early stage, underscoring the continued need for improved biomarkers. However, these data serve as a proof of concept for the potential value of biomarkers. Several other promising serum protein biomarkers such as des-γ-carboxyprothrombin (DCP) and lectin-bound AFP (AFP-L3) have been proposed and are undergoing phase 3 biomarker evaluation [27]. It should be noted neither test appears to perform better than AFP, much less AFP and US, in detection of early-stage HCC. Although not approved for surveillance or diagnosis, both biomarkers have been approved for use by the US Food and Drug Administration in determining future risk of HCC. Additional serum biomarkers, such as osteoponin, glypican-3, Golgi protein 73, and squamous cell carcinoma antigen have also been evaluated in phase 2 biomarker studies [28, 29].

Recently, advances in genomics have spurred identification of potential novel biomarkers, in the form of nucleic acids that are dysregulated in HCC. Given known aberrant microRNA (miRNA) expression identified in HCC and their reliable detection in body fluids (i.e., plasma, urine), circulating miRNAs, such as miR-21, miR-122, and miR-224, have been explored as early detection biomarkers [28]. Liquid biopsy describes the analysis of circulating tumor cells (CTCs) and circulating cell-free tumor DNA (cfDNA) shed from tumor sites, primary and metastatic, into the bloodstream [30]. Both DNA abnormalities isolated from CTCs as well as quantitative analysis of cfDNA have shown diagnostic value in HCC. For instance, quantitative analyses of cfDNA levels have shown moderate sensitivity and accuracy in detecting HCC [31, 32]. In addition, given recognition of aberrant DNA methylation patterns in tumorigenesis, analysis of CTC and cfDNA for such patterns, including differences in fragment length, is being evaluated for early HCC detection [33].Although the aforementioned biomarkers have shown promise in small phase 1 and 2 studies, further validation in large, multicenter longitudinal studies using standardized isolation and processing techniques as well as diagnostic cutoffs is needed. Hurdles of limited availability and prohibitive cost must also be overcome before practical clinical use.

Given the diverse etiologies and substantial heterogeneity between and even within HCC tumors, a single biomarker is unlikely to be sufficient—a panel of biomarkers may be necessary to achieve adequate sensitivity for early tumor detection. One of the best studied biomarker panels to date, GALAD, combines 3 biomarkers (AFP, AFP-L3, and DCP) with patient age and sex and has been shown to have high accuracy in large case-control studies. A validation phase 2 study with 6834 patients from the United Kingdom, Japan, and Germany (4404 with chronic liver disease and 2430 with HCC—1038 with early-stage HCC) reported sensitivities of 70% to 92% for HCC at any stage and specificities of 80% to 96% [34]. The sensitivity of the panel for early HCC remained high at 60% to 80%. A phase 2 study specifically in patients with MASH also demonstrated promising results, with a sensitivity of 68% for early HCC detection and specificity of 95%, with an area under the receiver operating characteristic (AUROC) of 0.91 [35]. An additional phase 2 study from the United States incorporating the multicenter Early Detection Research Network cohort confirmed excellent performance of the GALAD score in early-stage HCC (sensitivity of 76% and specificity of 86%, with an AUROC of 0.88) regardless of underlying etiology and including tumors with negative AFP [36]. Recent phase 3 evaluation in small cohort studies, as well as the EDRN Hepatocellular Carcinoma Early Detection Strategy (HEDS) Study, similarly report promising sensitivity for GALAD alone [37, 38], and this has now moved into a randomized phase 4 trial compared to ultrasound-based surveillance (NCT 06084234). Another biomarker panel showing promising early results is the Doylestown Plus algorithm, which includes log AFP, fucosylated kininogen, age, sex, ALT, and alkaline phosphatase. In a small phase 2 study with 93 cirrhosis patients and 69 patients with early-stage HCC, the Doylestown Plus algorithm had an AUROC of 0.97, compared to 0.80 for AFP alone [39]. Finally, Exact Sciences reported results of a phase 2 study evaluating 4 methylated DNA markers in combination with AFP and AFP-L3, which had a sensitivity of 71% for early-stage HCC and specificity of 90%, compared to sensitivity of 21% and specificity 98% for AFP alone (AUROC 0.91 vs 0.81) [40]. While phase 2 results for each of these biomarker panels are promising, they still require validation in large phase 3 and phase 4 (retrospective and prospective cohort) biomarker studies before routine use. Maturation of the HEDS Cohort and Texas HCC Consortium cohorts should allow validation of the biomarkers in the near future [41, 42]. While awaiting these data evaluating novel biomarkers, including biomarker panels, we should use the best of our currently validated surveillance tests.

In summary, ultrasound appears to be suboptimal for detecting HCC at an early stage when used alone. The addition of AFP to ultrasound appears to statistically significantly increase the sensitivity for HCC detection at an early stage, which can translate to increased curative treatment eligibility and improved survival. The increase in sensitivity is accompanied by a small, albeit statistically significant, decrease in specificity that can be mitigated with optimized AFP thresholds, longitudinal evaluation, and clinical interpretation. Additional evaluation is needed to determine whether the decrease in specificity relates to a clinical increase in physical, financial, or psychological harms. Until more optimal imaging modalities or biomarkers are available, we believe AFP should be used with ultrasound in at-risk patients undergoing HCC surveillance.

References

1. Singal AG, Kudo M, Bruix J. Breakthroughs in hepatocellular carcinoma therapies. Clin Gastroenterol Hepatol. 2023;21:2135–49.
2. Zhang BH, Yang BH, Tang ZY. Randomized controlled trial of screening for hepatocellular carcinoma. J Cancer Res Clin Oncol. 2004;130:417–22.
3. Singal AG, Zhang E, Narasimman M, et al. HCC surveillance improves early detection, curative treatment receipt, and survival in patients with cirrhosis: A meta-analysis. J Hepatol. 2022;77:128–39.
4. Singal AG, Llovet JM, Yarchoan M, et al. AASLD Practice Guidance on prevention, diagnosis, and treatment of hepatocellular carcinoma. Hepatology. 2023;78:1922–65.
5. Omata M, Cheng AL, Kokudo N, et al. Asia-Pacific clinical practice guidelines on the management of hepatocellular carcinoma: a 2017 update. Hepatol Int. 2017;11:317–70.
6. European Association for the Study of the Liver. Electronic address eee, European Association for the Study of the L. EASL Clinical practice guidelines: management of hepatocellular carcinoma. J Hepatol. 2018;69:182–236.
7. Simmons O, Fetzer DT, Yokoo T, et al. Predictors of adequate ultrasound quality for hepatocellular carcinoma surveillance in patients with cirrhosis. Aliment Pharmacol Ther. 2017;45:169–77.
8. Chong N, Schoenberger H, Yekkaluri S, et al. Association between ultrasound quality and test performance for HCC surveillance in patients with cirrhosis: a retrospective cohort study. Aliment Pharmacol Ther. 2022;55:683–90.
9. Schoenberger H, Chong N, Fetzer DT, et al. Dynamic changes in ultrasound quality for hepatocellular carcinoma screening in patients with cirrhosis. Clin Gastroenterol Hepatol. 2022;20:1561–1569 e4.
10. Fetzer DT, Browning T, Xi Y, et al. Associations of Ultrasound LI-RADS visualization score with examination, sonographer, and radiologist factors: retrospective assessment in over 10,000 examinations. AJR Am J Roentgenol. 2022;218:1010–20.
11. Nodine CF, Kundel HL, Lauver SC, et al. Nature of expertise in searching mammograms for breast masses. Acad Radiol. 1996;3:1000–6.
12. Atiq O, Tiro J, Yopp AC, et al. An assessment of benefits and harms of hepatocellular carcinoma surveillance in patients with cirrhosis. Hepatology. 2017;65:1196–205.
13. Huang DQ, Singal AG, Kono Y, et al. Changing global epidemiology of liver cancer from 2010 to 2019: NASH is the fastest growing cause of liver cancer. Cell Metab. 2022;34:969–977 e2.
14. Karim MA, Singal AG, Kum HC, et al. Prediagnostic CT or MRI utilization and outcomes in hepatocellular carcinoma: SEER-medicare database analysis. Cancer Res Commun. 2023;3:874–83.
15. Singal AG, Hoshida Y, Pinato DJ, et al. International Liver Cancer Association (ILCA) white paper on biomarker development for hepatocellular carcinoma. Gastroenterology. 2021;160:2572–84.
16. Gupta S, Bent S, Kohlwes J. Test characteristics of alpha-fetoprotein for detecting hepatocellular carcinoma in patients with hepatitis C. A systematic review and critical analysis. Ann Intern Med. 2003;139:46–50.
17. Trinchet JC, Chaffaut C, Bourcier V, et al. Ultrasonographic surveillance of hepatocellular carcinoma in cirrhosis: a randomized trial comparing 3- and 6-month periodicities. Hepatology. 2011;54:1987–97.
18. Tzartzeva K, Obi J, Rich NE, et al. Surveillance imaging and alpha fetoprotein for early detection of hepatocellular carcinoma in patients with cirrhosis: a meta-analysis. Gastroenterology. 2018;154:1706–1718 e1.
19. Gopal P, Yopp AC, Waljee AK, et al. Factors that affect accuracy of alpha-fetoprotein test in detection of hepatocellular carcinoma in patients with cirrhosis. Clin Gastroenterol Hepatol. 2014;12:870–7.
20. Sterling RK, Wright EC, Morgan TR, et al. Frequency of elevated hepatocellular carcinoma (HCC) biomarkers in patients with advanced hepatitis C. Am J Gastroenterol. 2012;107:64–74.
21. Vipani A, Lauzon M, Luu M, et al. Decreasing trend of serum alpha-fetoprotein level in hepatocellular carcinoma. Clin Gastroenterol Hepatol. 2022;20:1177–1179 e4.

22. Narasimman M, Hernaez R, Cerda V, et al. Hepatocellular carcinoma surveillance may be associated with potential psychological harms of in patients with cirrhosis. Hepatology. 2023;79:107.
23. Narasimman M, Hernaez R, Cerda V, et al. Financial burden of hepatocellular carcinoma screening in patients with cirrhosis. Clin Gastroenterol Hepatol. 2023.
24. Singal AG, Patibandla S, Obi J, et al. Benefits and harms of hepatocellular carcinoma surveillance in a prospective cohort of patients with cirrhosis. Clin Gastroenterol Hepatol. 2021;19:1925–1932 e1.
25. Tayob N, Stingo F, Do KA, et al. A Bayesian screening approach for hepatocellular carcinoma using multiple longitudinal biomarkers. Biometrics. 2018;74:249–59.
26. Tayob N, Christie I, Richardson P, et al. Validation of the Hepatocellular Carcinoma Early Detection Screening (HES) algorithm in a cohort of veterans with cirrhosis. Clin Gastroenterol Hepatol. 2019;17:1886–1893 e5.
27. Marrero JA, Feng Z, Wang Y, et al. Alpha-fetoprotein, des-gamma carboxyprothrombin, and lectin-bound alpha-fetoprotein in early hepatocellular carcinoma. Gastroenterology. 2009;137:110–8.
28. Parikh ND, Mehta AS, Singal AG, et al. Biomarkers for the early detection of hepatocellular carcinoma. Cancer Epidemiol Biomarkers Prev. 2020;29:2495–503.
29. Parikh ND, Tayob N, Singal AG. Blood-based biomarkers for hepatocellular carcinoma screening: Approaching the end of the ultrasound era? J Hepatol. 2022;78:207.
30. Manea I, Iacob R, Iacob S, et al. Liquid biopsy for early detection of hepatocellular carcinoma. Front Med (Lausanne). 2023;10:1218705.
31. Hu S, Liu Y, Yang Q, et al. Liquid biopsy using cell-free DNA in the early diagnosis of hepatocellular carcinoma. Investig New Drugs. 2023;41:532–8.
32. Arvind A, Singal AG. Emerging liquid biopsy techniques for early detection of hepatocellular carcinoma, prognostication, and disease monitoring. Clin Liver Dis (Hoboken). 2022;20:18–20.
33. Foda ZH, Annapragada AV, Boyapati K, et al. Detecting liver cancer using cell-free DNA fragmentomes. Cancer Discov. 2023;13:616–31.
34. Berhane S, Toyoda H, Tada T, et al. Role of the GALAD and BALAD-2 serologic models in diagnosis of hepatocellular carcinoma and prediction of survival in patients. Clin Gastroenterol Hepatol. 2016;14:875–886 e6.
35. Best J, Bechmann LP, Sowa JP, et al. GALAD score detects early hepatocellular carcinoma in an international cohort of patients with nonalcoholic steatohepatitis. Clin Gastroenterol Hepatol. 2020;18:728–735 e4.
36. Yang JD, Addissie BD, Mara KC, et al. GALAD score for hepatocellular carcinoma detection in comparison with liver ultrasound and proposal of GALADUS score. Cancer Epidemiol Biomarkers Prev. 2019;28:531–8.
37. Marrero J, Marsh T, Parikh N, et al. GALAD score improves early detection of hcc prior to the diagnosis of hcc: a phase 3 biomarker validation study. Hepatology. 2021;74:92A.
38. Singal AG, Tayob N, Mehta A, et al. GALAD demonstrates high sensitivity for HCC surveillance in a cohort of patients with cirrhosis. Hepatology. 2022;75:541–9.
39. Wang M, Devarajan K, Singal AG, et al. The doylestown algorithm: a test to improve the performance of AFP in the detection of hepatocellular carcinoma. Cancer Prev Res (Phila). 2016;9:172–9.
40. Chalasani NP, Porter K, Bhattacharya A, et al. Validation of a novel multitarget blood test shows high sensitivity to detect early stage hepatocellular carcinoma. Clin Gastroenterol Hepatol. 2022;20:173–182 e7.
41. Feng Z, Marrero JA, Khaderi S, et al. Design of the texas hepatocellular carcinoma consortium cohort study. Am J Gastroenterol. 2019;114:530–2.
42. Borges KA, Dai J, Parikh ND, et al. Rationale and design of the hepatocellular carcinoma early detection strategy study: a multicenter longitudinal initiative of the national cancer institute's early detection research network. Contemp Clin Trials. 2019;76:49–54.

Should Surveillance Be Performed Every 6 Months in All Patients with Cirrhosis?

10

Pierre Nahon and Olivier Seror

Main Points

1. Retrospective cohort studies have suggested that semiannual surveillance is associated with increased early detection and improved survival compared to annual surveillance.
2. A randomized controlled trial found no difference in early detection and curative treatment receipt between semiannual and quarterly surveillance.
3. There is heterogeneity in HCC risk and tumor growth patterns between patients, which raises the possibility of precision screening programs in the future.

The ideal time frame for the HCC surveillance interval can be estimated by the rate of tumor growth. Tumor-doubling time (TDT) is nevertheless heterogeneous and difficult to assess, but available data derived from older series obtained in hepatitis B (HBV)-infected populations in Asia reported median times of around 6 months [1, 2]. More recently, an American study performed in patients with HCC developed from cirrhosis, mostly infected by hepatitis C virus (HCV) or affected by nonviral causes of liver disease, described a median TDT of 229 days [3]. In line with these data, a 6-month interval represents a reasonable choice, and is endorsed by Western recommendations [4, 5].

P. Nahon (✉)
Liver unit, APHP, Bobigny, France
e-mail: pierre.nahon@aphp.fr

O. Seror
Radiology unit, APHP, Bobigny, France
e-mail: olivier.seror@aphp.fr

Clinical Justification of the Current Recommended 6-Month Surveillance Interval

During the last decade, several studies attempted to compare the performance of 3-month, 6-month, and 12-month intervals for screening. Except for one randomized trial, most evidence supporting the 6-month time frame is derived from retrospective analyses of longitudinal cohorts of patients included in surveillance programs or national registers. Most end points of these studies comprised HCC detection rates, access to curative procedures, and survival. However, because of their retrospective design, most reports did not consider pivotal factors such as compliance with screening, detection of nonmalignant lesions, futile recall procedures and other harms, or competing risks of death. Moreover, rigorous methods taking into account lead time or length-time biases were often not applied.

Studies Studying 6-Month vs 12-Month Screening Intervals

A retrospective study compared a semiannual and yearly ultrasound (US) time frame vs no screening in 269 patients with cirrhosis and HCC [6]. HCC in patients under the 6-month interval surveillance were more often diagnosed at an early stage (70% within Milan criteria) compared to 37% for those screened yearly. Similarly, an Italian study performed among 649 patients with cirrhosis confirmed the benefit of 6-monthly screening over 12-monthly (70% within Milan criteria vs 58%, respectively) [7]. Similar results were reported by observational cohorts [8]. In addition, a large retrospective registry study assessed the effect of different surveillance intervals in patients at risk of HCC [9]. Shorter US screening intervals were associated with reduced overall mortality in these patients, and as a whole provided additional arguments to support the 6-month time frame as the

N. S. Reau, A. G. Singal (eds.), *Controversies in Hepatocellular Carcinoma*, https://doi.org/10.1007/978-3-031-90225-3_10

optimal cutoff for HCC screening interval. More recently, the long prospective follow-up of patients with compensated viral cirrhosis showed that patients who respected the recommended 6-month screening interval had a higher proportion of HCC detected at an early stage [10]. Based on the longitudinal assessment of multiple end points and taking into account lead-time bias, this study showed that HCC early detection also translated into a survival benefit due to more frequent implementation of first-line curative procedures. Finally, a meta-regression analysis displayed an increased pooled sensitivity of US-based surveillance with the 6-month program reaching 70% compared to 50% with annual surveillance [11].

Is Reducing the Screening Interval to 3 Months Effective?

A French randomized controlled trial compared surveillance intervals of 3 and 6 months in more than 1300 cirrhotic patients to test whether shortening the surveillance time frame could result in greater early-stage HCC detection and improved survival [12]. With a mean follow-up of more than 40 months in both arms, most patients in both groups were similarly detected at an early HCC stage (79% vs 71%) and similar rates received first-line curative therapies (62% vs 58%). However, patients randomly assigned to the 3-month surveillance arm were more prone to be detected with smaller than 10-mm focal lesions that required further surveillance. This led to more investigations and recall procedures with no statistically significant overall increase in diagnosing HCC in comparison to the 6-monthly group. There was no difference in survival between both arms.

Factors that May Trigger Personalized Screening

There is broad patient variability in HCC incidence and doubling time. However, there are no data demonstrating that a higher liver cancer risk could be associated with faster tumor growth, both variables needing to be considered separately. However, these factors could be used to refine surveillance intervals and pave the way for personalized screening. Indeed, a 6-month period may be too long for some patients to be detected with an early-stage HCC, in particular in case of aggressive tumors, which are more likely to be missed by screening, or could be judged insufficient in high-risk populations. In contrast, semiannual surveillance programs may not be cost-effective in low-HCC risk patients [13] or, in case of indolent tumors, result in overtreatment and cause physical or psychological harm without benefit.

Heterogeneity in Hepatocellular Carcinoma Risk

All patients with cirrhosis do not have the same risk of developing HCC [14]. In this setting, Japanese guidelines have recommended performing surveillance every 3 months in "very high-risk" patients, such as HBV-infected cirrhotic patients with high viral replication, although the evidence on which this recommendation was based is scarce [15]. On the other hand, semiannual surveillance programs may not be cost-effective in certain subsets of cirrhotic patients because of their particularly low annual incidence of HCC, for example, in the case of cirrhotic patients whose HBV/HCV infection is controlled or eradicated [16, 17]. Beyond these extreme situations, personalized assessment of the individual risk of HCC and refinement of screening policies might be discussed according to HCC incidence. Various HCC scoring systems have been based on the combination of routine clinical features to stratify cirrhotic patients into different HCC risk classes [18] and might in the future be enriched by other features such as genomic traits. Based on such stratification, it is tempting to speculate that adapting HCC screening strategies might optimize both cost-effectiveness and the allocation of limited medical resources. In this context, the intensification of screening programs in intermediate- or high-risk groups is a timely challenge when considering the changing epidemiology of chronic liver disease [19].

Differences in Tumor Growth Patterns

The efficacy of cancer screening programs is affected by TDT. Indeed, surveillance is less effective in case of an aggressive tumor, which is usually detected at an advanced stage. Conversely, surveillance can detect indolent tumors leading to overdiagnosis and overtreatment.

Some specific subgroups of patients have been shown to develop aggressive tumors. This is for instance the case of HCV/HIV coinfected patients with cirrhosis, in whom infiltrating forms of HCC are commonly detected during surveillance programs [20]. Although such aggressive patterns should decline as HCV cure is achieved in these populations, [21] one can legitimately wonder if reducing the surveillance interval might improve early detection in these patients prone to develop such difficult-to-treat forms of liver cancers.

Regardless of specific subgroups, a recent retrospective multicenter study performed in the United States suggested heterogeneous growth patterns among 418 HCC patients with cirrhosis in whom sequential contrast-enhanced imaging procedures available before treatment were reviewed in 3 dimensions leading to TDT estimate [3]. Although the median TDT was around 6 months in the whole study population, 40% exhibited indolent growth patterns (TDT > 1 y), whereas rapid growth was observed in 25% (TDT < 3 mo). While patients with nonviral cirrhosis or larger tumors had a longer TDT, alpha fetoprotein levels were (not surprisingly) associated with rapid growth patterns. However, it must be acknowledged that the aforementioned randomized trial [12] testing 3-month vs 6-month intervals reported in both arms similar rates of patients diagnosed with advanced HCC (around 20%), suggesting that reducing the interval might not be sufficient to improve early-stage diagnosis in case of rapidly growing HCCs. Further studies are warranted to identify clinical situations associated with the development of aggressive tumors to provide refinement of HCC surveillance programs that should not exclusively concern screening interval, but also the methods of detection.

Conclusions

The recommended 6-month time frame, when respected, allows the detection of early-stage HCC in a majority of patients and has been shown to be cost-effective [22]. Since a broad heterogeneity exists in terms of HCC risk and tumor-aggressive behaviors among patients included in surveillance programs, precision screening should aim at adapting screening intervals according to these 2 parameters. It is tempting to speculate that high-risk patients and/or those more prone to develop tumors with a short TDT could benefit from a reduction to a 3-month interval. Conversely, applying a 12-month schedule in patients with low HCC risk and/or in whom indolent tumors might occur could also be discussed. In combination with more effective tools aimed at improving detection rates such as new imaging techniques or circulating biomarkers, personalized screening for HCC will contribute to the improvement of liver cancer prognosis. Nevertheless, such improvement will necessitate the accurate prediction both of incidence and TDT in patients with cirrhosis to redefine cost effective and safe guidelines appli cable for routine practice.

Grant Support National Agency for Research on HIV and Hepatitis (ANRS). Pierre Nahon's research is funded in part by the European Union (GENIAL, Grant agreement ID: 101096312) and French Agence Nationale de la Recherche (France 2030 DELIVER ANR-21-RHUS-0001).

References

1. An C, Choi YA, Choi D, et al. Growth rate of early-stage hepatocellular carcinoma in patients with chronic liver disease. Clin Mol Hepatol. 2015;21(3):279–86. https://doi.org/10.3350/cmh.2015.21.3.279.
2. Barbara L, Benzi G, Gaiani S, et al. Natural history of small untreated hepatocellular carcinoma in cirrhosis: a multivariate analysis of prognostic factors of tumor growth rate and patient survival. Hepatology. 1992;16(1):132–7. https://doi.org/10.1002/hep.1840160122.
3. Rich NE, John BV, Parikh ND, et al. Hepatocellular carcinoma demonstrates heterogeneous growth patterns in a multi-center cohort of patients with cirrhosis. Hepatology. 2020;72(5):1654–65. https://doi.org/10.1002/hep.31159.
4. European Association for the Study of the Liver. EASL clinical practice guidelines: management of hepatocellular carcinoma. J Hepatol. 2018;69(1):182–236. https://doi.org/10.1016/j.jhep.2018.03.019.
5. Heimbach JK, Kulik LM, Finn RS, Sirlin RS, et al. AASLD guidelines for the treatment of hepatocellular carcinoma. Hepatology. 2018;67(1):358–80. https://doi.org/10.1002/hep.29086.
6. Stravitz RT, Heuman DM, Chand N, Sterling RK, et al. Surveillance for hepatocellular carcinoma in patients with cirrhosis improves outcome. Am J Med. 2008;121(2):119–26. https://doi.org/10.1016/j.amjmed.2007.09.020.
7. Santi V, Trevisani F, Gramenzi A, et al. Italian liver cancer (ITA.LI.CA) group. Semiannual surveillance is superior to annual surveillance for the detection of early hepatocellular carcinoma and patient survival. J Hepatol. 2010;53(2):291–7. https://doi.org/10.1016/j.jhep.2010.03.010.
8. Santagostino E, Colombo M, Rivi M, Study Group of the Association of Italian Hemophilia Centers, et al. A 6-month versus a 12-month surveillance for hepatocellular carcinoma in 559 hemophiliacs infected with the hepatitis C virus. Blood. 2003;102(1):78–82.
9. Wu CY, Hsu YC, Ho HJ, Chen YJ, Lee TY, Lin JT. Association between ultrasonography screening and mortality in patients with hepatocellular carcinoma: a nationwide cohort study. Gut. 2016;65(4):693–701. https://doi.org/10.1136/gutjnl-2014-308786.
10. Costentin CE, Layese R, Bourcier V, et al. ANRS CO12 CirVir group. Compliance with hepatocellular carcinoma surveillance guidelines associated with increased lead-time adjusted survival of patients with compensated viral cirrhosis: a multi-center cohort study. Gastroenterology. 2018;155(2):431–442.e10.
11. Singal A, Volk ML, Waljee A, et al. Meta-analysis: surveillance with ultrasound for early-stage hepatocellular carcinoma in patients with cirrhosis. Aliment Pharmacol Ther. 2009;30(1):37–47. https://doi.org/10.1111/j.1365-2036.2009.04014.x.
12. Trinchet JC, Chaffaut C, Bourcier V, et al. Groupe d'Etude et de Traitement du Carcinome Hépatocellulaire (GRETCH). Ultrasonographic surveillance of hepatocellular carcinoma in cirrhosis: a randomized trial comparing 3- and 6-month periodicities. Hepatology. 2011;54(6):1987–97. https://doi.org/10.1002/hep.24545.
13. Andersson KL, Salomon JA, Goldie SJ, Chung RT. Cost effectiveness of alternative surveillance strategies for hepatocellular carcinoma in patients with cirrhosis. Clin Gastroenterol Hepatol. 2008;6(12):1418–24. https://doi.org/10.1016/j.cgh.2008.08.005.
14. Nahon P, Zucman-Rossi J. Single nucleotide polymorphisms and risk of hepatocellular carcinoma in cirrhosis. J Hepatol. 2012;57(3):663–74. https://doi.org/10.1016/j.jhep.2012.02.035.
15. Kokudo N, Hasegawa K, Akahane M, et al. Evidence-based clinical practice guidelines for hepatocellular carcinoma: the Japan Society of Hepatology 2013 update (3rd JSH-HCC guidelines). Hepatol Res. 2015;45(2) https://doi.org/10.1111/hepr.12464.
16. Nahon P, Bourcier V, Layese R, et al. ANRS CO12 CirVir group. Eradication of hepatitis C virus infection in patients with cirrhosis reduces risk of liver and non-liver complications. Gastroenterology. 2017;152(1):142–156.e2. https://doi.org/10.1053/j.gastro.2016.09.009.
17. Brichler S, Nahon P, Zoulim F, et al. ANRS CO12 CirVir study group. Non-virological factors are drivers of hepatocellular carcinoma in virosuppressed hepatitis B cirrhosis: results of ANRS CO12 CirVir cohort. J Viral Hepat. 2019;26(3):384–96. https://doi.org/10.1111/jvh.13029.

18. Sherman M. HCC risk scores: useful or not? Semin Liver Dis. 2017;37(4):287–95. https://doi.org/10.1055/s-0037-1607452.
19. Goossens N, Singal AG, King LY, et al. Cost-effectiveness of risk score-stratified hepatocellular carcinoma screening in patients with cirrhosis. Clin Transl Gastroenterol. 2017;8(6):e101. https://doi.org/10.1038/ctg.2017.26.
20. Gelu-Simeon M, Lewin M, Ostos M, et al. ANRS HC EP 25 Prethevic, ANRS CO13 HepaVih and ANRS CO12 CirVir study groups. Prognostic factors of survival in HIV/HCV CO-infected patients with hepatocellular carcinoma: the CARCINOVIC cohort. Liver Int. 2019;39(1):136–46. https://doi.org/10.1111/liv.13921.
21. Salmon-Ceron D, Nahon P, Layese R, et al. ANRS CO12 CirVir and ANRS CO13 HEPAVIH study groups. Human immunodeficiency virus/hepatitis C virus (HCV) CO-infected patients with cirrhosis are no longer at higher risk for hepatocellular carcinoma or end-stage liver disease as compared to HCV mono-infected patients. Hepatology. 2019;70(3):939–54. https://doi.org/10.1002/hep.30400.
22. Cadier B, Bulsei J, Nahon P, et al. ANRS CO12 CirVir and CHANGH groups early detection and curative treatment of hepatocellular carcinoma: a cost-effectiveness analysis in France and in the United States. Hepatology. 2017;65)4):1237-1248:1237. https://doi.org/10.1002/hep.28961.

Should Patients with Hepatitis C Virus Cirrhosis Undergo Hepatocellular Carcinoma Surveillance Indefinitely?

11

Adriaan J. van der Meer and Raoel Maan

Hepatocellular carcinoma (HCC) surveillance should be continued indefinitely after sustained virological response (SVR) in patients with cirrhosis.

Main Points

1. SVR statistically significantly reduces but does not eliminate the risk of HCC in patients with hepatitis C virus (HCV) cirrhosis.
2. Patients with HCV-related cirrhosis have an annual risk of HCC of 1.0% to 4.3% after SVR, with consistent risk even years after SVR.
3. At present, there are no validated risk models to identify low-risk patients with cirrhosis who do not require HCC surveillance.

Chronic HCV infection is a major global health problem as HCV affects an estimated 71,000,000 people worldwide and is associated with an impaired overall survival (OS) [1, 2]. Although patients with chronic HCV infection have an excess nonliver-related mortality, the predominant clinical sequelae of this disease are related to the progressive development of hepatic fibrosis [3]. At the time cirrhosis has developed, patients are at risk of liver failure and HCC. Older natural history studies showed that the annual risk of HCC among patients with cirrhosis and ongoing HCV infection is 3% to 7% [4, 5]. Even though the prevalence of HCV infection has been declining, the incidence of HCV-related HCC has only increased over recent decades [6]. Currently, HCC ranks sixth among all cancers worldwide, while it is the third most common cause of cancer-related death [7]. This highlights the poor outcome of HCC. Indeed, if not diagnosed in an early stage, HCC has a dismal prognosis with hardly any 5-year survival [8]. A recent Swedish national cohort including more than 3000 patients with HCC described median survival rates of 4.6 years following resection, 3.1 years after ablation, 1.4 years after transarterial chemoembolization, 0.5 years with sorafenib, and 0.3 years with best supportive care [9]. Those who underwent liver transplantation (LT) had the best outcome, with 75% survival at 5 years. Thus, HCC surveillance in patients with HCV-related cirrhosis aims to detect HCC early, when curative locoregional therapy (LRT) or LT is still possible [10]. Current guidelines support HCC surveillance in patients with cirrhosis [10, 11]. In this chapter we will discuss the efficacy of HCC surveillance among patients with HCV-related cirrhosis, also in relation to the recent development of highly effective antiviral treatment regimens.

A. J. van der Meer (✉) · R. Maan
Department of Gastroenterology and Hepatology, Erasmus MC University Medical Center, Rotterdam, The Netherlands
e-mail: a.vandermeer@erasmusmc.nl; r.maan@erasmusmc.nl

Sustained Virological Response and the Risk of Hepatocellular Carcinoma

While patients with HCV-related cirrhosis included in prior studies on HCC surveillance had ongoing infection, HCV is eradicated in almost all such patients in clinical care today. Even among those with compensated cirrhosis, SVR rates exceed 95% with direct-acting antivirals (DAAs). At present, however, there remain barriers to overcome for the universal treatment of the HCV-infected population [12].

From long-term followed cohorts of interferon (IFN)-treated patients with advanced hepatic fibrosis it became evident that achievement of SVR reduced their risk of HCC (~ four-fold) [13–16]. Yet, the risk of HCC was not eradicated on SVR and has been observed even long after virological cure. Among 842 patients with cirrhosis in Europe, followed for a median of almost 6 years, the annual risk of HCC after IFN-based SVR was 1.1% [17]. In this cohort the post-SVR HCC risk increased with age, presence of diabetes mellitus (a known risk factor of

N. S. Reau, A. G. Singal (eds.), *Controversies in Hepatocellular Carcinoma*, https://doi.org/10.1007/978-3-031-90225-3_11

HCC), and noninvasive markers of liver disease severity before antiviral therapy. A larger study from the United States described an annual HCC risk of 1.4% among patients with cirrhosis and IFN-induced SVR [18]. As this cohort concerns a predominantly male population of veterans, the higher risk may be caused by differences in patient characteristics and health behavior. Around the time of these reports, the first small and uncontrolled studies on the risk of HCC shortly after DAAs were presented [19–21]. Surprisingly, we were alarmed of an unexpected high rate of HCC recurrence and de novo HCC incidence after DAA therapy, with estimates of de novo HCC as high as 9.1% during the first 6 months of follow-up [20]. The sudden loss of HCV-stimulated immunological control over preexisting (undetectable) tumor nodules due to the rapid DAA-induced viral decline was a hypothesized explanation [21]. The clinical implications of these findings were heavily debated until larger cohort studies addressed the risk of HCC after DAA therapy. Reassuringly, all indicated that the higher observed HCC rate following DAAs was predominantly due to the ability of these drugs to treat and cure patients with more advanced liver disease [22–24]. After adjusting for the imbalance in confounding baseline characteristics, those treated with DAAs did not have an increased risk of HCC as compared to those treated with IFN therapy. In addition, the largest cohort study among 62,354 chronic HCV-infected patients confirmed that DAA-induced SVR was associated with HCC risk reduction (adjusted hazard ratio [HR] 0.29; 95% CI, 0.23–0.37), which was similar to IFN-induced SVR (adjusted HR 0.32; 95% CI, 0.28–0.37) [23]. Nevertheless, with the powerful DAAs curing patients who are more prone to HCC development, we expect to encounter HCC post SVR more frequently in the near future. While our experience with DAA therapy in patients with HCV-related cirrhosis will continue, the annual HCC risk after SVR ranges between 1.0% and 4.3% in current studies with short-term follow-up (Table 11.1) [12, 22, 24–41].

Table 11.1 Cohort studies on incidence of hepatocellular carcinoma among patients with cirrhosis and sustained virological response following direct-acting antiviral therapy

Author and publication y	No. of SVR patients with cirrhosis	Mean/median follow-up, y	No. of patients with HCC	Study design	(Calculated) annual HCC incidence rate[c], %	Reported risk factors for HCC
Cheung (2016) [27][a]	317	1.3	17	Prospective	4.3	NA
Kanwal (2017) [24]	7495	1.0	139	Retrospective	1.8	African American ethnicity
Mettke 2018 [34]	158	1.3	6	Prospective	2.9	Higher MELD score and higher AFP
Innes 2018 [22]	272	1.7	12	Retrospective	2.5	Higher age, CPS B/C, thrombocytopenia
Romano 2018 [38]	2497	1.4 (IQR, 1.0–1.9)	31	Prospective	1.0	Higher APRI score, HBsAg positive[d]
Ioannou 2018 [30]	7689	2.0	344	Retrospective	2.2	Higher age, Black race, lower platelet count, lower albumin, higher AST/√ALT
Calvaruso 2018 [25]	2140	1.2 (range, 6–24)	64	Prospective	2.6	Lower albumin and lower platelet count, CPS[d]
Kozbial 2018 [32]	393	1.3 (IQR, 0.3–3.0)	16	Retrospective	3.3	History of hepatic decompensation and low platelet count
Nahon 2018 [35]	274	1.8 (IQR, 1.1–2.2)	7	Retrospective	1.4	NA
Ioannou 2019 [29]	7533	3.0	619	Retrospective	0.5–1.4 in case of FIB-4 < 3.25 2.4–3.8 in case of FIB-4 ≥ 3.25	Baseline FIB-4 and change in FIB-4
Marino 2019 [33]	1070	1.6 (IQR, 1.4–1.9)	56	Retrospective	3.1	Noncharacterized nodules at baseline, liver function, alcohol intake, and hepatic decompensation
Park 2019 [12]	1218[b]	1.2 (SD = 0.7)	17[b]	Retrospective	1.2	Higher age, male sex, cirrhosis, diabetes, DAA therapy[d]
Degasperi 2019 [28]	546	2.1 (range, 0.3–3.3)	28[b]	Retrospective	3.4 (first year)	Male sex, diabetes, higher LSM, higher FIB-4 score

(continued)

Table 11.1 (continued)

Author and publication y	No. of SVR patients with cirrhosis	Mean/median follow-up, y	No. of patients with HCC	Study design	(Calculated) annual HCC incidence rate[c], %	Reported risk factors for HCC
Carrat 2019 [26]	2329	2.8 (IQR, 1.8–3.4)	166[b]	Prospective	2.2	Higher AFP, lower platelet count, lower albumin, hypertension, HCV genotype 3, higher age
Pinero 2019 [36]	653	1.3 (IQR, 0.8–1.9)	28	Prospective	2.8	Clinically significant portal hypertension
Shiha 2020 [39]	1734	2.0 (SD = 0.7)	101	Prospective	2.9	Higher age, male sex, higher, lower albumin
Tani 2020 [41]	191	1.2	10	Prospective	1.9 (first year)	Higher age and higher AFP
Kanwal 2020 [31]	6938	2.9 (SD = 0.6)	NA[e]	Retrospective	1.3–2.3	Higher age, African American ethnicity, higher MELD score
Pons 2020 [37]	572	2.9 (range, 0.3–3.8)	25	Prospective	1.5	Albumin before and after treatment, LSM after treatment

Abbreviations: *AFP* alpha fetoprotein, *ALT* alanine transaminase, *APRI* aspartate aminotransferase to platelet ratio, *AST* aspartate aminotransferase, *CPS* Child-Pugh score, *DAA* direct-acting antivirals, *FIB-4* fibrosis-4, *HBsAg* hepatitis B surface antigen, *HCC* hepatocellular carcinoma, *HCV* hepatitis C virus, *IQR* interquartile range, *LSM* liver stiffness measurement, *MELD* model for end-stage liver disease, *NA* not available, *SVR* sustained virological response

[a]Only patients with decompensated cirrhosis were included

[b]Reported number in all DAA-treated patients (not specifically those with SVR)

[c]When the annual HCC rate was not reported, this was calculated based on the presented data

[d]Analyses performed in all DAA-treated patients (not specifically those with SVR)

[e]In the entire cohort of 18,076 patients with DAA-induced SVR, there were 544 patients who were diagnosed with HCC. The adjusted hazard ratio of cirrhosis with respect to HCC was 4.13 (95% CI, 3.34–5.11)

Hepatocellular Carcinoma Surveillance

Surveillance Strategy in Light of Cirrhosis and Sustained Virological Response

Current guidelines recommend abdominal ultrasound (US) with or without alpha fetoprotein (AFP) measurement as HCC surveillance strategy in patients with cirrhosis, even in case of SVR [10, 11]. Low costs and the noninvasive nature of US are advantages in light of repeated applications. Downsides include its operator dependency, especially with respect to small nodules. Also, it may be difficult to distinguish a small malignant tumor from other benign histological changes in a nodular-transformed liver. A meta-analysis including 13,367 patients with cirrhosis indicated that the sensitivity of US for HCC of any stage was 84%, while the sensitivity declined to 47% for the detection of early HCC [42]. Despite the improved quality of the US machines we use today, the sensitivity of abdominal US for the detection of liver tumors is further challenged by an increase in fatty liver disease [43]. In selected patients magnetic resonance imaging may therefore be a more appropriate imaging modality [44]. Although more data are needed, and considering the additional costs, a recent prospective surveillance study indicated that magnetic resonance imaging had a higher HCC detection rate than US (86% vs 28%, respectively, $P < 0.001$) with fewer false-positive findings among cirrhotic patients with a high annual HCC risk [45].

There are no specific data on the accuracy of abdominal US for the detection of HCC in patients with HCV-related cirrhosis and SVR, but viral eradication may not be expected to influence its performance in the short term. This may, however, be different for the performance of AFP. In general, adding AFP (mostly studied at a cutoff of 20 ng/mL) to abdominal US increases the sensitivity (63%) for the detection of early-stage HCC as compared to US alone (45%, $P = 0.002$) [42]. In chronic HCV infection, however, AFP levels can be increased in the absence of HCC as well [42]. The combined strategy of US and AFP therefore results in more false-positive surveillance tests among HCV-infected patients, which led the European society to discourage the use of AFP [11, 42]. It may be questioned, however, whether these results should be extrapolated to patients with SVR. Successful antiviral therapy was shown to reduce AFP substantially, with hardly any patients remaining above 10 ng/mL [46]. Repeating prior studies on the performance of US and AFP for HCC detection following DAA therapy thus seems relevant.

Efficacy of Hepatocellular Carcinoma Surveillance in Cirrhosis

A large randomized controlled study on the efficacy of biannual abdominal US and AFP showed that HCCs detected through surveillance were more frequently treated with surgical resection and had a substantially better outcome compared to those diagnosed with HCC outside a surveillance program [47]. However, this study was performed more than 2 decades ago among Chinese patients with predominantly hepatitis B virus infection and a median age of approximately 40 years. These results can therefore not be extrapolated to an older and more obese population with HCV-related cirrhosis in Western countries today. First, steatosis and advanced hepatic fibrosis negatively affect the accuracy of abdominal US [43]. Second, in the case of cirrhosis, the possibilities to undergo curative LRT may be limited; this is the predominant reason why patients benefit from an early diagnosis [48]. Third, patients with cirrhosis are at an increased risk of liver-related mortality other than HCC, which may already limit their survival, even in case of SVR [17]. In the absence of randomized controlled trials in patients with cirrhosis, current practices are mainly based on the results of cohort studies in this population. A landmark meta-analysis included 15,158 patients with cirrhosis (of any etiology) and HCC from 47 studies performed in the United States, Europe, and Asia [49]. The proportion of early-stage HCC was 71% among patients in surveillance protocols vs 30% among those who were diagnosed incidentally or following symptoms. Surveillance-detected HCC was more frequently treated with curative intent (odds ratio 2.24), although curative treatment was received by only 52%. Nevertheless, the 3-year survival rate of 51% following surveillance-detected HCC was statistically significantly higher compared to the 3-year survival of 28% following HCC detected outside surveillance ($P < 0.001$). The suggested effect of HCC surveillance on OS remained in studies that adjusted for lead-time bias. In a more recent cohort of patients with cirrhosis, surveillance was associated with an earlier HCC diagnosis as well as a reduced mortality (adjusted HR 0.41; 95% CI, 0.26–0.65) as well, even though only 35% of patients with early HCC received curative therapy [50]. Increasing the uptake of curative LRT for early HCC may thus represent a route through which the benefit of surveillance may be maximized. Interestingly, in this study the positive association between surveillance and OS was present in those with HCC Barcelona Clinic Liver Cancer stage B to C as well, indicating that the relative survival benefit may not be restricted to patients receiving curative HCC treatment.

In contrast, a recent matched case-control study within the American Veterans Affairs health care system did not find US and/or AFP testing to be associated with a reduced mortality of HCC in cirrhosis [51]. However, in this study, the rate of curative treatment of early HCC was low and there were concerns regarding a bias due a poor compliance with the 6-month surveillance interval. Indeed, the median survival following HCC was higher among those compliant with the biannual surveillance recommendation in 2 European cohorts, while reducing the imaging interval to 3 months was not found to be beneficial [52–54]. Still, there remains controversy regarding the clinical benefit of HCC surveillance in patients with cirrhosis, which may explain part of the low uptake of the clear surveillance recommendations in society guidelines [55].

Cost-Efficacy of Hepatocellular Carcinoma Surveillance in Cirrhosis with Sustained Virological Response

At present, although based on older studies, HCC surveillance with biannual abdominal US with or without AFP is considered to be cost-effective in patient groups with an average HCC risk of 1.5% per year [10]. Patients with cirrhosis and ongoing HCV infection are well above this threshold. For those with cirrhosis and SVR, however, cost-efficacy should be assessed separately as was recently done in a Canadian modeling study [56]. A strong and exponential relationship between the annual HCC risk and the incremental cost-effectiveness ratios of biannual US was described. It was estimated that the incremental cost-effectiveness ratio was below the commonly suggested willingness-to-pay threshold of CAD $50,000 from an annual HCC risk of 1.3% onward. Thus, considering the currently available data on HCC incidence following SVR in patients with HCV-related cirrhosis, surveillance is likely to be cost-effective after successful DAA therapy (see Table 11.1) and possibly cost-effective after successful IFN therapy. Still, the assumptions driving the analyses in this study should be reviewed when interpreting its results in light of other health care systems in different populations.

The clinical efficacy of surveillance, however, may be hypothesized to have increased over time. Improved US quality could have eased the detection of HCC while there are more potential life-years to be gained following an early diagnosis due to better HCC management options [57]. In addition, 2 multicenter studies indicated that DAA therapy among patients with successfully treated early HCC was independently associated with a lower risk of death (adjusted HR 0.39–0.54) [58, 59]. Together, these developments could lower the risk cutoff for cost-effective HCC surveillance post SVR.

Unlike what one might assume, the annual risk of HCC did not show any reduction up to 10 years of follow-up after viral eradication with IFN-based therapy [17, 29]. During the first 4 years after DAA-induced SVR, the annual HCC risk remained above the suggested cost-efficacy threshold as well [29, 31]. As our experience with the course of HCV-related cirrhosis following successful DAA therapy continues, we will need to evaluate any dynamics in the incidence rate of HCC after SVR.

Hepatocellular Carcinoma Risk Stratification in Patients with Cirrhosis and Sustained Virological Response

Risk stratification could increase the cost-efficacy of HCC surveillance after SVR by selecting patients with the highest risk of HCC. The optimal surveillance protocol might also vary according to the HCC risk [60]. The major effect of risk stratification, however, would be the identification of patients with a negligible risk of HCC in whom surveillance can be stopped. In such patients surveillance is unlikely to be cost-effective, but perhaps more important is that their small chance of clinical benefit of surveillance may not outweigh its potential harms [56]. Downsides of surveillance have been insufficiently studied but include emotional distress, financial costs, and physical injuries as a result of invasive diagnostics or even treatment of false-positive nodules (which may thus not progress to clinically relevant HCC) [61].

Several parameters before antiviral therapy have been associated with the risk of HCC after SVR, among which age, ethnicity, features of metabolic syndrome, and noninvasive measures of liver disease severity are most frequently reported (see Table 11.1). Recently, a continuous risk model including such readily available clinical parameters before antiviral therapy was developed and validated among 7689 patients with HCV-related cirrhosis and SVR in the American Veterans Affairs national health care system [30]. Despite the relatively short mean follow-up of 2.2 years, as many as 344 patients were diagnosed with HCC and this contributed to the accuracy of the model. Still, external validation with longer follow-up in specifically those considered to have the lowest risk of HCC is needed before it can be implemented in daily clinical decision-making.

Next to pretreatment characteristics, the evolution of non-invasive parameters of hepatic fibrosis following successful antiviral therapy has been associated with HCC risk as well [29, 31, 37]. One of the largest studies included 7553 patients with cirrhosis and DAA-induced SVR. During a mean follow-up of 3.0 years, 619 patients were diagnosed with HCC [29]. Cirrhotic patients with a decline in their Fibrosis-4 (FIB-4) index from greater than or equal to 3.25 before treatment to less than 3.25 after SVR showed an HCC incidence of 2.45% per year, which was far above the threshold for cost-effective surveillance even though it was approximately half that of those patients with a FIB-4 that persisted at greater than or equal to 3.25 (5.08%/y). The lowest HCC incidence was observed in the group with a FIB-4 of less than 3.25 before and after DAA therapy. Their annual risk of HCC during the first years after SVR was just more than 1%, which is still just around the cutoff for cost-effective surveillance.

At present, however, there is no validated method to identify patients with HCV-related cirrhosis and SVR who will remain free of HCC. It has also been questioned whether the current clinical parameters have sufficient discriminative ability to exclude patients from surveillance [62]. Thus, novel molecular biomarkers related to HCC occurrence are actively explored through innovative translational research [63].

Should Patients with Hepatitis C Virus–Related Cirrhosis Undergo Hepatocellular Carcinoma Surveillance Indefinitely?

Patients with compensated cirrhosis and active HCV infection should continue HCC surveillance, although the primary focus in these patients is the achievement of SVR with DAA therapy. Viral eradication reduces the risk of HCC substantially. As our experience following DAA-induced SVR in patients with cirrhosis increases, we will learn whether surveillance is likely to indefinitely improve their prognosis. A precondition remains, however, that the patient be able to undergo HCC LRT with reasonable expectation of clinical benefit. This may not be the case in older patients, patients with substantial comorbidities, or patients with severely decompensated liver disease outside the trajectory of LT. Reasons to stop HCC surveillance can thus develop over time. A more favorable reason to refrain from surveillance in selected patients with cirrhosis and SVR would be their negligible risk of HCC. However, sufficiently validated prognostic tools to accurately identify these patients are not yet available. At present, therefore, all patients with HCV-related cirrhosis should remain included in long-term biannual liver imaging programs irrespective of successful antiviral therapy.

References

1. Lee MH, Yang HI, Lu SN, R.E.V.E.A.L.-HCV Study Group, et al. Chronic hepatitis C virus infection increases mortality from hepatic and extrahepatic diseases: a community-based long-term prospective study. J Infect Dis. 2012;206(4):469–77. https://doi.org/10.1093/infdis/jis385.
2. Polaris Observatory HCV Collaborators. Global prevalence and genotype distribution of hepatitis C virus infection in 2015: a modelling study. Lancet Gastroenterol Hepatol. 2017;2(3):161–76. https://doi.org/10.1016/S2468-1253(16)30181-9.
3. van der Meer AJ, Berenguer M. Reversion of disease manifestations after HCV eradication. J Hepatol. 2016;65(1 Suppl):S95–S108. https://doi.org/10.1016/j.jhep.2016.07.039.
4. Lok AS, Seeff LB, Morgan TR, HALT-C Trial Group, et al. Incidence of hepatocellular carcinoma and associated risk factors in hepatitis C-related advanced liver disease. Gastroenterology. 2009;136(1):138–48. https://doi.org/10.1053/j.gastro.2008.09.014.

5. Singal AG, Volk ML, Jensen D, Di Bisceglie AM, Schoenfeld PS. A sustained viral response is associated with reduced liver-related morbidity and mortality in patients with hepatitis C virus. Clin Gastroenterol Hepatol. 2010;8(3):280–8. https://doi.org/10.1016/j.cgh.2009.11.018.
6. Beste LA, Leipertz SL, Green PK, Dominitz JA, Ross D, Ioannou GN. Trends in burden of cirrhosis and hepatocellular carcinoma by underlying liver disease in US veterans, 2001–2013. Gastroenterology. 2015;149(6):1471–82. https://doi.org/10.1053/j.gastro.2015.07.056.
7. International Agency for Research on Cancer. Globocan. 2012. World Health Organization. Accessed 18 Jan 2020. http://gco.iarc.fr/
8. Llovet JM, Burroughs A, Bruix J. Hepatocellular carcinoma. Lancet. 2003;362(9399):1907–17. https://doi.org/10.1016/S0140-6736(03)14964-1.
9. Henriksson M, Björnsson B, Sternby Eilard M, et al. Treatment patterns and survival in patients with hepatocellular carcinoma in the Swedish national registry SweLiv. BJS Open. 2020;4(1):109–17. https://doi.org/10.1002/bjs5.50226.
10. Marrero JA, Kulik LM, Sirlin CB, et al. Diagnosis, staging, and management of hepatocellular carcinoma: 2018 practice guidance by the American Association for the Study of Liver Diseases. Hepatology. 2018;68(2):723–50. https://doi.org/10.1002/hep.29913.
11. European Association for the Study of the Liver. EASL clinical practice guidelines: management of hepatocellular carcinoma. J Hepatol. 2018;69(1):182–236. https://doi.org/10.1016/j.jhep.2018.03.019.
12. Park H, Wang W, Henry L, Nelson DR. Impact of all-oral direct-acting antivirals on clinical and economic outcomes in patients with chronic hepatitis C in the United States. Hepatology. 2019;69(3):1032–45. https://doi.org/10.1002/hep.30303.
13. Bruno S, Stroffolini T, Colombo M, et al. Italian Association of the Study of the Liver Disease (AISF). Sustained virological response to interferon-alpha is associated with improved outcome in HCV-related cirrhosis: a retrospective study. Hepatology. 2007;45(3):579–87. https://doi.org/10.1002/hep.21492.
14. Cardoso AC, Moucari R, Figueiredo-Mendes C, et al. Impact of peginterferon and ribavirin therapy on hepatocellular carcinoma: incidence and survival in hepatitis C patients with advanced fibrosis. J Hepatol. 2010;52(5):652–7. https://doi.org/10.1016/j.jhep.2009.12.028.
15. Morgan RL, Baack B, Smith BD, Yartel A, Pitasi M, Falck-Ytter Y. Eradication of hepatitis C virus infection and the development of hepatocellular carcinoma: a meta-analysis of observational studies. Ann Intern Med. 2013;158(5 PT 1):329–37. https://doi.org/10.7326/0003-4819-158-5-201303050-00005.
16. van der Meer AJ, Veldt BJ, Feld JJ, et al. Association between sustained virological response and all-cause mortality among patients with chronic hepatitis C and advanced hepatic fibrosis. JAMA. 2012;308(24):2584–93. https://doi.org/10.1001/jama.2012.144878.
17. van der Meer AJ, Feld JJ, Hofer H, et al. Risk of cirrhosis-related complications in patients with advanced fibrosis following hepatitis C virus eradication. J Hepatol. 2017;66(3):485–93. https://doi.org/10.1016/j.jhep.2016.10.017.
18. El-Serag HB, Kanwal F, Richardson P, Kramer J. Risk of hepatocellular carcinoma after sustained virological response in veterans with hepatitis C virus infection. Hepatology. 2016;64(1):130–7. https://doi.org/10.1002/hep.28535.
19. Cardoso H, Vale AM, Rodrigues S, et al. High incidence of hepatocellular carcinoma following successful interferon-free antiviral therapy for hepatitis C associated cirrhosis. J Hepatol. 2016;65(5):1070–1. https://doi.org/10.1016/j.jhep.2016.07.027.
20. Ravi S, Axley P, Jones D, et al. Unusually high rates of hepatocellular carcinoma after treatment with direct-acting antiviral therapy for hepatitis C related cirrhosis. Gastroenterology. 2017;152(4):911–2. https://doi.org/10.1053/j.gastro.2016.12.021.
21. Reig M, Mariño Z, Perelló C, et al. Unexpected high rate of early tumor recurrence in patients with HCV-related HCC undergoing interferon-free therapy. J Hepatol. 2016;65(4):719–26. https://doi.org/10.1016/j.jhep.2016.04.008.
22. Innes H, Barclay ST, Hayes PC, et al. The risk of hepatocellular carcinoma in cirrhotic patients with hepatitis C and sustained viral response: role of the treatment regimen. J Hepatol. 2018;68(4):646–54. https://doi.org/10.1016/j.jhep.2017.10.033.
23. Ioannou GN, Green PK, Berry K. HCV eradication induced by direct-acting antiviral agents reduces the risk of hepatocellular carcinoma. J Hepatol. 2018;68:25–32. https://doi.org/10.1016/j.jhep.2017.08.030.
24. Kanwal F, Kramer J, Asch SM, Chayanupatkul M, Cao Y, El-Serag HB. Risk of hepatocellular cancer in HCV patients treated with direct-acting antiviral agents. Gastroenterology. 2017;153(4):996–1005.e1. https://doi.org/10.1053/j.gastro.2017.06.012.
25. Calvaruso V, Cabibbo G, Cacciola I, et al. Rete Sicilia Selezione Terapia–HCV (RESIST-HCV). Incidence of hepatocellular carcinoma in patients with HCV-associated cirrhosis treated with direct-acting antiviral agents. Gastroenterology. 2018;155(2):411–421.e4. https://doi.org/10.1053/j.gastro.2018.04.008.
26. Carrat F, Fontaine H, Dorival C, et al. French ANRS CO22 Hepather Cohort. Clinical outcomes in patients with chronic hepatitis C after direct-acting antiviral treatment: a prospective cohort study. Lancet. 2019;393(10179):1453–64. https://doi.org/10.1016/S0140-6736(18)32111-1.
27. Cheung MCM, Walker AJ, Hudson BE. et al; HCV Research UK. Outcomes after successful direct-acting antiviral therapy for patients with chronic hepatitis C and decompensated cirrhosis. J Hepatol. 2016;65(4):741–7. https://doi.org/10.1016/j.jhep.2016.06.019.
28. Degasperi E, D'Ambrosio R, Iavarone M, et al. Factors associated with increased risk of de novo or recurrent hepatocellular carcinoma in patients with cirrhosis treated with direct-acting antivirals for HCV infection. Clin Gastroenterol Hepatol. 2019;17(6):1183–1191.e7. https://doi.org/10.1016/j.cgh.2018.10.038.
29. Ioannou GN, Beste LA, Green PK, et al. Increased risk for hepatocellular carcinoma persists up to 10 years after HCV eradication in patients with baseline cirrhosis or high FIB-4 scores. Gastroenterology. 2019;157(5):1264–78. https://doi.org/10.1053/j.gastro.2019.07.033.
30. Ioannou GN, Green PK, Beste LA, Mun EJ, Kerr KF, Berry K. Development of models estimating the risk of hepatocellular carcinoma after antiviral treatment for hepatitis C. J Hepatol. 2018;69(5):1088–98. https://doi.org/10.1016/j.jhep.2018.07.024.
31. Kanwal F, Kramer JR, Asch SM, Cao Y, Li L, El-Serag HB. Long-term risk of hepatocellular carcinoma in HCV patients treated with direct acting antiviral agents. Hepatology. 2020;71(1):44–55. https://doi.org/10.1002/hep.30823.
32. Kozbial K, Moser S, Al-Zoairy R, et al. Follow-up of sustained virological responders with hepatitis C and advanced liver disease after interferon/ribavirin-free treatment. Liver Int. 2018;38(6):1028–35. https://doi.org/10.1111/liv.13629.
33. Mariño Z, Darnell A, Lens S, et al. Time association between hepatitis C therapy and hepatocellular carcinoma emergence in cirrhosis: relevance of non-characterized nodules. J Hepatol. 2019;70(5):874–84. https://doi.org/10.1016/j.jhep.2019.01.005.
34. Mettke F, Schlevogt B, Deterding K, et al. Interferon-free therapy of chronic hepatitis C with direct-acting antivirals does not change the short-term risk for de novo hepatocellular carcinoma in patients with liver cirrhosis. Aliment Pharmacol Ther. 2018;47(4):516–25. https://doi.org/10.1111/apt.14427.

35. Nahon P, Layese R, Bourcier V, et al. ANRS CO12 CirVir Group. Incidence of hepatocellular carcinoma after direct antiviral therapy for HCV in patients with cirrhosis included in surveillance programs. Gastroenterology. 2018;155(5):1436–50. https://doi.org/10.1053/j.gastro.2018.07.015.
36. Pinero F, Mendizabal M, Ridruejo E, LALREAN, et al. Treatment with direct-acting antivirals for HCV decreases but does not eliminate the risk of hepatocellular carcinoma. Liver Int. 2019;39(6):1033–43. https://doi.org/10.1111/liv.14041.
37. Pons M, Rodríguez-Tajes S, Esteban JI, et al. Non-invasive prediction of liver-related events in patients with HCV-associated compensated advanced chronic liver disease after oral antivirals. J Hepatol. 2020;72(3):472–80. https://doi.org/10.1016/j.jhep.2019.10.005.
38. Romano A, Angeli P, Piovesan S, et al. Newly diagnosed hepatocellular carcinoma in patients with advanced hepatitis C treated with DAAs: a prospective population study. J Hepatol. 2018;69(2):345–52. https://doi.org/10.1016/j.jhep.2018.03.009.
39. Shiha G, Mousa N, Soliman R, Mikhail NN, Elbasiony MA, Khattab M. Incidence of HCC in chronic hepatitis C patients with advanced hepatic fibrosis who achieved SVR following DAAs: a prospective study. J Viral Hepat. 2020;27(1):671–9. https://doi.org/10.1111/jvh.13276.
40. Singal AG, Lim JK, Kanwal F. AGA clinical practice update on interaction between oral direct-acting antivirals for chronic hepatitis C infection and hepatocellular carcinoma: expert review. Gastroenterology. 2019;156(8):2149–57. https://doi.org/10.1053/j.gastro.2019.02.046.
41. Tani J, Morishita A, Sakamoto T, et al. Simple scoring system for prediction of hepatocellular carcinoma occurrence after hepatitis C virus eradication by direct-acting antiviral treatment: All Kagawa Liver Disease Group Study. Oncol Lett. 2020;19(3):2205–12. https://doi.org/10.3892/ol.2020.11341.
42. Tzartzeva K, Obi J, Rich NE, et al. Surveillance imaging and alpha fetoprotein for early detection of hepatocellular carcinoma in patients with cirrhosis: a meta-analysis. Gastroenterology. 2018;154(6):1706–1718.e1. https://doi.org/10.1053/j.gastro.2018.01.064.
43. Simmons O, Fetzer DT, Yokoo T, et al. Predictors of adequate ultrasound quality for hepatocellular carcinoma surveillance in patients with cirrhosis. Aliment Pharmacol Ther. 2017;45(1):169–77. https://doi.org/10.1111/apt.13841.
44. Expert Panel on Gastrointestinal Imaging, Horowitz JM, Kamel IR, Arif-Tiwari H, et al. ACR Appropriateness Criteria chronic liver disease. J Am Coll Radiol. 2017;14(11S):S391–405. https://doi.org/10.1016/j.jacr.2017.08.045.
45. Kim SY, An J, Lim YS, et al. MRI with liver-specific contrast for surveillance of patients with cirrhosis at high risk of hepatocellular carcinoma. JAMA Oncol. 2017;3(4):456–63. https://doi.org/10.1001/jamaoncol.2016.3147.
46. Oze T, Hiramatsu N, Yakushijin T, et al. Osaka Liver Forum. Post-treatment levels of alpha-fetoprotein predict incidence of hepatocellular carcinoma after interferon therapy. Clin Gastroenterol Hepatol. 2014;12(7):1186–95. https://doi.org/10.1016/j.cgh.2013.11.033.
47. Zhang BH, Yang BH, Tang ZY. Randomized controlled trial of screening for hepatocellular carcinoma. J Cancer Res Clin Oncol. 2004;130(7):417–22. https://doi.org/10.1007/s00432-004-0552-0.
48. Forner A, Reig M, Bruix J. Hepatocellular carcinoma. Lancet. 2018;391(10127):1301–14. https://doi.org/10.1016/S0140-6736(18)30010-2.
49. Singal AG, Pillai A, Tiro J. Early detection, curative treatment, and survival rates for hepatocellular carcinoma surveillance in patients with cirrhosis: a meta-analysis. PLoS Med. 2014;11(4):e1001624. https://doi.org/10.1371/journal.pmed.1001624.
50. Singal AG, Mittal S, Yerokun OA, et al. Hepatocellular carcinoma screening associated with early tumor detection and improved survival among patients with cirrhosis in the US. Am J Med. 2017;130(9):1099–1106.e1. https://doi.org/10.1016/j.amjmed.2017.01.021.
51. Moon AM, Weiss NS, Beste LA, et al. No association between screening for hepatocellular carcinoma and reduced cancer-related mortality in patients with cirrhosis. Gastroenterology. 2018;155(4):1128–1139.e6. https://doi.org/10.1053/j.gastro.2018.06.079.
52. Costentin CE, Layese R, Bourcier V, et al. ANRS CO12 CirVir Group. Compliance with hepatocellular carcinoma surveillance guidelines associated with increased lead-time adjusted survival of patients with compensated viral cirrhosis: a multi-center cohort study. Gastroenterology. 2018;155(2):431–442.e10. https://doi.org/10.1053/j.gastro.2018.04.027.
53. Santi V, Trevisani F, Gramenzi A, et al. Italian Liver Cancer (ITA.LI.CA) Group. Semiannual surveillance is superior to annual surveillance for the detection of early hepatocellular carcinoma and patient survival. J Hepatol. 2010;53(2):291–7. https://doi.org/10.1016/j.jhep.2010.03.010.
54. Trinchet JC, Chaffaut C, Bourcier V, et al. Groupe d'Etude et de Traitement du Carcinome Hépatocellulaire (GRETCH). Ultrasonographic surveillance of hepatocellular carcinoma in cirrhosis: a randomized trial comparing 3- and 6-month periodicities. Hepatology. 2011;54(6):1987–97. https://doi.org/10.1002/hep.24545.
55. Singal AG, Tiro JA, Murphy CC, et al. Mailed outreach invitations significantly improve HCC surveillance rates in patients with cirrhosis: a randomized clinical trial. Hepatology. 2019;69(1):121–30. https://doi.org/10.1002/hep.30129.
56. Farhang Zangneh H, Wong WWL, Sander B, et al. Cost effectiveness of hepatocellular carcinoma surveillance after a sustained virologic response to therapy in patients with hepatitis C virus infection and advanced fibrosis. Clin Gastroenterol Hepatol. 2019;17(9):1840–1849.e16. https://doi.org/10.1016/j.cgh.2018.12.018.
57. Noone A, Howlader N, Krapcho M, et al., editors. SEER cancer statistics review, 1975–2015. National Cancer Institute; 2018.
58. Cabibbo G, Celsa C, Calvaruso V, et al. Rete Sicilia Selezione Terapia–HCV (RESIST-HCV) and Italian Liver Cancer (ITA.LI.CA.) Group. Direct-acting antivirals after successful treatment of early hepatocellular carcinoma improve survival in HCV-cirrhotic patients. J Hepatol. 2019;71(2):265–73. https://doi.org/10.1016/j.jhep.2019.03.027.
59. Singal AG, Rich NE, Mehta N, et al. Direct-acting antiviral therapy for hepatitis C virus infection is associated with increased survival in patients with a history of hepatocellular carcinoma. Gastroenterology. 2019;157(5):1253–1263.e2. https://doi.org/10.1053/j.gastro.2019.07.040.
60. Kim HL, An J, Park JA, et al. Magnetic resonance imaging is cost-effective for hepatocellular carcinoma surveillance in high-risk patients with cirrhosis. Hepatology. 2019;69(4):1599–613. https://doi.org/10.1002/hep.30330.
61. Rich NE, Parikh ND, Singal AG. Overdiagnosis: an understudied issue in hepatocellular carcinoma surveillance. Semin Liver Dis. 2017;37(4):296–304. https://doi.org/10.1055/s-0037-1608775.
62. Calvaruso V, Bruix J. Towards personalized screening for hepatocellular carcinoma: still not there. J Hepatol. 2020;73(6):1319–21. https://doi.org/10.1016/j.jhep.2020.06.032.
63. Fujiwara N, Friedman SL, Goossens N, Hoshida Y. Risk factors and prevention of hepatocellular carcinoma in the era of precision medicine. J Hepatol. 2018;68(3):526–49. https://doi.org/10.1016/j.jhep.2017.09.016.

Part III

Diagnosis

Should All Liver Lesions in Patients with Cirrhosis Be Characterized and Reported Using Liver Imaging Reporting and Data System Criteria?

12

Kathryn J. Fowler, Guilherme M. Cunha, Victoria Chernyak, and Claude B. Sirlin

All liver lesions in patients with cirrhosis undergoing hepatocellular carcinoma (HCC) surveillance should be characterized and reported using Liver Imaging Reporting and Data System (LI-RADS) criteria.

Main Points

1. The LI-RADS algorithm provides an accurate diagnosis of HCC, with the LR-5 category having a 95% positive predictive value for the diagnosis of HCC. LI-RADS also offers a probabilistic approach to indeterminate liver lesions, with LR-4 observations having a substantially higher risk of HCC than LR-3 observations.
2. The LI-RADS algorithm is based on a standardized, precise lexicon of imaging findings that are regularly reviewed and updated. This facilitates clear communication between providers as well as facilitates large-scale, multisite studies of imaging findings.
3. LI-RADS has been endorsed by the American College of Radiology and the American Association for the Study of Liver Diseases (AASLD), and is now aligned with the Organ Procurement and Transplantation Network (OPTN) for transplant-eligible patients.
4. There are increasing data evaluating the interrater reliability and accuracy of LI-RADS classifications, although prospective multicenter studies are still needed.

K. J. Fowler
Diagnostic Radiology, Scripps Medical Group, San Diego, CA, USA

G. M. Cunha
Liver Imaging Group, Department of Radiology, UC San Diego, San Diego, CA, USA
e-mail: gcunha@uw.edu

V. Chernyak
Department of Radiology, University of Washington, Seattle, WA, USA

C. B. Sirlin (✉)
Department of Radiology, Columbia University Irving Medical Center, New York, NY, USA
e-mail: csirlin@health.ucsd.edu

The LI-RADS provides a comprehensive and integrated approach to diagnosing HCC in at-risk patients. LI-RADS includes 4 separate algorithms: ultrasound (US) screening/surveillance, contrast-enhanced US diagnosis, computed tomography/magnetic resonance imaging (CT/MRI) diagnosis, and CT/MRI treatment response. LI-RADS algorithms are based on a lexicon of standardized terms and definitions, relevant to liver imaging. The algorithms are summarized in core documents, which provide the essential material and instructions for applying each of the algorithms. The LI-RADS lexicon is provided as a standalone compendium of terms and definitions, supplemented by comments to clarify usage and fields to enable searching and sorting. A LI-RADS manual provides a complete reference for the entire spectrum of liver imaging in at-risk patients, and it includes background content, illustrative examples, and schematic drawings, scientific rationale, emerging evidence, usage pitfalls, and references. All LI-RADS materials are available free of charge on the American College of Radiology website (1).

There are multiple established benefits to using LI-RADS, yet some challenges remain to be solved. This chapter provides an overview of the pros and cons of LI-RADS use in practice, along with a final recommendation based on best available evidence and expert opinion.

Pros

Minimum Requirements

When considering diagnostic approaches to HCC in at-risk patients, it is important to first address the minimum requirements of any system: acceptable accuracy and reliability. As

N. S. Reau, A. G. Singal (eds.), *Controversies in Hepatocellular Carcinoma*, https://doi.org/10.1007/978-3-031-90225-3_12

reviewed in the following sections, several systems, including LI-RADS, meet these minimum requirements.

However, LI-RADS is unique in its breadth, completeness, rigor, international perspective, and capability to update its algorithms and other materials.

Liver Imaging Reporting and Data System Provides an Accurate Diagnosis of Hepatocellular Carcinoma

Many imaging algorithms exist for diagnosis of HCC (2). The philosophy behind these algorithms—that is, the relative priority on sensitivity vs specificity for the detection and diagnosis of HCC—may differ depending on geographic factors, including regional epidemiology, available and preferred treatment options, and cost considerations. In the United States, the most common risk factor for HCC is cirrhosis, which often precludes resection for early-stage HCC because of diminished liver function reserve (3). This is unlike other parts of the world where hepatitis B viral infection prevails, and resection is a commonly used treatment option for HCC (4). Hence, in the United States, the best option for cure and long-term survival of patients with early-stage HCC is liver transplantation. As such, exception points and priority on the waitlist are given to HCC patients who meet criteria. To achieve equipoise in the allocation system for livers, and avoid misallocation for false-positive diagnoses, the determination of HCC must be made with near 100% certainty. Hence, near-perfect diagnostic certainty (i.e., ~ 95% or greater positive predictive value) for LR-5 category (definite HCC) is the goal of LI-RADS criteria. In a systematic review of 17 studies by van der Pol et al. (5), an LR-5 category assignment indicated HCC in 94% (95% CI, 92%–96%) and malignancy in 97% (95% CI, 95%–99%) of individual liver observations. More recent meta-analyses have reported similar results (6, 7). In studies comparing systems, including AASLD, OPTN, National Comprehensive Cancer Network, and international guidelines, LI-RADS demonstrated the highest overall accuracy (8). When an observation meets LR-5 criteria, the diagnosis of HCC can be made definitively by imaging alone, obviating the need for histological confirmation before treatment. LI-RADS achieves this high specificity through a hierarchical decision tree that requires users to stepwise consider or reject other diagnostic options before considering the possibility of HCC, as well as by specifying stringent imaging criteria delineated in a diagnostic table. Among these criteria are strict requirements for size greater than or equal to 10 mm and presence of nonrim arterial-phase hyperenhancement (APHE).

By comparison, other guidelines, particularly those in Asia, prioritize sensitivity by having less stringent criteria (9). While these more relaxed criteria improve the sensitivity for HCC, they also increase the frequency of false-positive interpretations, which could lead to inappropriate choice of treatment if confirmatory biopsy is not performed.

Liver Imaging Reporting and Data System Provides Reliable Diagnostic Features and Criteria for Hepatocellular Carcinoma

Given the primary aim of standardization, the interreader agreement of LI-RADS features and categories has been a major area of investigation. The agreement for LI-RADS major features and categories has been shown to be moderate to substantial (10, 11), but importantly, the system can be applied by experts and nonexperts alike without a substantial effect on reliability (12). LI-RADS compares favorably with other systems like the former OPTN criteria (13).

Beyond the Bare Minimum

Liver Imaging Reporting and Data System Provides Standardized, Precise Terminology that Is Regularly Updated and Improved

The diagnosis of HCC has suffered historically from inconsistent and ambiguous terms, definitions, and diagnostic certainties (14). Before the development of LI-RADS, phrases such as "concerning for" or "suspicious of" littered radiology reports for HCC. These ambiguous phrases challenge the ability of the multidisciplinary team to appropriately direct care, especially for high-stakes decisions such as transplant eligibility (15).

In recognition of this challenge, LI-RADS first issued a diagnostic algorithm and lexicon in 2011, which subsequently has been regularly updated and improved. The currently published LI-RADS algorithms are the most comprehensive to date, with the latest update in 2018 heralding the unification of LI-RADS and the AASLD and, more recently, alignment with OPTN. Similarly, the LI-RADS lexicon is the most complete dictionary of liver imaging terms available (1). The lexicon has gone through a substantial vetting process to ensure that the terms are precise, easy to understand by native and nonnative English speakers, operationally sound, and accurate (16). While other HCC imaging guidelines provide criteria for diagnosis of HCC, the LI-RADS approach is singular by precisely defining the lexicon to help ensure standardization in the application. Other approaches use undefined terms, which impose unavoidable ambiguity and inconsistency.

Uniquely, the LI-RADS algorithms and lexicon are considered "living" documents, which are regularly improved by active, standing committees as new evidence emerges, new technology develops, and in response to user feedback.

The Application of Standardized, Precise Terminology in the Literature and in Clinical Practice Has Important Downstream Benefits

The use of templated reporting and standardized terminology in radiology can improve compliance with diagnostic criteria and management recommendations (17), allows clear communication, and is increasingly endorsed by major international societies as essential for providing quality in the practice of radiology (18). When considering which standardized system to use, it is important to consider the downstream effects on patient care. Furthermore, LI-RADS' templated reporting and standardized terminology allows routine clinical radiology reports to contain discrete and well-defined data elements with the potential to accelerate progress. This labeled clinical data can easily enable large-scale data mining for quality improvement or research purposes, such as training and development of deep learning algorithms.

Liver Imaging Reporting and Data System Is Based on Best Available Evidence, Expert Consensus, and Has Been Extensively Investigated

Best available evidence is critically reviewed and applied in the selection, definition, and incorporation of imaging features into the categorization algorithms. Tang et al. (19) performed a systematic review of available evidence supporting LI-RADS major features and LR-M features and provided rationale and biological basis for each feature. For instance, APHE is described regarding the process of angiogenesis in the context of carcinogenesis and citing histological evidence of unpaired arteries in progressed HCC.

LI-RADS is supported by the American College of Radiology. The infrastructure of LI-RADS is regulated to include a steering committee of expert radiologists and clinical liaisons, overseeing numerous working groups, each dedicated to a focused topic related to imaging of high-risk patients. Working group charges range from lexicon development, defining imaging technique requirements, developing reporting instructions and templates, and clarifying management implications to international considerations. The overall membership of LI-RADS comprises representatives from around the world.

Since its inception in 2011, LI-RADS has benefitted from an exponential growth in publications and scientific presentations, and LI-RADS is now the most investigated current diagnostic algorithm for HCC in the world. A PubMed search August 7, 2024, of English-language articles on HCC published since January 1, 2017, found 419 articles with "LI-RADS" in the title, 9 with "AASLD," 19 with "EASL" (European Association for the Study of the Liver), and 3 with "APASL" (Asian Pacific Association for the Study of the Liver).

Liver Imaging Reporting and Data System Is a Granular, Probabilistic Approach that Permits Individualized Management

Instead of a simplistic binary outcome of HCC or not—as supported by OPTN, EASL, and other systems—LI-RADS categories communicate unambiguously the relative probability of malignancy and HCC (Fig. 12.1). This allows for individualized management of patients. As discussed earlier, the LR-5 category communicates near 100% certainty for the diagnosis of HCC to protect against false-positive transplantation allocation. Because the availability of organs is far lower than the number of patients with HCC who could benefit, it is important to offer alternative treatment modalities to patients with lower but still substantial probabilities of

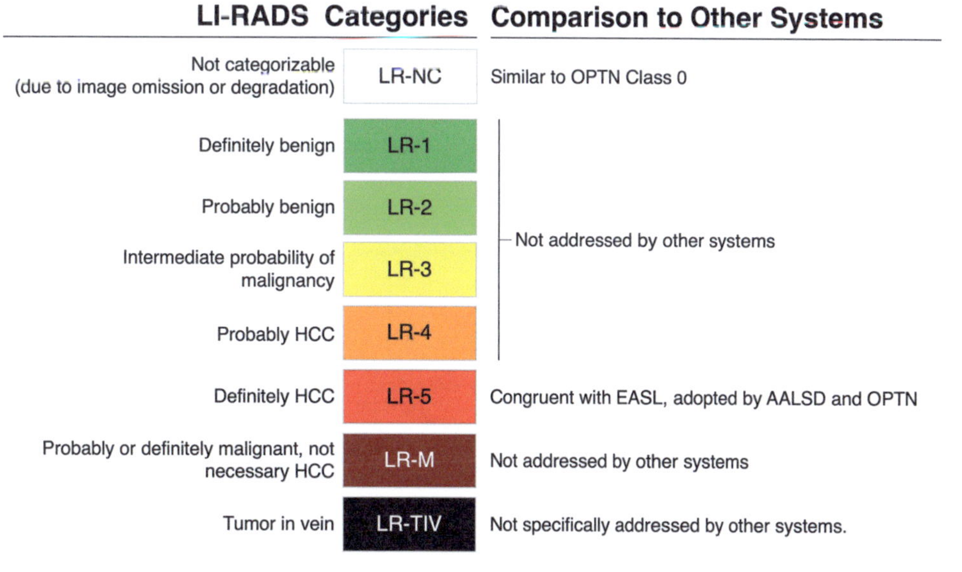

Fig. 12.1 Liver Imaging Reporting and Data System (LI-RADS) categories and correspondence to other diagnostic systems. Note the granular approach of LI-RADS opposed to the binary approach of most systems. *AASLD* American Association for the Study of Liver Diseases, *APHE* nonrim arterial-phase hyperenhancement, *EASL* European Association for the Study of the Liver

having HCC. Accordingly, the probabilistic algorithm includes categories such as LR-4 that provide opportunity in nontransplant candidates for early treatment of lesions that are likely but not definitely HCC. In the systematic review by van der Pol et al. (5), LR-4 observations represented malignancy 80% (95% CI, 75%–85%) and HCC 74% (95% CI, 67%–80%) of the time. Hence, in some clinical contexts, it may be appropriate to treat LR-4 observations as presumptive HCC without biopsy confirmation, especially in patients who are not eligible for transplantation but who could benefit from other intent-to-cure therapies, such as percutaneous ablation or, if it could be performed safely, resection. The LI-RADS granular approach also permits differentiation of HCC from non-HCC malignancy and tailored management for other categories, as discussed next.

Liver Imaging Reporting and Data System Accurately Differentiates Hepatocellular Carcinoma from Non-HCC Malignancies

The stepwise LI-RADS approach to arriving at a specific diagnosis of HCC entails first identifying nonhepatocellular malignancies. Non-HCC malignancies that arise in patients at risk of HCC most frequently are intrahepatic cholangiocarcinomas (iCCAs) and less often combined hepatocellular cholangiocarcinomas (cHCC-CCAs). The management for non-HCC malignancy differs from HCC, especially when regarding transplant eligibility. In recognition of this important distinction, LI-RADS created the LR-M diagnostic category, which is defined as a probable or definite malignancy that does not show specific HCC imaging features. Operationally, LR-M observations are characterized by targetoid features on dynamic postcontrast imaging (e.g., rim APHE, peripheral washout), diffusion imaging, and hepatobiliary phase imaging, or—in lesions that do not meet LR-5 criteria—by necrosis, infiltrative appearance, or marked diffusion restriction. The literature on the LR-M category shows robust ability to differentiate, other malignancies from HCC in patients with cirrhosis (20). Most other diagnostic systems do not specify imaging criteria to differentiate HCC from other malignancies and so do not safeguard against these diagnostic errors with important mismanagement implications.

Liver Imaging Reporting and Data System Provides Tailored Management Recommendations for each Category

Management of patients with or at risk for HCC is complex, with patient comorbidity, support network, and preferences often altering the intended course of action. However, management of indeterminate lesions is particularly complex and controversial. With a probabilistic approach and individual management recommendations for each category, LI-RADS provides guidance on best practice. Figure 12.2 shows LI-RADS categories and the corresponding management recommendations. For instance, for an LR-3 (intermediate probability of malignancy) observation on MRI, the main recommendation is for short-interval follow-up imaging, whereas for LR-M observations, which are highly likely to be malignant, a biopsy is a primary consideration. In other systems, like prior versions of AASLD and the current EASL system, the suggested action for all indeterminate lesions is biopsy. The recommendation to biopsy all indeterminate lesions is problematic on several grounds. One, it is unrealistic; anecdotally, many indeterminate lesions are not biopsied

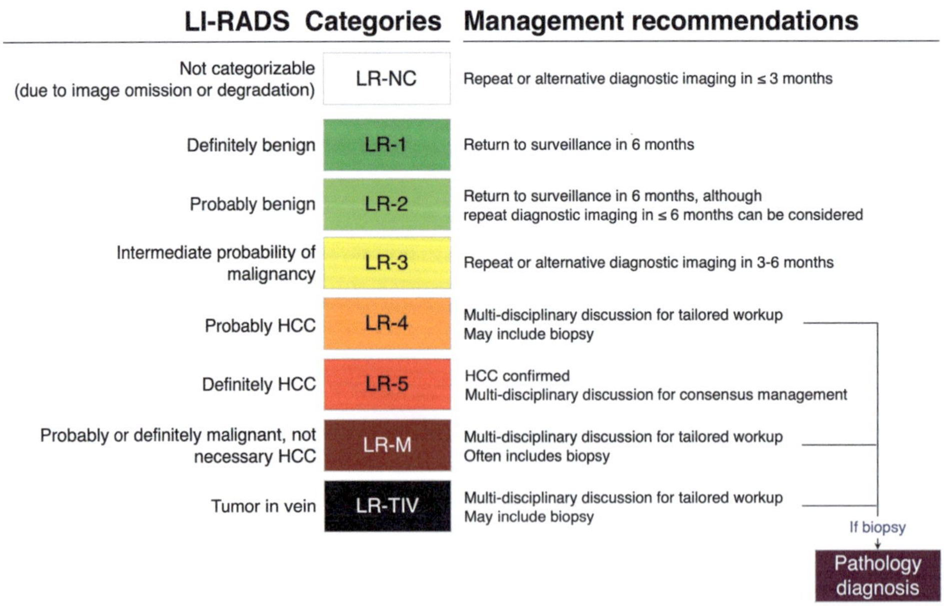

Fig. 12.2 Liver Imaging Reporting and Data System (LI-RADS) categories and management recommendations. *HCC* hepatocellular carcinoma

in clinical practice but instead followed or treated presumptively. Two, there is no scientific evidence that routine biopsy of such lesions portents any outcome benefit (21). Three, it exposes patients to additional morbidity.

Liver Imaging Reporting and Data System Guides the Entire Spectrum of Hepatocellular Carcinoma Imaging from Surveillance to Treatment Response

As previously described, beyond the diagnostic algorithms, LI-RADS covers additional imaging contexts in patients at risk for HCC: US screening/surveillance (LI-RADS US Surveillance v2024), treatment response assessment (TRA) after nonradiation-based locoregional therapy or resection (LI-RADS CT/MRI Nonradiation TRA v2024, LI-RADS CEUS Nonradiation TRA v2024) on CT and MRI or on CEUS, respectively, and TRA after radiation-based locoregional therapy (LI-RADS CT/MRI Radiation TRA v2024) on CT and MRI (1). This breadth of materials is exceptional among imaging systems, and guides standardization for each clinical context.

Some advantages of LI-RADS US Surveillance are worth mentioning: i) It is the first system to provide guidance on image interpretation, reporting, and management in patients undergoing US for HCC screening and surveillance; ii) it provides a description of examination completeness and quality (e.g., visualization score), similar to breast imaging, to supplement clinical decision making and help inform referring physicians of the potential need for alternative surveillance strategy in cases for which the examination may be inadequate (1); and iii) it is aligned with the 2023 AASLD clinical practice guidance (3).

The LI-RADS TRA algorithms also come with presumed advantages. They provide a lesion-level assessment of viability, which is essential in the context of bridging to transplantation and/or qualifying for exception status by downstaging (22). Likewise, a lesion-level approach allows individualized management, helping to direct locoregional therapy decision such as determining the need to retreat a lesion. Importantly, LI-RADS provides separate algorithms for TRA after nonradiation- and radiation-based therapy, reflecting each therapy's unique mechanisms of action and incorporating the latest advances in knowledge, as well as modality-tailored algorithms for CT/MRI and for CEUS (1).

Cons

While LI-RADS has many benefits and exceeds all other HCC algorithms in comprehensiveness, it has noteworthy drawbacks. As explained next, most of these drawbacks are shared by other systems, but some are particular to LI-RADS.

Liver Imaging Reporting and Data System Is Complex

It is inarguable that a simplistic binary approach is easier for the radiologist. With a total of 8 independent diagnostic categories, 5 major features, 21 ancillary features, 5 targetoid LR-M feature, 3 nontargetoid LR-M features, and several rules, familiarity and confidence in generating LI-RADS reports come with a learning curve. The rate of error in applying LI-RADS features, lexicon, and categorization in routine practice is likely higher than that reported in the literature.

Liver Imaging Reporting and Data System Is Subjective

Furthermore, virtually all LI-RADS imaging features are subjective and often defined in relation to the background liver. For instance, washout appearance is defined as the visually assessed temporal reduction in enhancement from earlier to later phase relative to the composite background liver. Imaging features that require comparison to background liver become problematic when the background liver is abnormal or heterogeneous because of the presence of fibrosis, iron, or fat. The latter may be especially problematic as the obesity and steatotic liver disease epidemics progress. The challenge of the background liver is compounded by the different effects of iron or fat on MRI vs CT, potentially introducing categorization variability depending on the applied modality. It is important to note that subjectivity is not confined to LI-RADS—all HCC imaging guidelines rely on subjective features—but remains a challenge to the liver imaging community at large.

Liver Imaging Reporting and Data System Interreader Agreement Is Imperfect

While the literature suggests that interreader agreement using LI-RADS is substantial for LI-RADS major features (12, 23), agreement for ancillary features is variable, with some features showing only fair agreement (23). Similar to errors in application arising from its complexity, variability between different radiologists arising from subjectivity in practice may be higher than reported in the literature. Although anecdotal, poor agreement for LI-RADS and other systems in practice is supported by the not infrequent experience of substantial overreads during multidisciplinary conferences.

While the LI-RADS treatment response algorithm provides guidance on a lesion-by-lesion level for purposes of transplantation listing, the categories do not yet translate easily for downstaging purposes. For instance, the LR-TR equivocal category could be interpreted as either viable tumor that counts toward listing or as not definitely viable and hence does not count toward listing. Early evidence suggests that up to 70% of treated lesions with equivocal viability according to LI-RADS contain microscopic foci of nonnecrotic cancer cells (24, 25). Further research is needed to understand the clinical relevance of microscopic foci of residual tumor. Such data will inform future refinements of the LI-RADS treatment response algorithm to communicate response based on clinically meaningful and actionable thresholds of residual viability.

Liver Imaging Reporting and Data System Categories May Create Confusion with Regard to Tumor Burden

The reporting of individual observations may not be the best way to communicate the overall tumor burden in the liver in all cases. For instance, a report stating that there are "5 LR-4 observations" may be less clear to the receiving clinical team than a report indicating "high probability for multifocal HCC." Figure 12.3 shows an example of multifocal HCC as a potential source of confusion. As of this writing, the LI-RADS algorithms provide few rules for when and how to transition from individual to aggregate reporting. How to best report multiple lesions is a problem for all diagnostic systems, not just LI-RADS.

The Presence of Liver Imaging Reporting and Data System 3 and LI-RADS 4 Lesions in Transplant Candidates May Create Confusion for the Multidisciplinary Team

LR-3 and LR-4 observations frequently coexist with LR-5 lesions in transplant candidates. The presence of these lesions may create confusion for the multidisciplinary team and potentially could negatively affect a patient's treatment options. While many LR-3 and LR-4 observations are HCC (5), their presence does not necessarily indicate metastatic spread beyond the liver and so may portend little if any risk in the setting of transplantation. To our knowledge, there is currently no evidence that LR-3 or LR-4 observations when present in addition to T2-stage HCC have any effect on outcomes or posttransplant recurrence. The initial impetus for OPTN criteria adoption in 2013 was to reduce false-positive rates for HCC diagnosis in the context of transplantation. Subsequent publications suggest this goal has been

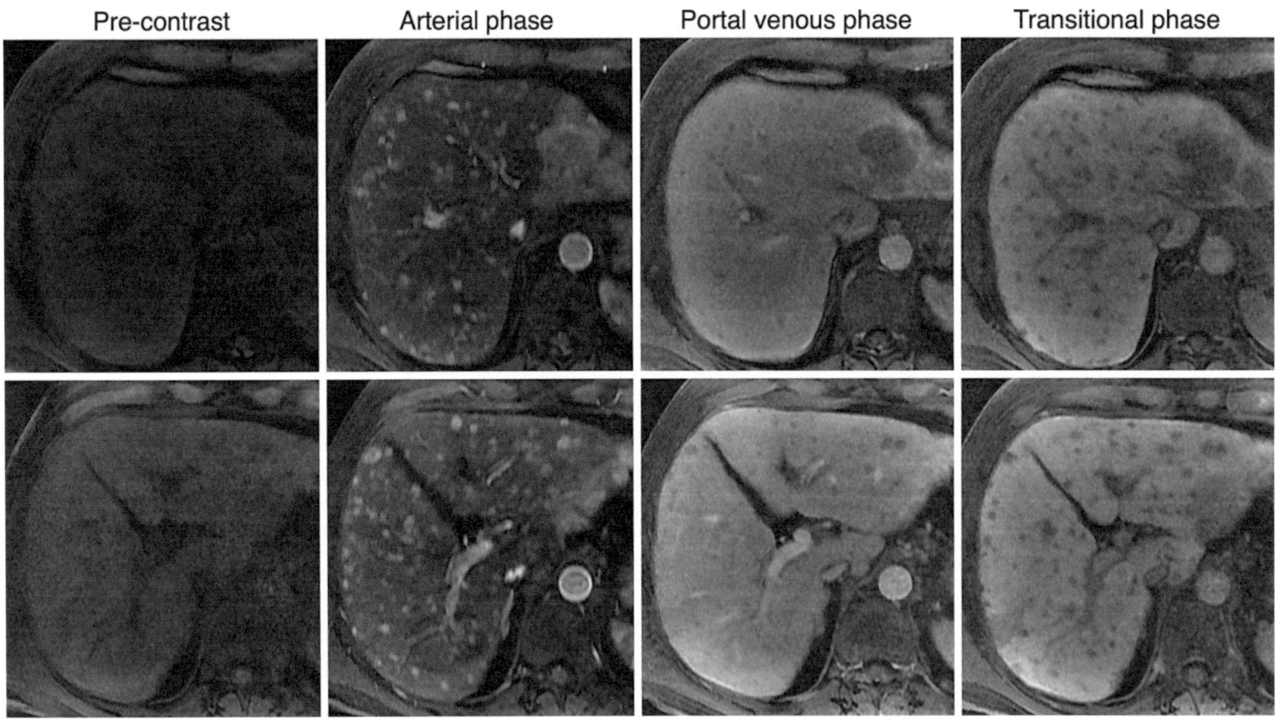

Fig. 12.3 A 65-year-old man, chronic hepatitis C cirrhosis, with multifocal HCC. A dominant mass is seen in the left lobe, measuring 28 mm, with nonrim arterial-phase hyperenhancement and washout, indicating Liver Imaging Reporting and Data System (LR-5; definite hepatocellular carcinoma [HCC]). Multiple additional small observations are seen throughout the liver, most measuring less than 10 mm. If reported individually, these observations would be categorized as LR-4. However, these should be reported in aggregate and in the context of the dominant mass, indicating multifocal HCC

accomplished (26). Whether and how an ordinal probabilistic approach benefits the management of transplant patients is an area that merits additional research.

Liver Imaging Reporting and Data System Can be Challenging to Apply in Longitudinal Follow-Up

LI-RADS considers some aspects of longitudinal follow-up, such as subthreshold and threshold growth, and on the flip side spontaneous resolution. However, the application of ancillary and major features over time can be confusing, especially if they come and go based on differences in dynamic contrast timing or technique. If applied stringently at each time point, fluctuations in the "callability" of imaging features may cause temporal variability in lesion categorization. For instance, a 15-mm lesion with APHE and washout appearance (categorized LR-5) might fail to show APHE when reimaged 3 months later because of a technical factor such as arterial phase mistiming rather than a biological change, potentially leading to a spurious recategorization as LR-4. Fluctuations in imaging features and categorization, while not solely a LI-RADS problem, can lead to confusion.

Evidence for Liver Imaging Reporting and Data System Is Not High Level

Despite the abundance of research articles focused on validating and describing LI-RADS' performance, most of these papers are single-center, retrospective, observational cohorts (5–7), with unavoidable selection or verification bias. The current paradigm of imaging as the definitive diagnosis often precludes the opportunity for a true independent reference standard, that is, histopathology. Many studies of LI-RADS have focused on enriched populations, not reflective of natural consecutive cohorts, which affects the ability to truly measure the positive predictive value of the system. Low-level evidence is a problem for all diagnostic systems. The few prospective studies in the field have also been single-center experiences. Multicenter prospective registries with data both from expert and community centers are needed.

Likewise, the benefits of standardized reporting for US surveillance examinations and use of alternative surveillance strategies in the event of an inadequate examination have not been studied and so the effect on patient-level outcomes is not known. In addition, the LR-TR algorithm categories may require further validation and refinement to determine what might represent a clinically meaningful threshold for determining residual viable tumor. Unlike more established patient-level response assessment systems, the lesion-level LR-TR categories do not by themselves provide patient-level prognostic data, and it is anticipated that integration of patient-level categorizations will be required to provide this capability.

Conclusion

Many HCC diagnostic systems are in use. The main difference between LI-RADS and other systems (e.g., prior versions of AASLD, current EASL system) is its ordinal probabilistic approach vs a binary approach. The LI-RADS probabilistic approach may introduce some complexity in clinical practice, particularly in the setting of transplantation. However, most HCC patients encountered in clinical practice ultimately will not receive a liver transplantation and recognition of high-risk lesions in these patients can allow for earlier access to treatment as well as individualized management. Furthermore, many of the challenges discussed in this review are not unique to LI-RADS. What is unique is that LI-RADS has standing committees aimed to solve these issues by incorporating emerging evidence and expert consensus to continuously improve the LI-RADS approach. The LI-RADS lexicon is the most vetted and comprehensive dictionary of liver imaging terms. The LI-RADS algorithm is accurate and reliable for the diagnosis of HCC. LI-RADS is currently the only approach that covers the entire spectrum of imaging in patients at risk for HCC, the only one that differentiates HCC from non-HCC malignancies, and the only one to provide a probabilistic approach to indeterminate lesions to allow for tailored management.

Hence, it is our recommendation that "all liver lesions in patients with cirrhosis undergoing HCC surveillance should be characterized and reported using LI-RADS criteria."

Conflict of Interest Information

Claude B. Sirlin

Financial disclosures:

- Educational materials royalties: Wolters Kluwer
- Payment for educational lectures on liver imaging: Medscape
- Research grants: ACR, Foundation of NIH, Bayer, GE, Gilead, NIH, Philips, Siemens, V Foundation
- Consulting: Altimmune, Ascelia Pharma AB, Blade, Boehringer, Epigenomics, Guerbet, Livivos
- Consulting under auspices of UC: AMRA, BMS, Exact Sciences, GE Digital, IBM-Watson
- Core lab services: Enanta, Gilead, ICON, Intercept, Nusirt, Orsobio, Shire, Synageva, Takeda
- Stock and stock options: Livivos
- Executive position in company: Livivos (Chief Medical Officer, unsalaried position with stock options and

stock) through June 28, 2023; Principal Scientific Advisor to Livivos (unsalaried position with stock options and stock) since June 28, 2023
- Equipment loans to institution: Butterfly, GE, Siemens, Mayo
- Provision of contrast material to institution: Bayer

Non-Financial disclosures:

- Founder, Past Chair, and current Member, LI-RADS Steering Committee
- Co-Chair, LI-RADS International Working Group
- Co-Chair, LI-RADS Lexicon & Writing Group
- Member, numerous other LI-RADS Working Groups
- Member, Data Safety Monitoring board for National Cancer Institute funded Early Detection Research Network

Guilherme Moura Cunha
Financial disclosures

- Research grants: Radiologic Society of North America, Microsoft corporation, Society of Radiologists in Ultrasound

Non-Financial disclosures

- Co-Chair, LI-RADS Reporting Working Group
- Member, LI-RADS international working group, guarantor of Portuguese translation.
- Member, LI-RADS Steering Committee.

Victoria Chernyak
Financial disclosures

- None

Non-Financial disclosures

- Co-chair LI-RADS Steering Committee
- Member, numerous other LI-RADS Working Groups

Kathryn Fowler
Financial disclosures

- Institutional grant support: Bayer, Median, Pfizer, Cannon, Siemens, GE, Echosense
- Personal consulting: Bayer, Ascelia, Guerbet
- CME talks: VRAD, CME science, Bayer
- Speaker: Medscape
- Expert Witness
- Founding advisory board: Quantix bio

Non-Financial disclosures

- ACR Panel Chair
- ACR LI-RADS Steering Committee Co-Chair and working group chair/member
- SAR Portfolio Director

Conflict of Interest Claude B. Sirlin payment for educational lectures on liver imaging: Health Pro Match.

References

1. https://www.acr.org/Clinical-Resources/Reporting-and-Data-Systems/LI-RADS. 2020. Accessed on 11 Aug 2024.
2. Tang A, Cruite I, Mitchell DG, Sirlin CB. Hepatocellular carcinoma imaging systems: why they exist, how they have evolved, and how they differ. Abdom Radiol (NY). 2018;43(1):3–12.
3. Singal AG, Llovet JM, Yarchoan M, Mehta N, Heimbach JK, Dawson LA, Jou JH, Kulik LM, Agopian VG, Marrero JA, Mendiratta-Lala M, Brown DB, Rilling WS, Goyal L, Wei AC, Taddei TH. AASLD practice guidance on prevention, diagnosis, and treatment of hepatocellular carcinoma. Hepatology. 2023;78(6):1922–65.
4. Omata M, Cheng AL, Kokudo N, Kudo M, Lee JM, Jia J, et al. Asia-Pacific clinical practice guidelines on the management of hepatocellular carcinoma: a 2017 update. Hepatol Int. 2017;11(4):317–70.
5. van der Pol CB, Lim CS, Sirlin CB, McGrath TA, Salameh JP, Bashir MR, et al. Accuracy of the liver imaging reporting and data system in computed tomography and magnetic resonance image analysis of hepatocellular carcinoma or overall malignancy-a systematic review. Gastroenterology. 2019;156(4):976–86.
6. Lee S, Kim YY, Shin J, Son WJ, Roh YH, Choi JY, Sirlin CB, Chernyak V. Percentages of hepatocellular carcinoma in LI-RADS categories with CT and MRI: a systematic review and meta-analysis. Radiology. 2023;307(1):e220646.
7. Liang Y, Xu F, Guo Y, Lai L, Jiang X, Wei X, Wu H, Wang J. Diagnostic performance of LI-RADS for MRI and CT detection of HCC: a systematic review and diagnostic meta-analysis. Eur J Radiol. 2021 Jan;134:109404.
8. Erkan B, Meier J, Clark TJ, Kaplan J, Lambert JR, Chang S. Non-invasive diagnostic criteria of hepatocellular carcinoma: comparison of diagnostic accuracy of updated LI-RADS with clinical practice guidelines of OPTN-UNOS, AASLD, NCCN, EASL-EORTC, and KLSCG-NCC. PLoS One. 2019;14(12):e0226291.
9. Kim TH, Kim SY, Tang A, Lee JM. Comparison of international guidelines for noninvasive diagnosis of hepatocellular carcinoma: 2018 update. Clin Mol Hepatol. 2019;25(3):245–63.
10. Kang JH, Choi SH, Lee JS, Park SH, Kim KW, Kim SY, Lee SS, Byun JH. Interreader agreement of liver imaging reporting and data system on MRI: a systematic review and meta-analysis. J Magn Reson Imaging. 2020;52(3):795–804.
11. Hong CW, Chernyak V, Choi JY, Lee S, Potu C, Delgado T, Wolfson T, Gamst A, Birnbaum J, Kampalath R, Lall C, Lee JT, Owen JW, Aguirre DA, Mendiratta-Lala M, Davenport MS, Masch W, Roudenko A, Lewis SC, Kierans AS, Hecht EM, Bashir MR, Brancatelli G, Douek ML, Ohliger MA, Tang A, Cerny M, Fung A, Costa EA, Corwin MT, McGahan JP, Kalb B, Elsayes KM, Surabhi VR, Blair K, Marks RM, Horvat N, Best S, Ash R, Ganesan K, Kagay CR, Kambadakone A, Wang J, Cruite I, Bijan B, Goodwin M, Moura Cunha G, Tamayo-Murillo D,

Fowler KJ, Sirlin CB. A multicenter assessment of Interreader reliability of LI-RADS version 2018 for MRI and CT. Radiology. 2023;307(5):e222855.

12. Fowler KJ, Tang A, Santillan C, Bhargavan-Chatfield M, Heiken J, Jha RC, et al. Interreader reliability of LI-RADS version 2014 algorithm and imaging features for diagnosis of hepatocellular carcinoma: a large international multireader study. Radiology. 2018;286(1):173–85.
13. Davenport MS, Khalatbari S, Liu PS, Maturen KE, Kaza RK, Wasnik AP, et al. Repeatability of diagnostic features and scoring systems for hepatocellular carcinoma by using MR imaging. Radiology. 2014;272(1):132–42.
14. Corwin MT, Lee AY, Fananapazir G, Loehfelm TW, Sarkar S, Sirlin CB. Nonstandardized terminology to describe focal liver lesions in patients at risk for hepatocellular carcinoma: implications regarding clinical communication. AJR Am J Roentgenol. 2018;210(1):85–90.
15. Poullos PD, Tseng JJ, Melcher ML, Concepcion W, Loening AM, Rosenberg J, et al. Structured reporting of multiphasic CT for hepatocellular carcinoma: effect on staging and suitability for transplant. AJR Am J Roentgenol. 2018;210(4):766–74.
16. Chernyak V, Tang A, Do RKG, Kamaya A, Kono Y, Santillan CS, Fowler KJ, Bashir MR, Cunha GM, Fetzer DT, Kielar A, Lee JT, Mendiratta-Lalla M, Sirlin CB, LI-RADS. Steering committee and guarantors of translation integrity listed in the acknowledgements. Liver imaging: it is time to adopt standardized terminology. Eur Radiol. 2022;32(9):6291–301.
17. Aase A, Fabbrini AE, White KM, Averill S, Gravely A, Melzer AC. Implementation of a standardized template for reporting of incidental pulmonary nodules: feasibility, acceptability, and outcomes. J Am Coll Radiol. 2020;17(2):216–23.
18. European Society of Radiology (ESR). ESR paper on structured reporting in radiology. Insights Imaging. 2018;9(1):1–7. https://doi.org/10.1007/s13244-017-0588-8. Epub 2018 Feb 19. PMID: 29460129; PMCID: PMC5825315
19. Tang A, Bashir MR, Corwin MT, Cruite I, Dietrich CF, Do RKG, et al. Evidence supporting LI-RADS major features for CT- and MR imaging-based diagnosis of hepatocellular carcinoma: a systematic review. Radiology. 2018;286(1):29–48.
20. Kim YY, Kim MJ, Kim EH, Roh YH, An C. Hepatocellular carcinoma versus other hepatic malignancy in cirrhosis: performance of LI-RADS version 2018. Radiology. 2019 Apr;291(1):72–80.
21. Heimbach JK, Kulik LM, Finn RS, Sirlin CB, Abecassis MM, Roberts LR, Zhu AX, Murad MH, Marrero JA. AASLD guidelines for the treatment of hepatocellular carcinoma. Hepatology. 2018 Jan;67(1):358–80.
22. Cunha GM, Tamayo-Murillo DE, Fowler KJ. LI-RADS and transplantation: challenges and controversies. Abdom Radiol (NY). 2021;46(1):29–42.
23. Alhasan A, Cerny M, Olivié D, Billiard JS, Bergeron C, Brown K, Bodson-Clermont P, Castel H, Turcotte S, Perreault P, Tang A. LI-RADS for CT diagnosis of hepatocellular carcinoma: performance of major and ancillary features. Abdom Radiol (NY). 2019;44(2):517–28.
24. Chaudhry M, McGinty KA, Mervak B, Lerebours R, Li C, Shropshire E, et al. The LI-RADS version 2018 MRI treatment response algorithm: evaluation of ablated hepatocellular carcinoma. Radiology. 2020;294(2):320–6.
25. Shropshire EL, Chaudhry M, Miller CM, Allen BC, Bozdogan E, Cardona DM, et al. LI-RADS treatment response algorithm: performance and diagnostic accuracy. Radiology. 2019;292(1):226–34.
26. Fowler KJ, Karimova EJ, Arauz AR, Saad NE, Brunt EM, Chapman WC, Heiken JP. Validation of organ procurement and transplant network (OPTN)/united network for organ sharing (UNOS) criteria for imaging diagnosis of hepatocellular carcinoma. Transplantation. 2013;95(12):1506–11.

13 Can Imaging Findings of Arterial Enhancement and Washout Still Represent a Benign Liver Lesion?

William R. Masch and Mishal Mendiratta-Lala

Nonrim arterial-phase enhancement and delayed washout are highly specific in at-risk patients and should be considered diagnostic of hepatocellular carcinoma (HCC).

Main Points

1. Arterial-phase hyperenhancement (APHE) is defined as enhancement in the arterial phase resulting in brightness that is greater than the background liver. Washout is defined as a temporal reduction of enhancement within an observation from an earlier to later phase and the observation must become hypoenhancing with respect to the background liver.
2. The computed tomography/magnetic resonance imaging (CT/MRI) diagnostic algorithm of the Liver Imaging Reporting and Data System (LI-RADS) applies only to patients with known or prior HCC, chronic hepatitis B virus (HBV) infection, or cirrhosis. Similarly, LI-RADS criteria have not been validated in patients with extrahepatic malignancies and diagnostic performance may be lower in this patient population.
3. The presence of nonrim APHE and delayed washout have a high specificity (95–100%) and positive predictive value (PPV; 91–100%) for diagnosis of HCC in at-risk patients. Diagnostic performance is higher in lesions larger than 2 cm than in those 1–2 cm.
4. Features such as targetoid appearance or rim hyperenhancement raise concern for non-HCC malignancies.

All major international guidelines for imaging-based diagnosis of HCC consider the combination of APHE and "washout" at multiphasic contrast-enhanced CT or MRI to be a hallmark of definite HCC [1]. In certain patient populations, this combination of imaging features in observations greater than or equal to 10 mm in size is considered equivalent to a histologic diagnosis and precludes the need for percutaneous biopsy [2]. Often overlooked is the fact that such diagnostic algorithms allowing for noninvasive HCC diagnosis apply only to relatively small, well-defined patient populations at high risk of the development of HCC. The CT/MRI Diagnostic Algorithm of the LI-RADS [3] applies only to patients with known or prior HCC, chronic HBV, and/or cirrhosis, excluding pediatric patients and patients with cirrhosis from a vascular disorder or congenital hepatic fibrosis. According to the guidelines published by the American Association for the Study of Liver Disease (AASLD), [2, 4] the Asian Pacific Association for the Study of the Liver (APASL), [5] the European Association for the Study of the Liver (EASL), [6] and the Korean Liver Cancer Association-National Cancer Center (KLCA-NCC), [7] the diagnostic imaging algorithms incorporating APHE and "washout" apply only to patients at high risk of HCC with a positive screening or surveillance test, and in the case of the AASLD and the APASL, to patients in whom there is clinical suspicion of HCC. While all international guidelines have slight differences in what constitutes the high-risk population, they generally include most patients with cirrhosis and/or chronic HBV infection, similarly to LI-RADS.

Of all patients undergoing cross-sectional imaging of the liver, the proportion of patients meeting the criteria for the application of an HCC imaging algorithm is small. In the United States, the prevalence of chronic liver disease is estimated between 4% and 7%, [8] and the prevalence of cirrhosis is less than 0.5% [9]. Even in enriched populations of patients undergoing liver imaging, only a small minority are known to have chronic liver disease. One study examining inpatient individuals undergoing liver imaging for elevated liver function tests found that only 14% of patients had cirrhosis [10] and would meet the criteria for application of an imaging-based algorithm for HCC diagnosis. Conversely, incidental liver lesions are common and found in more than

W. R. Masch · M. Mendiratta-Lala (✉)
Division of Abdominal Imaging, Department of Radiology, University of Michigan, Ann Arbor, MI, USA
e-mail: mascwill@med.umich.edu; mmendira@med.umich.edu

N. S. Reau, A. G. Singal (eds.), *Controversies in Hepatocellular Carcinoma*, https://doi.org/10.1007/978-3-031-90225-3_13

30% of patients undergoing liver imaging, and it has been reported that up to 50% of the population has a benign liver lesion [11, 12]. Thus, the proportion of liver observations identified at cross-sectional imaging not meeting the criteria for the application of an HCC imaging algorithm is large.

The appropriate incorporation of clinical data is a requisite for the accurate characterization of liver lesions. The likelihood that a liver observation with APHE and "washout" reflects HCC is highly dependent on clinical context and varies with the pretest probability of HCC. In a young patient without liver disease and a low pretest probability of HCC, said observation is statistically likely to be benign, probably a focal nodular hyperplasia [FNH] or a hepatocellular adenoma, as these lesions are known to intermittently exhibit combination APHE and "washout" [13–16]. However, in a patient with hepatitis C virus–related cirrhosis, said observation is likely to be HCC because potential HCC mimics (e.g., FNH, hepatocellular adenoma, and hypervascular liver metastasis) are uncommon in the cirrhotic milieu.

In addition to clinical context and the presence or absence of APHE and "washout," there are many additional imaging factors, such as pattern of APHE, observation size, presence or absence of capsule appearance, and growth rate, that further influence likelihood of HCC, and that should be taken into account at the time of image interpretation [17]. Rapid growth rate and lesion size are both nonspecific markers suggestive of malignancy; benign regenerative nodules do not tend to rapidly enlarge and rarely exceed 1.5 cm in size, [18–20] Capsule appearance is considered a relatively specific finding for HCC as HCC is the only tumor occurring in a cirrhotic liver known to have a fibrous tumor capsule at histologic examination [21, 22]. Because the purpose of this chapter is to focus on a comprehensive review of liver lesions exhibiting APHE and "washout," the other major features of HCC (observation size, threshold growth, and capsule appearance) will not be further discussed in depth.

One challenge to the interpretation and reporting of liver imaging is that the definitions of APHE and "washout" differ, albeit slightly, among the various imaging algorithms for HCC diagnosis. These differences largely relate to regional variations in HCC epidemiology and management patterns [23]. To standardize and enhance communication regarding diagnosis and management of patients with liver disease, as well as improve the quality of literature, standardized terminology is critical. Importantly, the AASLD and LI-RADS harmonized recommendations in 2018 when the AASLD incorporated the LI-RADS categories into its guidance document for HCC diagnosis, staging, and management [4]. In addition, in 2020, the LI-RADS Steering Committee in conjunction with the LI-RADS International Work Group created a standardized lexicon for describing liver observations [Tang A, et al. Liver cancer: It is time to adopt a standardized imaging terminology for primary liver cancer]. The definitions discussed in this chapter will be those elaborated by LI-RADS, unless otherwise noted, because these definitions are the most widely accepted in North America. The subsequent sections of this chapter will provide an in-depth discussion of the rationale and evidence of APHE and "washout" as major imaging features for HCC in at-risk populations, and will highlight various situations that may alter the diagnostic accuracy of these findings.

Arterial Phase Hyperenhancement—Definition, Physiologic Basis, and Evidence

APHE is defined as enhancement in the arterial phase of CT or MRI that is unequivocally greater (in whole or in part) than background liver resulting in brightness that is greater than liver. Thus, an observation with APHE must be higher in attenuation (CT) or intensity (MRI) than liver in the arterial phase. In addition to simple presence or absence, the pattern of APHE is an important consideration for liver observation characterization. Consequently, LI-RADS distinguishes between "nonrim APHE" (APHE associated with HCC) and "rim APHE" (APHE suggestive of a non-HCC malignancy). "Nonrim APHE" is considered a major feature of HCC, and the LI-RADS lexicon specifically defines it as APHE that is not most pronounced at the periphery of an observation [Tang A, et al. Liver cancer: It is time to adopt a standardized imaging terminology for primary liver cancer]. Conversely, "rim APHE" is the pattern of APHE associated with non-HCC malignancies, including intrahepatic cholangiocarcinoma and liver metastasis [24, 25]. Yet another pattern of APHE is that which is peripheral, discontinuous, and nodular or flame-shaped in morphology; such an enhancement pattern is considered diagnostic of hemangioma and does not qualify as suspicious for HCC [26, 27]. Thus, the pattern of APHE, not simply its presence, is important in the characterization of a liver nodule.

During multistep hepatocarcinogenesis, a liver nodule undergoes multiple pathophysiologic changes as it transforms from a benign regenerative nodule, through serially higher-risk HCC precursor nodules (i.e., low-grade dysplastic nodule, high-grade dysplastic nodule, early HCC) into progressed, encapsulated HCC [19]. Notably, there is progressive accumulation of unpaired, nontriadal neoarteries at each step of the process [28, 29]. This accumulation of neoarteries is initially offset by the serial loss of portal tracts containing blood flow from native hepatic arteries and portal veins, and consequently most HCC precursor nodules remain hypoenhancing or isoenhancing in the arterial phase. Typically, it is not until a nodule transforms into a progressed (overtly malignant) HCC that it accumulates a sufficient number of neoarteries to offset the loss of arterial flow from the native hepatic artery to manifest APHE at multiphasic CT or MRI, thus allowing it to be distinguished from nonmalignant liver nodules [18].

Traditionally, APHE has been considered highly sensitive for the detection of HCCs, with reported sensitivities up to 96% [30–32]. However, many of the studies touting such high sensitivities were retrospective and prone to incorporation and verification bias [17]. Studies using explant pathology report lower sensitivities ranging from 76% to 83% [33, 34]. Importantly, the sensitivity of APHE for the detection of HCC increases with observation size. For large HCCs greater than 20 mm, sensitivities of up to 100% have been reported, [33] and for HCCs smaller than 10 mm, sensitivities as low as 10% have been reported [34]. While APHE is very sensitive for progressed HCC, it is not sensitive for early, well-differentiated HCCs [35]. Last, APHE, as an imaging finding by itself, lacks specificity for HCC as many benign (e.g., hemangioma, arterioportal shunt) and malignant (e.g., intrahepatic cholangiocarcinoma) lesions may exhibit APHE in a cirrhotic liver.

"Washout"—Definition, Biologic Basis, and Evidence

"Washout" has a 2-pronged definition [Tang A, et al. Liver cancer: It is time to adopt a standardized imaging terminology for primary liver cancer]. First, there must be a temporal reduction of enhancement (in whole or in part) within an observation relative to composite liver tissue from an earlier to later phase following some degree of arterial-phase enhancement. Second, the observation must be hypoenhancing with respect to composite liver tissue on said later phase. At contrast-enhanced CT and MRI using extracellular agents, said hypoenhancement may be identified in the portal venous phase or any delayed phase thereafter. However, at MRI performed with the hepatobiliary contrast agent gadoxetate disodium, the hypoenhancement indicative of "washout" may be identified only in the portal venous phase (according to the current CT/MRI LI-RADS v 2018 diagnostic algorithm) [3].

There are two main theories for the biologic mechanism of "washout" in HCC. During multistep hepatocarcinogenesis, there is a progressive decline in the number of portal tracts perfusing a nodule [36]. By the time a nodule evolves into progressed HCC with a fibrous capsule, venous drainage of the nodule has transitioned from hepatic veins to portal veins, and there is no longer a portal venous blood supply [37]. Consequently, many theorize that progressed HCCs receive relatively less contrast-enhanced blood in comparison to background liver parenchyma because of the diminution of portal flow, and this results in hypoenhancement in the portal venous and delayed phases. Alternatively, others theorize that "washout" is due to rapid venous drainage of HCC nodules after the arterial phase [38]. Regardless of the mechanism, the imaging manifestation of "washout" is the result of these aforementioned vascular changes occurring to HCC nodules at the terminal stages of multistep hepatocarcinogenesis.

Most large international societies worldwide share a similar definition of "washout" for contrast-enhanced CT and MRI with extracellular agents. However, there are important regional differences in the definition of "washout" for gadoxetate disodium–enhanced MRI. Guidelines in Europe and North America restrict washout appearance to the portal venous phase only [3, 4, 6]. After the portal-venous phase of a gadoxetate-enhanced MRI, an appreciable component of liver parenchymal enhancement is due to intracellular accumulation of the contrast molecules within hepatocytes expressing OATP transport proteins in their cell membranes [39]. Consequently, when using gadoxetate disodium, it is not possible to discern whether transitional or hepatobiliary phase hypointensity within an observation is due to true washout kinetics or due to low OATP expression and diminished intracellular accumulation of contrast molecules. In comparison to Western definitions, guidelines elaborated by the KLCA-NCC and APASL liberalize the definition of "washout" for gadoxetate disodium–enhanced MRI to include hepatobiliary phase hypointensity within their "washout" definitions as long as steps are taken to exclude hemangioma as a diagnosis [5, 7]. Differences in the definition of "washout" among various guidelines worldwide largely relate to regional differences in HCC epidemiology and treatment strategy beyond the scope of this chapter. However, it should be noted that the definition of "washout" on gadoxetate disodium–enhanced MRI is an active area of scientific investigation [40, 41].

Most studies examining the diagnostic performance of "washout" for the diagnosis of HCC examine the imaging feature in combination with APHE. The few published studies comparing washout appearance alone to combination APHE and washout appearance show washout appearance alone to have a higher sensitivity (59–68%) and a lower specificity (87.5–95%) compared to combination APHE and washout appearance (sensitivity: 44–58.2%; specificity 96.4–100%) [31, 42]. It is important to realize that the perception of washout appearance at multiphasic CT or MRI requires adequate perfusion and enhancement of background liver parenchyma. In patients with altered hepatic perfusion (e.g., advanced cirrhosis, chronic portal vein thrombosis, or indwelling transjugular intrahepatic portosystemic shunt), washout appearance may be less apparent or absent within progressed HCCs; however, this is an area that is not well studied [17].

Combination Arterial-Phase Hyperenhancement and "Washout" in High-Risk Populations

In patients at risk of HCC meeting criteria for application of the LI-RADS CT/MRI diagnostic algorithm, neither APHE nor "washout" alone have specificity for the diagnosis of HCC. However, these two imaging features in combination are highly specific for HCC. Most studies report a high specificity (95–100%) and a high PPV (91–100% [31, 33, 34, 42–44]) for the diagnosis of HCC in high-risk patients when nodules exhibit both APHE and "washout." The performance data are size-dependent, with higher reported specificities approaching 100% for larger nodules greater than or equal to 20 mm in size, and slightly lower reported specificities ranging from approximately 90% to 95% for smaller nodules of 10–19 mm. Importantly, the combination of APHE and "washout" is rarely observed in intrahepatic cholangiocarcinoma [24].

The high specificity of APHE and "washout" for HCC diagnosis comes at the cost of a lowered sensitivity, which has a widely reported range of 44–98% [31, 33, 34, 42–44]. However, since the imaging-based diagnosis of HCC may be used as surrogate for histologic diagnosis, highly specific criteria for HCC diagnosis are a necessity since therapeutic and transplant allocation decisions are frequently made on the basis of imaging findings alone. In the LI-RADS CT/MRI diagnostic algorithm, the combination of APHE and "washout" is the most important criterion for the designation of definite HCC, LR-5, for liver observations greater than or equal to 10 mm. However, many HCCs will not exhibit both APHE and "washout" together, lowering the sensitivity of this combination of findings (Fig. 13.1). Given this, LI-RADS categorization of liver observations in patients at risk of HCC is helpful to determine the likelihood of HCC when the rigid criteria for definite diagnosis are not present. A recent large systematic review including 3556 liver observations in 2760 patients from 17 retrospective studies found increasing LI-RADS categories connoted increasing likelihood of HCC; LR-3 observations were 38% likely to be HCC, LR-4 observations were 74% likely to be HCC, and LR-5 observations were 94% likely to be HCC [45]. The LI-RADS score assigned to a liver observation in a patient at high risk of HCC is based on the presence or absence of the major features of HCC (APHE, washout appearance, capsule appearance, size, and threshold growth) and is further modified by the presence of ancillary features suggestive of malignancy, HCC in particular, or benignity (see Fig. 13.1).

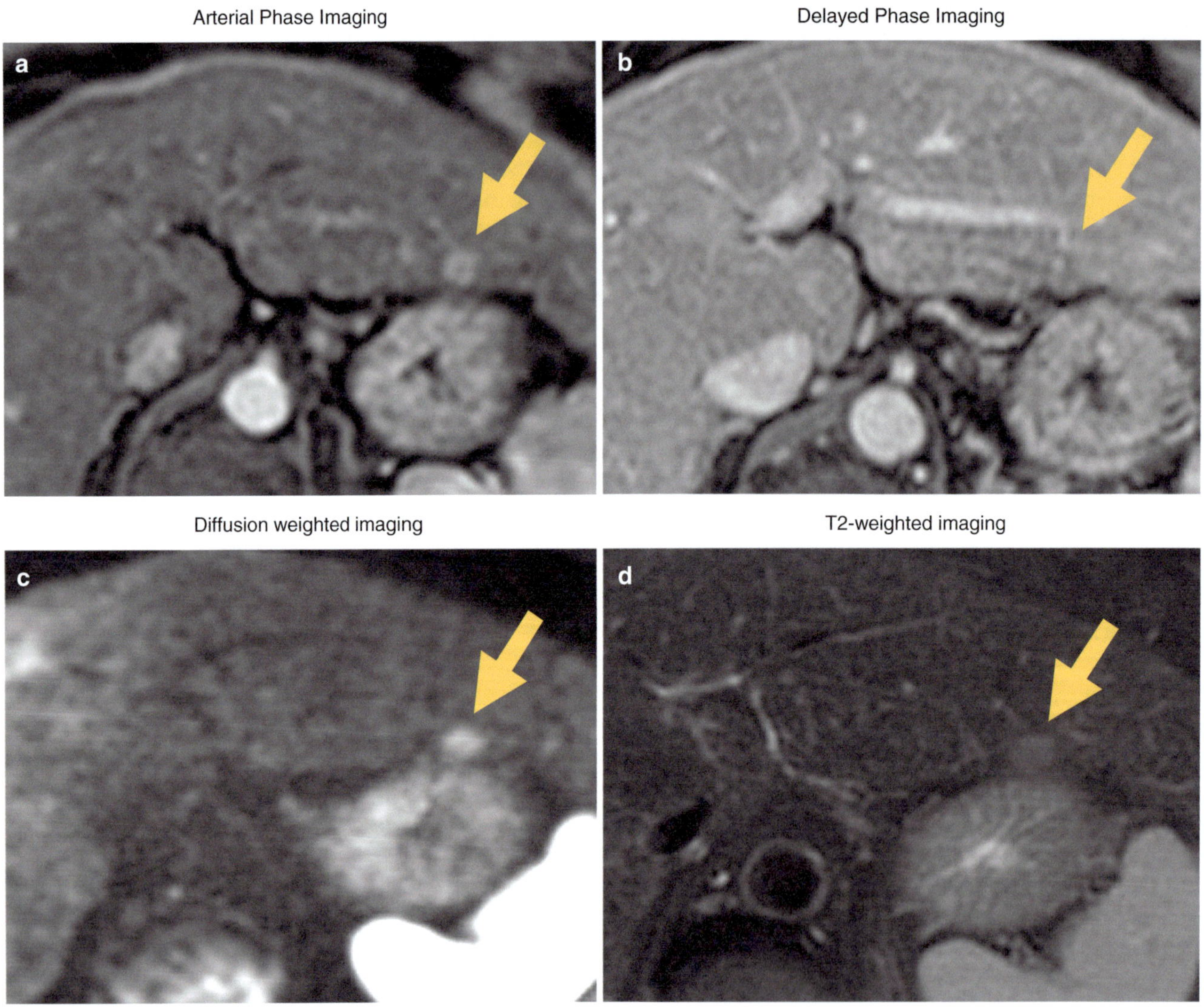

Fig. 13.1 Biopsy-proven hepatocellular carcinoma in a patient with hepatitis B–related cirrhosis. The nodule exhibits arterial-phase hyperenhancement in (**a**) the arterial phase, but no washout appearance or capsule appearance in (**b**) the delayed phases. (**c**) The observation restricts diffusion on the diffusion-weighted image, suggesting increased cellularity; (**d**) this is considered an ancillary feature suggesting malignancy in the computed tomography/magnetic resonance imaging Liver Imaging Reporting and Data System (LI-RADS) diagnostic algorithm. Given this spectrum of findings, this observation would be characterized LI-RADS 4

An important consideration is the performance of combination APHE and "washout" for HCC diagnosis in at-risk patients with an active extrahepatic malignancy. There is accumulating evidence that targetoid features (e.g., rim APHE) are common in non-HCC malignancies and rare in HCC, [46, 47] and if present, such features should preclude the imaging-based diagnosis of HCC. However, most studies evaluating these imaging features focus on the discrimination of HCC from intrahepatic cholangiocarcinoma and hepatocholangiocarcinoma, not from metastatic deposits. HCC imaging algorithms such as LI-RADS have not been validated in this specific patient population, and it is possible that the PPV of LR-5 will be lowered in patients with active extrahepatic cancer. Thus, caution should be taken when entertaining the diagnosis of HCC on the basis of APHE and "washout" in a patient with cirrhosis and an active extrahepatic malignancy; such patients may warrant a multidisciplinary tumor board discussion and consideration for biopsy.

Combination Arterial-Phase Hyperenhancement and "Washout" in Lower-Risk Populations

There are many patients at increased risk of HCC who meet the criteria for HCC screening programs but who do not meet the rigid inclusion criteria for the CT/MRI diagnostic algorithm. These include patients with noncirrhotic hepatitis C virus and noncirrhotic metabolic dysfunction-associated steatotic liver disease. However, the diagnostic performance of APHE and "washout" in this patient population has not been well established. A recent retrospective study found that the LI-RADS 5 designation was highly specific (97–100%) for differentiating HCC from non-HCC liver cancer in these patients (most LR-5 designations in the study were made on the basis of APHE and "washout" and not on the basis of other features) [48]. However, this study did not examine the ability of APHE and "washout" to differentiate HCC from benign liver lesions because all individuals enrolled in the study had known liver malignancy.

Conversely, the ability of combination APHE and "washout" to diagnose HCC in patients with noncirrhotic livers with no risk factors for HCC is felt to be poor. It is well documented that APHE and "washout" may be seen in some benign hepatocellular liver lesions (i.e., some hepatocellular adenomas and FNH) as well as in some hypervascular liver metastases. One study examining the enhancement characteristics of benign hepatocellular lesions in patients without liver disease found the presence of APHE and "washout" in 58.1% (25/43) of hepatocellular adenomas [13]. While a less common occurrence in FNH, these lesions have been shown to occasionally exhibit APHE and "washout" as well; one study comparing imaging characteristics of FNH to HCC found APHE and "washout" within 22.2% (4/18) of FNHs at dynamic contrast-enhanced CT [14]. Washout appearance within arterial-phase hypervascular liver metastases (e.g., neuroendocrine tumor, thyroid carcinoma, melanoma, renal cell carcinoma metastases) has also been well described [15, 16]. Thus, the combination of APHE and "washout" should not be used to diagnose HCC in patients without risk factors for HCC when the pretest probability of HCC is low (Fig. 13.2) in accordance with the LI-RADS diagnostic algorithm.

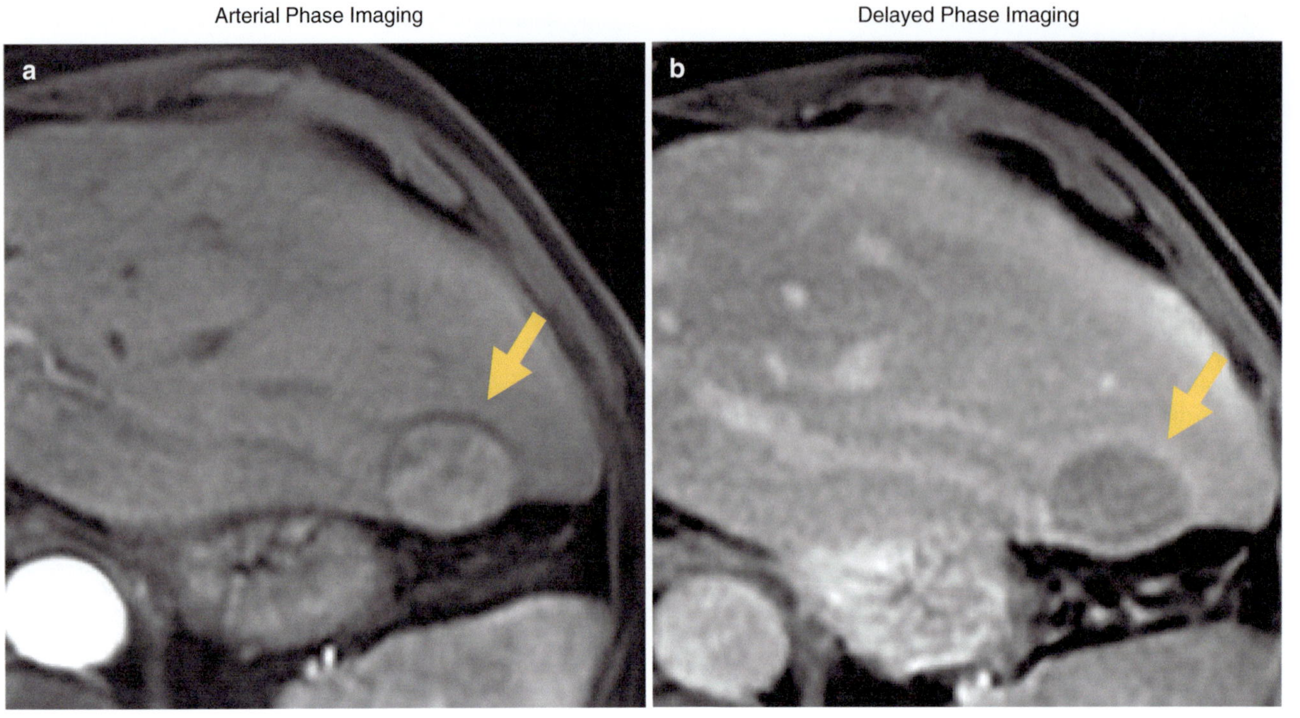

Fig. 13.2 Resection proven focal nodular hyperplasia in a young patient with no known liver disease. The nodule exhibits arterial phase hyperenhancement (APHE) in (**a**) the arterial phase and washout appearance in (**b**) the delayed phases. The combination of APHE and washout should not be used to diagnosis hepatocellular carcinoma (HCC) in patients without risk factors for HCC (i.e., not meeting criteria for application of the Liver Imaging Reporting and Data System diagnostic algorithm) as it may result in misdiagnosis and patient harm

Conclusion

The specificity of APHE and "washout" for the diagnosis of HCC on multiphasic CT or MRI is dependent on clinical history, observation size, pattern of APHE, and presence or absence of features suggestive of non-HCC malignancy. The combination of APHE and "washout" should be considered diagnostic of HCC only in patients at high risk of HCC (e.g., patients with cirrhosis from a nonvascular cause, noncirrhotic HBV, known prior HCC) and for observations greater than or equal to 10 mm in size; in such a setting, these imaging features have been shown to have exceedingly high diagnostic specificity. Furthermore, the APHE must be nonperipheral in morphology and not in a pattern suggestive of hemangioma or other malignancy. In patients without risk factors for HCC, APHE and "washout" should not be used in isolation to diagnose HCC because of the risk of inappropriate HCC diagnosis.

References

1. Kim TH, Kim SY, Tang A, Lee JM. Comparison of international guidelines for noninvasive diagnosis of hepatocellular carcinoma: 2018 update. Clin Mol Hepatol. 2019;25(3):245–63. https://doi.org/10.3350/cmh.2018.0090.
2. Heimbach JK, Kulik LM, Finn RS, et al. AASLD guidelines for the treatment of hepatocellular carcinoma. Hepatology. 2018;67(1):358–80. https://doi.org/10.1002/hep.29086.
3. American College of Radiology (ACR). Liver imaging reporting and data system version 2018. Accessed 30 Jan 2020. https://www.acr.org/Clinical-Resources/Reporting-and-Data-Systems/LI-RADS
4. Marrero JA, Kulik LM, Sirlin CB, et al. Diagnosis, staging, and management of hepatocellular carcinoma: 2018 practice guidance by the American Association for the Study of Liver Diseases. Hepatology. 2018;68(2):723–50. https://doi.org/10.1002/hep.29913.
5. Omata M, Cheng AL, Kokudo N, et al. Asia-Pacific clinical practice guidelines on the management of hepatocellular carcinoma: a 2017 update. Hepatol Int. 2017;11(4):317–70. https://doi.org/10.1007/s12072-017-9799-9.
6. European Association for the Study of the Liver. EASL clinical practice guidelines: management of hepatocellular carcinoma. J Hepatol. 2018;69(1):182–236. https://doi.org/10.1016/j.jhep.2018.03.019.
7. Korean Liver Cancer Association; National Cancer Center. 2018 Korean Liver Cancer Association–National Cancer Center Korea practice guidelines for the management of hepatocellular carcinoma. Gut Liver. 2019;13(3):227–99. https://doi.org/10.5009/gnl19024.
8. Setiawan VW, Stram DO, Porcel J, Lu SC, Le Marchand L, Noureddin M. Prevalence of chronic liver disease and cirrhosis by underlying cause in understudied ethnic groups: the multiethnic cohort. Hepatology. 2016;64(6):1969–77. https://doi.org/10.1002/hep.28677.
9. Scaglione S, Kliethermes S, Cao G, et al. The epidemiology of cirrhosis in the United States: a population-based study. J Clin Gastroenterol. 2015;49(8):690–6. https://doi.org/10.1097/MCG.0000000000000208.
10. Rothschild JM, Khorasani R, Silverman SG, Hanson RW, Fiskio JM, Bates DW. Abdominal cross-sectional imaging for inpatients with abnormal liver function test results: yield and usefulness. Arch Intern Med. 2001;161(4):583–8. https://doi.org/10.1001/archinte.161.4.583.

11. Gore RM, Pickhardt PJ, Mortele KJ, et al. Management of incidental liver lesions on CT: a white paper of the ACR incidental findings committee. J Am Coll Radiol. 2017;14(11):1429–37. https://doi.org/10.1016/j.jacr.2017.07.018.
12. Gore RM, Thakrar KH, Wenzke DR, Newmark GM, Mehta UK, Berlin JW. That liver lesion on MDCT in the oncology patient: is it important? Cancer Imaging. 2012;12(2):373–84. https://doi.org/10.1102/1470-7330.2012.9028.
13. Grazioli L, Bondioni MP, Haradome H, et al. Hepatocellular adenoma and focal nodular hyperplasia: value of gadoxetic acid-enhanced MR imaging in differential diagnosis. Radiology. 2012;262(2):520–9. https://doi.org/10.1148/radiol.11101742.
14. Kitao A, Matsui O, Yoneda N, et al. Differentiation between hepatocellular carcinoma showing hyperintensity on the hepatobiliary phase of gadoxetic acid-enhanced MRI and focal nodular hyperplasia by CT and MRI. AJR Am J Roentgenol. 2018;211(2):347–57. https://doi.org/10.2214/AJR.17.19341.
15. Namasivayam S, Martin DR, Saini S. Imaging of liver metastases: MRI. Cancer Imaging. 2007;7(1):2–9. https://doi.org/10.1102/1470-7330.2007.0002.
16. Reimer P, Rummeny EJ, Daldrup HE, et al. Enhancement characteristics of liver metastases, hepatocellular carcinomas, and hemangiomas with Gd-EOB-DTPA: preliminary results with dynamic MR imaging. Eur Radiol. 1997;7(2):275–80. https://doi.org/10.1007/s003300050150.
17. Tang A, Bashir MR, Corwin MT, LI-RADS Evidence Working Group, et al. Evidence supporting LI-RADS major features for CT- and MR imaging-based diagnosis of hepatocellular carcinoma: a systematic review. Radiology. 2018;286(1):29–48. https://doi.org/10.1148/radiol.2017170554.
18. Choi JY, Lee JM, Sirlin CB. CT and MR imaging diagnosis and staging of hepatocellular carcinoma: part I. Development, growth, and spread: key pathologic and imaging aspects. Radiology. 2014;272(3):635–54. https://doi.org/10.1148/radiol.14132361.
19. Park YN. Update on precursor and early lesions of hepatocellular carcinomas. Arch Pathol Lab Med. 2011;135(6):704–15. https://doi.org/10.5858/2010-0524-RA.1.
20. Tanaka Y, Sasaki Y, Katayama K, et al. Probability of hepatocellular carcinoma of small hepatocellular nodules undetectable by computed tomography during arterial portography. Hepatology. 2000;31(4):890–8. https://doi.org/10.1053/he.2000.5979.
21. International Consensus Group for Hepatocellular Neoplasia. Pathologic diagnosis of early hepatocellular carcinoma: a report of the International Consensus Group for Hepatocellular Neoplasia. Hepatology. 2009;49(2):658–64. https://doi.org/10.1002/hep.22709.
22. Kojiro M. Histopathology of liver cancers. Best Pract Res Clin Gastroenterol. 2005;19(1):39–62. https://doi.org/10.1016/j.bpg.2004.10.007.
23. Tang A, Cruite I, Mitchell DG, Sirlin CB. Hepatocellular carcinoma imaging systems: why they exist, how they have evolved, and how they differ. Abdom Radiol (NY). 2018;43(1):3–12. https://doi.org/10.1007/s00261-017-1292-3.
24. Rimola J, Forner A, Reig M, et al. Cholangiocarcinoma in cirrhosis: absence of contrast washout in delayed phases by magnetic resonance imaging avoids misdiagnosis of hepatocellular carcinoma. Hepatology. 2009;50(3):791–8. https://doi.org/10.1002/hep.23071.
25. Kang Y, Lee JM, Kim SH, Han JK, Choi BI. Intrahepatic mass-forming cholangiocarcinoma: enhancement patterns on gadoxetic acid-enhanced MR images. Radiology. 2012;264(3):751–60. https://doi.org/10.1148/radiol.12112308.
26. Cogley JR, Miller FH. MR imaging of benign focal liver lesions. Radiol Clin North Am. 2014;52(4):657–82. https://doi.org/10.1016/j.rcl.2014.02.005.
27. Kamaya A, Maturen KE, Tye GA, Liu YI, Parti NN, Desser TS. Hypervascular liver lesions. Semin Ultrasound CT MR. 2009;30(5):387–407. https://doi.org/10.1053/j.sult.2009.06.001.
28. Hayashi M, Matsui O, Ueda K, et al. Correlation between the blood supply and grade of malignancy of hepatocellular nodules associated with liver cirrhosis: evaluation by CT during intraarterial injection of contrast medium. AJR Am J Roentgenol. 1999;172(4):969–76. https://doi.org/10.2214/ajr.172.4.10587130.
29. Matsui O, Kadoya M, Kameyama T, et al. Benign and malignant nodules in cirrhotic livers: distinction based on blood supply. Radiology. 1991;178(2):493–7. https://doi.org/10.1148/radiology.178.2.1846240.
30. Kim TK, Lee KH, Jang HJ, et al. Analysis of gadobenate dimeglumine-enhanced MR findings for characterizing small (1-2-cm) hepatic nodules in patients at high risk for hepatocellular carcinoma. Radiology. 2011;259(3):730–8. https://doi.org/10.1148/radiol.11101549.
31. Rimola J, Forner A, Tremosini S, et al. Non-invasive diagnosis of hepatocellular carcinoma ≤ 2 cm in cirrhosis. Diagnostic accuracy assessing fat, capsule and signal intensity at dynamic MRI. J Hepatol. 2012;56(6):1317–23. https://doi.org/10.1016/j.jhep.2012.01.004.
32. Lee KHY, O'Malley ME, Haider MA, Hanbidge A. Triple-phase MDCT of hepatocellular carcinoma. AJR Am J Roentgenol. 2004;182(3):643–9. https://doi.org/10.2214/ajr.182.3.1820643.
33. Burrel M, Llovet JM, Ayuso C, Barcelona Clínic Liver Cancer Group, et al. MRI angiography is superior to helical CT for detection of HCC prior to liver transplantation: an explant correlation. Hepatology. 2003;38(4):1034–42. https://doi.org/10.1053/jhep.2003.50409.
34. Denecke T, Grieser C, Fröling V, et al. Multislice computed tomography using a triple-phase contrast protocol for preoperative assessment of hepatic tumor load in patients with hepatocellular carcinoma before liver transplantation. Transpl Int. 2009;22(4):395–402. https://doi.org/10.1111/j.1432-2277.2008.00793.x.
35. Sano K, Ichikawa T, Motosugi U, et al. Imaging study of early hepatocellular carcinoma: usefulness of gadoxetic acid-enhanced MR imaging. Radiology. 2011;261(3):834–44. https://doi.org/10.1148/radiol.11101840.
36. Kojiro M, Roskams T. Early hepatocellular carcinoma and dysplastic nodules. Semin Liver Dis. 2005;25(2):133–42. https://doi.org/10.1055/s-2005-871193.
37. Kitao A, Zen Y, Matsui O, Gabata T, Nakanuma Y. Hepatocarcinogenesis: multistep changes of drainage vessels at CT during arterial portography and hepatic arteriography – radiologic-pathologic correlation. Radiology. 2009;252(2):605–14. https://doi.org/10.1148/radiol.2522081414.
38. Santillan C, Fowler K, Kono Y, Chernyak V. LI-RADS major features: CT, MRI with extracellular agents, and MRI with hepatobiliary agents. Abdom Radiol (NY). 2018;43(1):75–81. https://doi.org/10.1007/s00261-017-1291-4.
39. Kitao A, Matsui O, Yoneda N, et al. The uptake transporter OATP8 expression decreases during multistep hepatocarcinogenesis: correlation with gadoxetic acid enhanced MR imaging. Eur Radiol. 2011;21(10):2056–66. https://doi.org/10.1007/s00330-011-2165-8.
40. Vernuccio F, Cannella R, Meyer M, et al. LI-RADS: diagnostic performance of hepatobiliary phase hypointensity and major imaging features of LR-3 and LR-4 lesions measuring 10-19 mm with arterial phase hyperenhancement. AJR Am J Roentgenol. 2019;213(2):W57–65. https://doi.org/10.2214/AJR.18.20979.
41. Kim DH, Choi SH, Kim SY, Kim MJ, Lee SS, Byun JH. Gadoxetic acid-enhanced MRI of hepatocellular carcinoma: value of washout in transitional and hepatobiliary phases. Radiology. 2019;291(3):651–7. https://doi.org/10.1148/radiol.2019182587.
42. Sangiovanni A, Manini MA, Iavarone M, et al. The diagnostic and economic impact of contrast imaging techniques in the

diagnosis of small hepatocellular carcinoma in cirrhosis. Gut. 2010;59(5):638–44. https://doi.org/10.1136/gut.2009.187286.
43. Jang HJ, Kim TK, Khalili K, et al. Characterization of 1-to 2-cm liver nodules detected on HCC surveillance ultrasound according to the criteria of the American Association for the study of liver disease: is quadriphasic CT necessary? AJR Am J Roentgenol. 2013;201(2):314–21. https://doi.org/10.2214/AJR.12.9341.
44. Marrero JA, Hussain HK, Nghiem HV, Umar R, Fontana RJ, Lok AS. Improving the prediction of hepatocellular carcinoma in cirrhotic patients with an arterially-enhancing liver mass. Liver Transpl. 2005;11(3):281–9. https://doi.org/10.1002/lt.20357.
45. van der Pol CB, Lim CS, Sirlin CB, et al. Accuracy of the liver imaging reporting and data system in computed tomography and magnetic resonance image analysis of hepatocellular carcinoma or overall malignancy—a systematic review. Gastroenterology. 2019;156(4):976–86. https://doi.org/10.1053/j.gastro.2018.11.020.
46. Fraum TJ, Tsai R, Rohe E, et al. Differentiation of hepatocellular carcinoma from other hepatic malignancies in patients at risk: diagnostic performance of the liver imaging reporting and data system version 2014. Radiology. 2018;286(1):158–72. https://doi.org/10.1148/radiol.2017170114.
47. Kim YY, Kim MJ, Kim EH, Roh YH, An C. Hepatocellular carcinoma versus other hepatic malignancy in cirrhosis: performance of LI-RADS version 2018. Radiology. 2019;291(1):72–80. https://doi.org/10.1148/radiol.2019181995.
48. Ludwig DR, Fraum TJ, Cannella R, et al. Expanding the Liver Imaging Reporting and Data System (LI-RADS) v2018 diagnostic population: performance and reliability of LI-RADS for distinguishing hepatocellular carcinoma (HCC) from non-HCC primary liver carcinoma in patients who do not meet strict LI-RADS high-risk criteria. HPB (Oxford). 2019;21(12):1697–706. https://doi.org/10.1016/j.hpb.2019.04.007.

14 Should Diagnostic Liver Biopsy Be Performed in Patients with Cirrhosis Who Have Characteristic Imaging Features of Hepatocellular Carcinoma?

Joseph C. Ahn and Ju Dong Yang

Highly specific imaging algorithms have obviated the need for liver biopsy in most cases, and biopsy should be reserved for cases in which imaging is not characteristic for hepatocellular carcinoma (HCC).

Main Points

1. The Liver Imaging Reporting and Data System (LI-RADS) diagnostic algorithm allows for noninvasive diagnosis of HCC in most cases among at-risk patients, including those with cirrhosis or chronic hepatitis B virus infection. LI-RADS 5 features have a 94% positive predictive value for the diagnosis of HCC.
2. A selective biopsy strategy, that is, performing biopsy in patients only without characteristic imaging, avoids rare, albeit potentially substantial, risks such as bleeding and needle-track seeding.
3. The disadvantage of a selective biopsy strategy is the lack of tumor tissue for molecular characterization, which has slowed personalized treatment strategies.

HCC is the most common type of primary liver cancer, occurring mostly in patients with chronic liver diseases, and it is among the leading causes of cancer death globally [1]. Except for a select group of patients with small, localized early-stage HCC who may be candidates for curative therapies, the majority of HCC patients present with advanced-stage cancer, have an extremely poor prognosis, and are palliatively managed with systemic therapy and supportive care. Therefore, HCC surveillance with biannual liver ultrasound with or without serum alpha fetoprotein is recommended for high-risk patients who have cirrhosis or chronic hepatitis B infection to achieve early detection of cancer [2–4]. Detection of a liver nodule in a high-risk individual should raise high suspicion for HCC and trigger recall procedures to establish or exclude the diagnosis of HCC. In contrast to most other solid cancers that require invasive biopsy for diagnosis, HCC is frequently diagnosed noninvasively based on characteristic imaging findings. In fact, clinical practice guidelines do not mandate routine liver biopsy for pathologic confirmation of HCC before treatment if the lesion meets specific imaging criteria for a definitive diagnosis of HCC [2–4]. Currently, a diagnostic liver biopsy is conducted selectively for patients whose HCC diagnosis remains indeterminate on contrast-enhanced imaging, and in patients with advanced HCC for whom molecular analyses of the tumor tissue could dictate specific therapy [5]. The rationale for a selective biopsy strategy is primarily based on the high diagnostic accuracy of contrast-enhanced imaging modalities, which obviates the need for an invasive biopsy procedure with complication risks. However, there also exist arguments for a more routine biopsy strategy to implement an individualized treatment approach for patients based on the molecular characteristics of the tumor.

Selective Biopsy Strategy Using Liver Imaging Reporting and Data System Criteria

In 2008, a joint committee consisting of radiologists, hepatologists, surgeons, and pathologists developed the standardized LI-RADS, which to date serves as a standardized lexicon and comprehensive imaging algorithm with high specificity for HCC [6]. With the most recent update in 2018, the LI-RADS is now integrated into the American Association

J. C. Ahn
Division of Gastroenterology and Hepatology, Mayo Clinic, Rochester, MN, USA
e-mail: ahn.joseph@mayo.edu

J. D. Yang (✉)
Division of Digestive and Liver Diseases, Department of Medicine, Comprehensive Transplant Center, Samuel Oschin Comprehensive Cancer Institute, Cedars Sinai Medical Center, Los Angeles, CA, USA
e-mail: JuDong.Yang@cshs.org

N. S. Reau, A. G. Singal (eds.), *Controversies in Hepatocellular Carcinoma*, https://doi.org/10.1007/978-3-031-90225-3_14

for the Study of Liver Diseases (AASLD) 2018 HCC clinical practice guidance [3, 7]. At-risk patients with abnormal results on surveillance (lesions ≥ 10 mm or alpha fetoprotein ≥ 20 ng/mL) or clinical suspicion of HCC should undergo multiphase computed tomography or magnetic resonance imaging for diagnostic testing [3]. The major imaging features suggestive of HCC include nonrim arterial-phase hyperenhancement (APHE), nonperipheral portal venous or delayed-phase "washout" appearance, enhancing "capsule" appearance, size of at least 10 mm, and growth by 50% or more in 6 months or less [7]. Based on the presence/absence of the major features as well as several additional ancillary features, the lesion can be divided into 8 unique LI-RADS categories that reflect the probability of HCC. The LI-RADS categories suggestive of malignancy include LR-4 (probably HCC), LR-5 (definitely HCC), LR-M (probably or definitely malignant, not specific for HCC), and LR-TIV (malignancy with tumor in vein) [7]. A lesion is categorized LR-5 if it is greater than or equal to 20 mm in size with APHE and has one or more of the other major features such as nonperipheral "washout," enhancing capsule, and threshold growth by 50% or more in 6 months or less. For lesions between 10 and 19 mm in size, APHE and one or more of the other major features, either nonperipheral "washout" or threshold growth by 50% or more in 6 months or less would qualify for LR-5 [7]. An LR-5 lesion has 94% positive predictive value of being HCC, and liver biopsy is not needed in these cases to confirm the diagnosis of HCC [8]. On the other hand, an LR-4 lesion may be greater than or equal to 20 mm in size with APHE but without the other major features. An LR-4 lesion has an average of 74% probability of HCC and needs pathologic confirmation to establish the diagnosis [8]. An LR-M lesion is almost certainly malignant, but may not be another type of cancer that is not HCC. As for LR-TIV lesions, while HCC is the most common hepatic tumor associated with macrovascular invasion, the differential includes intrahepatic cholangiocarcinoma and rarely other malignancies. Therefore, AASLD recommends selective liver biopsy strategies for patients with LR-4, LR-M, and LR-TIV lesions using an individualized, multidisciplinary approach. While AASLD recommends repeat imaging for LR-2 (probably benign) and LR-3 (intermediate) lesions, [3] the European Association for the Study of the Liver endorses a more aggressive approach and recommends performing a liver biopsy for LR-2 and LR-3 lesions as well, whenever technically feasible, at least in the dominant nodule [2].

Advantages of Selective Biopsy Strategy

The main advantage of a selective biopsy strategy is to avoid the risk of liver biopsy when the diagnosis of HCC can be established by radiologic test and additional histologic examination gives no additional information that can affect the selection of treatment or prognostication. While liver biopsy can provide a pathologic confirmation of HCC, it is an invasive procedure with known risks. Bleeding is a risk with all biopsy procedures, but HCC patients tend to be at higher risk because of the hypervascular nature of HCC tumors and the presence of thrombocytopenia and coagulopathy related to their underlying liver diseases. Severe bleeding occurs between 1 in 2500 and 1 in 10,000 biopsies, while less severe bleeding happens in 1 in 500 biopsies [9]. Patients with severe bleeding can bleed intraperitoneally and may require hospitalization for hemodynamic resuscitation, transfusion, radiological intervention, or surgery. Death from uncontrolled hemorrhage is a possibility and has been reported to occur in less than or equal to 1 in 10,000 liver biopsies [9]. Another concern for liver biopsy in the context of HCC tissue diagnosis is tumor seeding of the biopsy needle track. A meta-analysis of 8 studies (N = 1340) showed that tumor seeding occurred in about 2.7% (0%–5.8%) of cases, but can be treated by excision or radiation and has not been shown to affect the outcome of oncological treatment [10, 11]. Incidence of needle-track seeding was much lower in studies conducted after 2000 (0.6%), suggesting that the risk has decreased recently. Another large retrospective study of 5092 HCC patients found only a 0.14% incidence of needle-track seeding [12]. Reduced risk of needle-track tumor seeding could be in part attributed to the employment of the coaxial technique during needle passes, which protects normal tissue along the tract with a needle introducer, thus reducing the risk of seeding [10, 13]. The theoretical risk of tumor seeding becomes insignificant in patients with advanced HCC because the patient is already deemed to have metastatic disease requiring systemic therapy [5]. Overall, the risks of liver biopsy should not be considered as reasons to withhold liver biopsy when there is a clear clinical indication. However, an invasive liver biopsy may not be needed if there is a less-invasive alternative that will provide the necessary diagnostic information with high accuracy.

Disadvantages of Selective Biopsy Strategy

The biggest disadvantage of relying on noninvasive imaging modalities to diagnose HCC is the lack of tumor tissue sampling, which can be used for molecular characterization of the tumor and guide treatment decisions. Molecular-targeted agent sorafenib has been the mainstay of systemic therapy for advanced-stage HCC, and several additional molecular-targeted agents such as levatinib, regorafenib, and cabozantinib, as well as immune checkpoint inhibitors nivolumab and pembrolizumab, have recently been approved as first- and second-line agents [14]. However, response rates to these agents have been limited to about 20% or less and patients typically receive no more than a few months of survival benefit. Given the poor prognosis associated with advanced-stage HCC and the suboptimal outcomes with systemic therapy, there exists an unmet need in identifying patients who are likely to benefit from specific treatments [15]. Over the last decade, high-throughput molecular technologies have discovered that HCC is a highly heterogeneous entity with various genetic alterations [16]. Genome-wide molecular profiling of clinical HCC tissues has identified statistically significant molecular heterogeneity across HCC tumors, [17] and a meta-analysis of transcriptome profiles from diverse cohorts of patients identified various aggressive and indolent molecular subtypes of HCC tumors [18]. For example, fibrolamellar carcinoma is a subtype of HCC that arises in young patients without cirrhosis, and is characterized by the presence of fibrous bands in a unique "lamellar" pattern under the microscope [19]. The chromophobe HCC with abrupt anaplasia is another distinct subtype with a set of characteristic histological features and uses a telomerase-independent mechanism of telomere maintenance, in contrast to telomerase activation seen in most HCCs [20]. While it is not yet established how the subtypes of HCC should be treated differently, there are emerging data showing that certain subtypes may be associated with prognosis and varying response to specific treatments. A recently described subtype called the *macrotrabecullar-massive HCC (MTM-HCC)* was shown to be an independent predictor of early and overall recurrence on multivariate analysis [21]. At a molecular level, the MTM-HCC was characterized by overexpression of angiopoietin 2 and vascular endothelial growth factor-A, suggesting that it could potentially be targeted with antiangiopoietin 2 and anti-vascular endothelial growth factor-A antibodies [21]. Recently, implementation of prospective next-generation sequencing for patients with HCC showed that such analyses could provide meaningful predictive and prognostic information used to guide systemic therapy and match patients to genome-directed targeted therapies [22]. In this study, activating mutations in the PI3k-mTOR pathway were associated with poor outcomes in sorafenib-treated patients, while activating mutations in the WNT pathway were associated with innate resistance to immune checkpoint blockade. Furthermore, the authors identified potentially actionable mutations in a small subset of patients that allowed use of targeted therapy based on their genomic reports [22].

Thus, some experts advocate for a routine biopsy strategy by which all patients with suspected advanced HCC, even those with LR-5 lesions where the diagnosis of HCC is not in question, should undergo a liver biopsy for molecular characterization of the tumor for better prognostication and prediction of treatment response. Currently, pretreatment liver biopsy is often a prerequisite to participate in clinical trials to identify molecular biomarkers predicting treatment response to novel therapeutic agents.

In addition, it must be noted that the LI-RADS can misclassify lesions in a small proportion of cases, resulting in some non-HCC tumors receiving false-positive LR-5 designation. Notably, there have been increasing reports of combined hepatocellular cholangiocarcinoma (cHCC-CCA) with LR-5 designation in up to 25.7% (18/70) of cases, meaning up to one-quarter of cHCC-CCAs can be falsely diagnosed as HCC on radiologic assessment [23].

Conclusion

Development of standardized and highly specific imaging algorithms such as the LI-RADS has enabled noninvasive diagnosis of HCC in patients whose lesions meet the criteria for LR-5, and liver biopsy is usually reserved as an option for patients whose diagnoses remain uncertain after appropriate multiphase contrast-enhanced imaging. Such a selective biopsy strategy has the advantage of being able to establish the diagnosis of HCC in many patients without subjecting them to the risks of invasive liver biopsy. Having said that, there is still no role for a routine biopsy strategy in patients with early-stage HCC when the diagnosis could be confirmed based on characteristic radiologic findings. However, even for patients with LR-5 lesions whose HCC diagnoses are not in question, tumor tissue sampling can still offer valuable information regarding the phenotypic and molecular characteristics of the HCC tumor. HCC is a highly heterogeneous entity with different subtypes and genetic alterations, which explains the suboptimal response rate to the currently employed "one-size fits all" approach to systemic therapy. Advancements in high-throughput next-generation sequencing will enable clinicians to identify specific mutations in patients' HCC tumor samples that will help guide the choice of systemic therapy. Thus, we still advocate for a selective liver biopsy strategy in patients with HCC, reserving it primarily for patients with diagnostic uncertainty on imaging or patients with advanced HCC who are being considered for systemic therapy or participation in therapeutic clinical trials (Table 14.1). Finally, "liquid biopsy" techniques such as circulating tumor cells (CTCs) and circulating tumor DNA (ctDNA) have recently emerged as noninvasive biomarkers that can provide important diagnostic and prognostic information in patients with HCC [24, 25]. CTCs are malignant cells derived from the primary tumor or sites of metastases that migrate into the systemic circulation, [24] while ctDNA is the fraction of cell-free DNA that originates from dying tumor cells [25]. CTCs and ctDNA can be analyzed for specific mutations and epigenetic changes that can be used to guide targeted therapy. Profound heterogeneities exist between tumor cells within the primary tumors and those within sites of metastases, and a single biopsy specimen containing a small amount of tumor tissue may not be representative of the whole HCC tumor [15].

Table 14.1 Advantages and disadvantages of selective vs routine biopsy strategies

	Selective biopsy strategy	Routine biopsy strategy
Advantages	High accuracy of LR-5 designation Avoidance of liver biopsy and its complications (bleeding, tumor seeding)	Ability to determine molecular subtypes of HCC and identify specific mutations to guide targeted therapy
Disadvantages	Lack of tissue confirmation Lack of molecular analysis	Risks of liver biopsy and its complications (bleeding, tumor seeding)

Abbreviations: *HCC* hepatocellular carcinoma, *LR-5* Liver Imaging Reporting and Data System category 5 (definitely HCC)

Therefore, liquid biopsy is not merely a less-invasive alternative to an actual biopsy, but may represent a novel approach to obtaining an even deeper understanding of the heterogeneous tumor cells throughout the body, and may obviate the need for routine tissue biopsy in patients with advanced-stage HCC in the future.

References

1. Global Burden of Disease Liver Cancer Collaboration, Akinyemiju T, Abera S, Ahmed M, et al. The burden of primary liver cancer and underlying etiologies from 1990 to 2015 at the global, regional, and national level: results from the global burden of disease study 2015. JAMA Oncologia. 2017;3(12):1683–91. https://doi.org/10.1001/jamaoncol.2017.3055.
2. European Association for the Study of the Liver. EASL clinical practice guidelines: management of hepatocellular carcinoma. J Hepatol. 2018;69(1):182–236. https://doi.org/10.1016/j.jhep.2018.03.019.
3. Marrero JA, Kulik LM, Sirlin CB, et al. Diagnosis, staging, and management of hepatocellular carcinoma: 2018 practice guidance by the American Association for the Study of Liver Diseases. Hepatology. 2018;68(2):723–50. https://doi.org/10.1002/hep.29913.
4. Omata M, Cheng AL, Kokudo N, et al. Asia-Pacific clinical practice guidelines on the management of hepatocellular carcinoma: a 2017 update. Hepatol Int. 2017;11(4):317–70. https://doi.org/10.1007/s12072-017-9799-9.
5. Finn RS. The role of liver biopsy in hepatocellular carcinoma. Gastroenterol Hepatol (N Y). 2016;12(10):628–30.
6. Tang A, Bashir MR, Corwin MT, et al. LI-RADS evidence working group. Evidence supporting LI-RADS major features for CT- and MR imaging–based diagnosis of hepatocellular carcinoma: a systematic review. Radiology. 2017;286(1):29–48. https://doi.org/10.1148/radiol.2017170554.
7. Chernyak V, Fowler KJ, Kamaya A, et al. Liver imaging reporting and data system (LI-RADS) version 2018: imaging of hepatocellular carcinoma in at-risk patients. Radiology. 2018;289(3):816–30. https://doi.org/10.1148/radiol.2018181494.
8. van der Pol CB, Lim CS, Sirlin CB, et al. Accuracy of the liver imaging reporting and data system in computed tomography and magnetic resonance image analysis of hepatocellular carcinoma or overall malignancy—a systematic review. Gastroenterology. 2019;156(4):976–86. https://doi.org/10.1053/j.gastro.2018.11.020.
9. Rockey DC, Caldwell SH, Goodman ZD, Nelson RC, Smith AD. American Association for the Study of Liver Diseases. Liver biopsy Hepatology. 2009;49(3):1017–44. https://doi.org/10.1002/hep.22742.
10. Silva MA, Hegab B, Hyde C, Guo B, Buckels JAC, Mirza DF. Needle track seeding following biopsy of liver lesions in the diagnosis of hepatocellular cancer: a systematic review and meta-analysis. Gut. 2008;57(11):1592–6. https://doi.org/10.1136/gut.2008.149062.
11. Fuks D, Cauchy F, Fusco G, Paradis V, Durand F, Belghiti J. Preoperative tumour biopsy does not affect the oncologic course of patients with transplantable HCC. J Hepatol. 2014;61(3):589–93. https://doi.org/10.1016/j.jhep.2014.04.046.
12. Ahn DW, Shim JH, Yoon JH, et al. Treatment and clinical outcome of needle-track seeding from hepatocellular carcinoma. Korean J Hepatol. 2011;17(2):106–12. https://doi.org/10.3350/kjhep.2011.17.2.106.
13. Maturen KE, Nghiem HV, Marrero JA, et al. Lack of tumor seeding of hepatocellular carcinoma after percutaneous needle biopsy using coaxial cutting needle technique. AJR Am J Roentgenol. 2006;187(5):1184–7. https://doi.org/10.2214/AJR.05.1347.
14. Kudo M. Systemic therapy for hepatocellular carcinoma: latest advances. Cancers (Basel). 2018;10(11):412. https://doi.org/10.3390/cancers10110412.
15. Zhu S, Hoshida Y. Molecular heterogeneity in hepatocellular carcinoma. Hepat Oncologia. 2018;5(1):HEP10. https://doi.org/10.2217/hep-2018-0005.
16. Sparchez Z, Mocan T. Contemporary role of liver biopsy in hepatocellular carcinoma. World J Hepatol. 2018;10(7):452–61. https://doi.org/10.4254/wjh.v10.i7.452.
17. Goossens N, Sun X, Hoshida Y. Molecular classification of hepatocellular carcinoma: potential therapeutic implications. Hepat Oncol. 2015;2(4):371–9. https://doi.org/10.2217/hep.15.26.
18. Hoshida Y, Nijman SMB, Kobayashi M, et al. Integrative transcriptome analysis reveals common molecular subclasses of human hepatocellular carcinoma. Cancer Res. 2009;69(18):7385–92. https://doi.org/10.1158/0008-5472.CAN-09-1089.
19. Rieber A. Fibrolamellar carcinoma [article in German]. Z Gastroenterol. 1994;32(11):651–3.
20. Wood LD, Heaphy CM, Daniel HD, et al. Chromophobe hepatocellular carcinoma with abrupt anaplasia: a proposal for a new subtype of hepatocellular carcinoma with unique morphological and molecular features. Mod Pathol. 2013;26(12):1586–93. https://doi.org/10.1038/modpathol.2013.68.
21. Ziol M, Poté N, Amaddeo G, et al. Macrotrabecular-massive hepatocellular carcinoma: a distinctive histological subtype with clinical relevance. Hepatology. 2018;68(1):103–12. https://doi.org/10.1002/hep.29762.
22. Harding JJ, Nandakumar S, Armenia J, et al. Prospective genotyping of hepatocellular carcinoma: clinical implications of next-generation sequencing for matching patients to targeted and immune therapies. Clin Cancer Res. 2019;25(7):2116–26. https://doi.org/10.1158/1078-0432.CCR-18-2293.
23. Jeon SK, Joo I, Lee DH, et al. Combined hepatocellular cholangiocarcinoma: LI-RADS v2017 categorisation for differential diagnosis and prognostication on gadoxetic acid-enhanced MR imaging. Eur Radiol. 2019;29(1):373–82. https://doi.org/10.1007/s00330-018-5605-x.
24. Ahn JC, Teng PC, Chen PJ, et al. Detection of circulating tumor cells and their implications as a novel biomarker for diagnosis, prognostication, and therapeutic monitoring in hepatocellular carcinoma. Hepatology. 2021;73(1):422–36. https://doi.org/10.1002/hep.31165.
25. Yang JD, Liu MC, Kisiel JB. Circulating tumor DNA and hepatocellular carcinoma. Semin Liver Dis. 2019;39(4):452–62. https://doi.org/10.1055/s-0039-1688503.

Is Contrast-Enhanced Ultrasound Sufficiently Accurate to Be Used for Noninvasive Diagnosis of Hepatocellular Carcinoma?

15

Yuko Kono and Stephanie R. Wilson

Contrast-enhanced ultrasound (CEUS) has increasing data showing high accuracy for hepatocellular carcinoma (HCC) diagnosis and it can be used as an alternative diagnostic modality.

Main Points

1. CEUS has several advantages including a high safety profile and real-time imaging capability.
2. Diagnostic criteria of HCC on CEUS include arterial-phase hyperenhancement (APHE) followed by late, mild washout. Non-HCC malignancies such as intrahepatic cholangiocarcinoma or metastases typically have rim enhancement followed by early or marked washout.
3. Increasing data show that CEUS can be highly accurate for diagnosis of HCC including differentiation between HCC and intrahepatic cholangiocarcinoma in high-volume expert centers.
4. Unlike computed tomography (CT) or magnetic resonance imaging (MRI), CEUS provides only diagnostic information and does not provide tumor staging.
5. The European Association for the Study of Liver (EASL) and American Association for the Study of Liver Diseases (AASLD) both recommend CEUS as an alternative diagnostic modality if CT or MRI are inconclusive or contraindicated. Some centers have also started to use CEUS reflexively when a patient is found to have a liver lesion on surveillance US.

Y. Kono (✉)
Division of Gastroenterology and Hepatology, Department of Medicine, University of California San Diego, San Diego, CA, USA

Department of Radiology, University of California San Diego, San Diego, CA, USA
e-mail: ykono@health.ucsd.edu

S. R. Wilson
Department of Radiology, University of Calgary, Calgary, AB, Canada

Department of Medicine, University of Calgary, Calgary, AB, Canada
e-mail: stephanie.wilson@albertahealthservices.ca

CEUS was introduced in the late 1990s and has been widely used for many applications in Europe, Asia, and Canada. In the United States, while the use of US contrast agents has been approved for cardiology for more than 20 years, liver and other abdominal applications in adults and children were approved only in April 2016. Therefore, its use in the United States is still developmental, and its broad acceptance is incomplete.

In patients at risk for HCC, including those with cirrhosis and chronic hepatitis B virus, HCC can be diagnosed noninvasively by imaging without histological confirmation as endorsed by major liver societies [1, 2]. In the United States, HCC is diagnosed by dynamic contrast CT or MRI, and CEUS is marginally used. In this chapter, we introduce CEUS in the evaluation of focal liver disease with emphasis on the accuracy in HCC diagnosis. The advantages and limitations of CEUS, with comparison to CT and MRI, are included.

Contrast-Enhanced Ultrasound

CEUS is an advanced form of US using microbubble contrast agents, pure blood pool agents composed of tiny gas-filled microbubbles with a size of 2–4 μm. Visualization of signals from microbubbles requires contrast-specific imaging techniques available on the majority of high-end US machines today. Following intravenous injection of miniscule volumes of microbubble agents (0.2–2 mL), the liver and any other visible organs will show contrast enhancement. The background tissues exhibit only linear response to US signals. The

N. S. Reau, A. G. Singal (eds.), *Controversies in Hepatocellular Carcinoma*, https://doi.org/10.1007/978-3-031-90225-3_15

microbubbles, by comparison, resonate and exhibit augmented harmonic signals under very weak US exposure. Using these differences, contrast-specific imaging cancels the linear tissue signal, and shows a contrast-enhanced microbubble-only image. This is highly sensitive, and even a single bubble can be detected within the field of view.

CEUS permits dynamic characterization of liver nodules, and allows assessment of blood flow, including large visible vessels and also tissue perfusion at the capillary level.

It has been shown that CEUS is extremely safe, with less than 0.009% serious adverse events in more than 23,000 abdominal examinations [3]. There is no kidney or liver toxicity. Therefore, no pre-examination blood test is required. There is no ionizing radiation. CEUS can be performed portably and repeated as needed. Pediatric exams are equally safe and require no sedation [4].

Advantages of Contrast-Enhanced Ultrasound

CEUS offers several advantages.

Dynamic Real-Time Imaging

All grayscale and CEUS examinations are performed in dynamic real time, usually performed at a rate of 10–15 frames per second, compared to CT scan, taken once in several minutes. This provides US with superior temporal resolution as compared with CT and MRI. Therefore, CEUS will show enhancement regardless of its timing or duration. This is valuable in the identification of the rapidly changing enhancement patterns shown in all benign tumors and especially flash-filling hemangiomas, for example, and also for resolution of arterial phase mistiming, commonly encountered both on CT and MRI.

Contrast-Specific Subtraction Technique

Specialized software for performance of CEUS is a subtraction technique rendering the image virtually black before the injection of the contrast agent. The technique greatly enhances the signal from blood. The contrast-specific mode, therefore, provides images composed of signals from microbubbles only, providing a highly sensitive technique to confirm subtle arterial-phase enhancement when it cannot be seen on CT/MRI.

Use of a Purely Intravascular Contrast Agent

Microbubble contrast agents are purely intravascular and therefore CEUS always reflects the blood flow characteristics of both the liver and any observed nodule, including both enhancement and then washout. Contrast agents for CT and MRI, by comparison, have a well-recognized interstitial phase whereby the contrast agent may pass through the vascular endothelium into the interstitium of a tumor. This is pseudoenhancement and its occurrence will always be discordant with true vascular imaging from CEUS [5].

Focus on a Real Nodule at Baseline

Quality grayscale US is an essential first step for performance of CEUS as the examination is performed focused on an identified nodule. This differs from CT/MRI, in which many contrast observations are related to perfusion changes including arterial-portal shunting (APS). APS is commonly seen in a cirrhotic liver on CT/MRI, showing as a single or multiple foci of APHE without other imaging features including washout. APS are not true lesions and are not associated with nodules on US. Additionally, CEUS does not show these shunts and is therefore invaluable for their differentiation from HCC without washout. By comparison, suspect APS from CT or MRI related to HCC without washout will show a nodule on US that will raise a high suspicion for HCC.

Limitations of Contrast-Enhanced Ultrasound

CEUS carries the limitations of US. If US quality is poor because of patient body habitus or other factors, such as bowel gas, CEUS may be of limited quality. Today, most US machines are limited for lesion depth for CEUS at greater than 10 cm, although newer machines can provide good enhancement images up to a depth of 15 cm. Some patients have blind spots in the subdiaphragmatic liver. Further, severe fatty liver attenuates US signals, and CEUS may be of poor quality.

CEUS is a focused examination, and usually one nodule is evaluated with one injection. Because microbubbles stay in the bloodstream for several minutes only, repeated injections can be given as needed. CEUS may be performed to evaluate a few nodules per examination. Nonetheless, CEUS may not be optimal for staging the entire liver.

Slow adoption and lack of expertise in CEUS in the United States remain important limitations.

Contrast-Enhanced Ultrasound for Focal Liver Lesion Characterization

Focal liver lesion characterization on CEUS is based on an established algorithm showing the enhancement of a nodule relative to the adjacent liver parenchyma in the arterial and the portal venous phase. Not surprisingly, these algorithms are similar to those for CT/MRI performed for the same indication [6]. The details of nodule characterization are beyond the scope of this chapter. However, simple rules are invaluable:

Arterial-Phase Enhancement

- Benign tumors, hemangioma, and focal nodular hyperplasia show highly specific arterial-phase enhancement patterns that are most easily assessed with dynamic real-time imaging on CEUS.
- APHE is otherwise frequently nonspecific.
- HCC APHE is most often globular.
- Non-HCC malignancy may show arterial-phase globular or rim enhancement.

Washout

- Washout is the most valuable indicator of malignancy.
- CEUS documents the presence of washout as well as its timing and intensity.
- Classic HCC is characterized by APHE followed by late (> 1 min) and mild washout.
- Nonhepatocellular malignancy, including metastasis and intrahepatic cholangiocarcinoma (ICC), is characterized by early (< 1 min) onset and marked washout.

Although many imaging features/algorithms are similar between CEUS and CT/MRI, it is also known that there are distinct differences between CEUS and CT/MRI enhancement patterns in certain liver lesions [7]. Particularly well recognized is discordance in portal venous-phase enhancement of ICC between CEUS and CT/MRI due to the difference of behavior of contrast agents [5]. US contrast agents are purely intravascular microbubbles 2–4 μm in size, close to the size of red blood cells, which do not leak into the interstitial space. Contrast agents for CT/MRI by comparison are much smaller, and leak into the interstitial space after first circulation. Lesions with fibrous tissue such as ICC often show progressive enhancement on CT/MRI due to the leakage of contrast agents into the interstitial fibrous tissue. Further, on rare occasions, ICC can be misdiagnosed as a benign lesion on CT/MRI if the progressive enhancement pattern is misinterpreted as that of an "atypical hemangioma." CEUS not only can differentiate HCC from ICC, but also increases the confidence of the malignancy that can be misdiagnosed by CT/MRI as stated earlier. Knowledge of this difference between CT/MRI and CEUS is a valuable addition for ICC diagnosis, in particular.

CEUS generally shows true washout in all malignant tumors, including HCC and nonhepatocellular (ICC, metastasis) malignancy. Their differentiation is accurately achieved by evaluating the timing and degree of washout. CT/MRI on the other hand evaluate only the presence or absence of washout.

Contrast-Enhanced Ultrasound for Hepatocellular Carcinoma Diagnosis

CEUS has been endorsed for HCC diagnosis by society guidelines and recommendations, including the Asian Pacific Association for the Study of the Liver, [8] the Japanese Society of Hepatology, [9] the Korean Liver Cancer Study Group–National Cancer Center, [10] the Canadian Association for the Study of the Liver, [11] the Italian Association for the Study of the Liver, [12] and the World Federation for Ultrasound in Medicine and Biology–European Federation of Society for Ultrasound in Medicine and Biology [13]. However, CEUS was removed from the HCC diagnosis algorithm in the previous AASLD 2011 [14] and EASL 2012 guidelines [15] out of concern of misdiagnosis of ICC as HCC. This concern was based on a small retrospective study with 21 ICC cases [16] that reported 10 of 21 (47.6%) cases of ICC showed homogeneous hyperenhancement followed by washout on CEUS, with a high risk of misdiagnosis as HCC. However, subsequent experience and publication do not support the interpretation of this small retrospective study [17–19].

This has led to the refinement of the diagnostic criteria for HCC and ICC on CEUS. HCC is diagnosed with APHE followed by late (> 60 s) washout of mild degree (Fig. 15.1). On the other hand, nonhepatocellular malignancy such as ICC and metastasis shows rim APHE followed by early (< 60 seconds) or marked washout (Fig. 15.2).

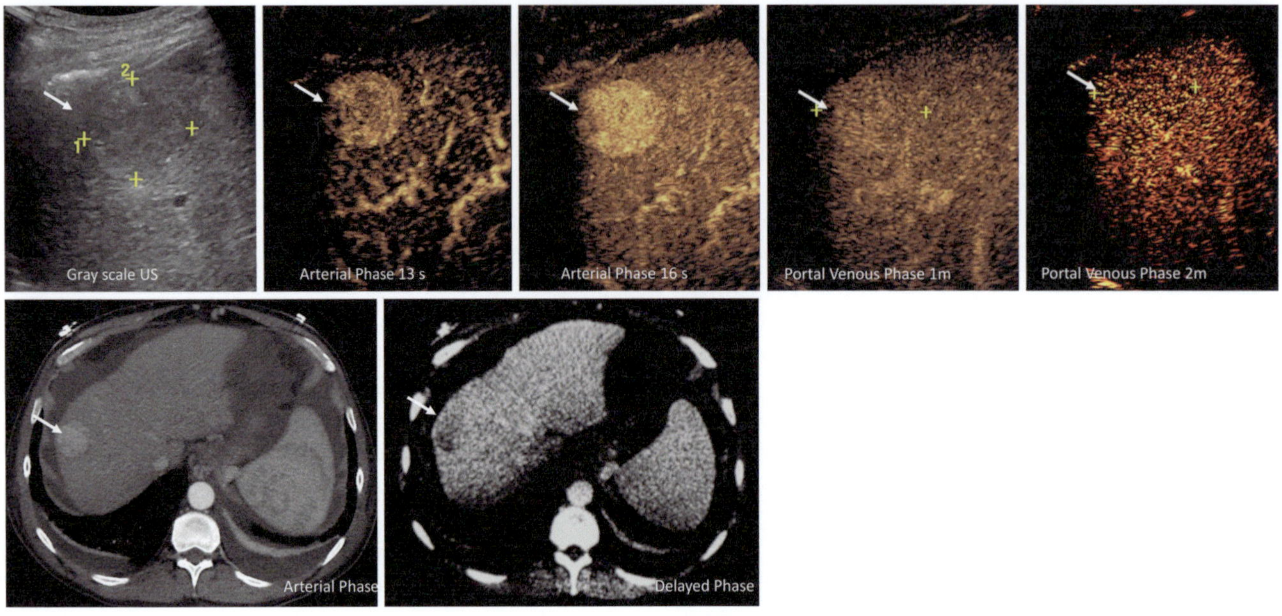

Fig. 15.1 Contrast-enhanced ultrasound Liver Imaging Reporting and Data System 5 (CEUS LR-5). A 54-year-old man with hepatitis C virus/alcohol cirrhosis and a CEUS LR-5 lesion. Grayscale US shows a 3.8-cm slightly hypoechoic nodule. On CEUS images, there is arterial phase hyperenhancement (APHE) and late (>60 s) and mild washout. It meets all major criteria for CEUS LR-5: size (≥10 mm), APHE and late and mild washout. Computed tomography (CT) scan also shows APHE and late washout, LR-5. CT and CEUS findings are concordant

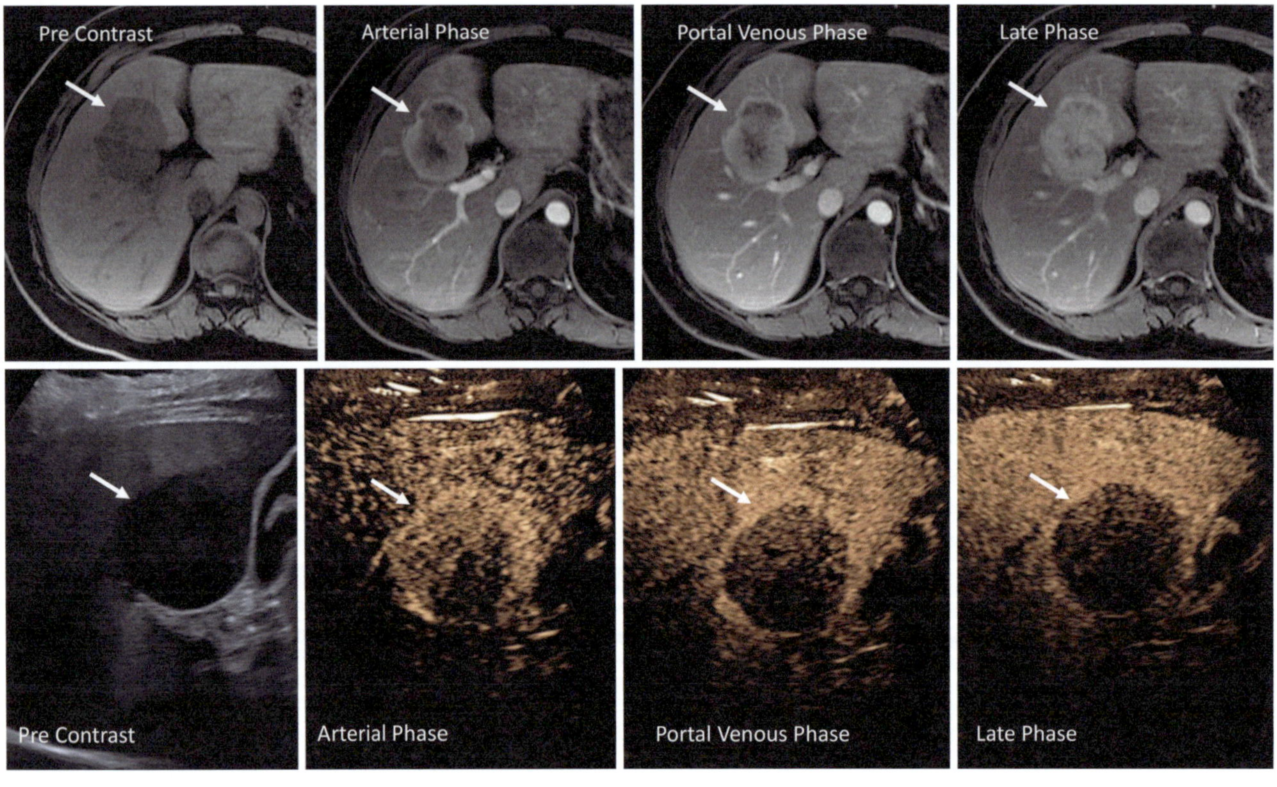

Fig. 15.2 Contrast-enhanced ultrasound Liver Imaging Reporting and Data System M (CEUS LR-M). A 69-year-old woman with metabolic dysfunction-associated steatohepatitis and biopsy-proven intrahepatic cholangiocarcinoma (ICC) in the right lobe. Magnetic resonance imaging shows rim arterial-phase hyperenhancement (APHE) with progressive enhancement, consistent with LR-M. CEUS shows rim APHE and early (< 60 s) and marked washout. Rim APHE and early, marked washout are major features for CEUS LR-M with a typical washout pattern for nonhepatocellular malignancy. Discordance of washout between MRI and CEUS is a frequent observation in ICC

Contrast-Enhanced Ultrasound Liver Imaging Reporting and Data System and Accuracy of CEUS in Hepatocellular Carcinoma Diagnosis

Accurate imaging criteria are critical for HCC diagnosis as the majority of HCC are diagnosed by imaging only. The Liver Imaging Reporting and Data System (LI-RADS) was created to meet this need, initially for CT/MRI, and expanded for CEUS in 2016. LI-RADS is endorsed by the American College of Radiology (https://www.acr.org/Clinical-Resources/Reporting-and-Data-Systems/LI-RADS). LI-RADS standardizes terminology, technique, reporting, and data collection for liver imaging on high-risk patients for HCC.

LI-RADS was developed by multidisciplinary, international consortiums of diagnostic and interventional radiologists, hepatobiliary surgeons, hepatologists, and hepatopathologists. LI-RADS includes five categories with different probabilities of HCC or malignancy, from LR-1 (definitely benign) to LR-5 (definitely HCC). LR-5 allows noninvasive diagnosis of HCC. However, additional HCCs may be found in adjacent categories in the diagnostic algorithm such as LR-4, LR-3, and LR-TIV (tumor in vein) (Table 15.1). LR-3 and LR-4 may also include multiple other regenerative and dysplastic nodules. To ensure detection of all nonhepatocellular malignancies without reduction of specificity for LR-5, LI-RADS also includes LR-M (probably or definitely malignant but not specific for HCC). LR-M includes ICC, metastasis, and also some HCC.

Focal liver lesion characterization is one of the most widely used applications of CEUS [13]. In a prospective multicenter trial with 1349 patients, there was no statistically significant difference between CEUS and spiral CT in liver tumor differentiation (malignant or benign) and tumor specification (entity) [20]. In a meta-analysis of 21 studies, for CEUS, sensitivity was 88% (95% CI, 87–90%), specificity was 81% (95% CI, 79–84%), and 38.62 (95% CI, 13.64–109.35%) for diagnostic odds ratio (DOR); all were similar and there was no statistical difference compared with contrast-enhanced CT or contrast-enhanced MRI [21].

Table 15.1 LI-RADS classification

LI-RADS category	
LR-1	Definitely benign
LR-2	Probably benign
LR-3	Intermediate probability for malignancy
LR-4	Probably HCC
LR-5	Definitely HCC
LR-M	Probably or definitely malignant, not necessarily HCC
LR-TIV	Tumor in vein
LR-NC	Not categorizable

HCC hepatocellular carcinoma, *LI-RADS* Liver Imaging Reporting and Data System, *LR* LI-RADS

The assessment of washout is the most important feature of CEUS when differentiating HCC from nonhepatocellular malignancy. Washout on CT/MRI in malignant tumors is interpreted as present or absent. Washout on CEUS by comparison is additionally assessed on its timing and intensity as described in the prior section. It has been shown the onset of washout takes place earlier than 60 s after contrast injection in the majority of ICCs (87–100%), [17, 18, 22–24] and washout is marked (50–78%), [18, 24] different from typical HCCs that show late (onset later than 60 s) and mild washout.

Several retrospective validation studies have been published for CEUS LI-RADS [24–27]. An Italian multicenter study with 1006 liver nodules in 848 patients showed a positive predictive value of CEUS LR-5 for HCC was 98.5% overall and 98% for lesions smaller than or equal to 20 mm [25], supporting the high specificity of CEUS LR-5, the major objective of LI-RADS. A Chinese study with 175 nodules all 2 cm or smaller showed 97.1% (68/70) of LR-5 were HCCs [24]. A Canadian study with 196 nodules in 184 patients showed a specificity of 96% (55/57) and positive predictive value of 98% (119/121) for the LR-5 category for the diagnosis of HCC [27]. Timing of washout is most important in determining hepatocellular and nonhepatocellular malignancy [26]. Overall, the data support high specificity of CEUS LR-5 for HCC, equivalent to the specificity of CT/MRI LR-5 [28].

A multicenter international prospective validation study involving 545 nodules confirmed the high accuracy of CEUS, with a specificity of 95.1% for CEUS LR-5 [29]. A subanalysis of the same study demonstrated the significant clinical impact of CEUS on indeterminate lesions detected by CT and MRI [5]. CEUS led to changes in management recommendations in 33.3% (25/75) of cases, with 96% (24/25) of these changes being correct. Current EASL guidelines now endorse the use of CEUS as a second-line modality for HCC diagnosis [2]. AASLD guidance for HCC management published in 2018 and in 2023 fully endorsed LI-RADS for CT/MRI for HCC diagnosis, and currently CEUS is acknowledged as an alternate imaging modality for HCC diagnosis in expert centers [30, 31]. An increasing number of publications with favorable data on CEUS diagnosis of HCC should support its greater inclusion in the future.

When and How Contrast-Enhanced Ultrasound Should Be Used for Hepatocellular Carcinoma Diagnosis

While data suggest CEUS can provide accurate diagnosis of HCC with differentiation from ICC, what are the advantages and disadvantages of using CEUS instead of or in addition to CT/MRI?

The advantages of CEUS include (1) a high safety profile, (2) no nephrotoxicity, (3) no ionizing radiation, (4) provision of additional/different information from CT/MRI, which can be complementary, and (5) repeatability as needed.

In the United States, CEUS is currently most widely used for HCC diagnosis when CT/MRI are not conclusive or contraindicated. It is a well-recognized problem-solving tool for indeterminate lesions on CT/MR [32]. CEUS is an excellent choice for resolution of indeterminate MRI in those at risk of HCC, resolving scans with HCC, showing APHE only and no washout, and reducing the time to diagnosis [29]. In addition, for LR-3, LR-4, or LR-M lesions on CT/MRI, CEUS may provide different diagnostic information.

In some practices, CEUS is performed as first-line diagnostic imaging: If there is a nodule suspected on the surveillance US, immediate performance of CEUS will facilitate management of those with malignancy, including expedited performance of staging CT/MRI. Following CEUS, CT/MRI are recommended for all LR-4, LR-M, and LR-5 nodules. Following multidisciplinary discussion, CT/MRI may also be recommended for LR-2/LR-3. Further, if CEUS shows LR-1, there is no need to perform CT or MRI, and patients can return to routine surveillance US.

CEUS is an excellent choice for resolution of indeterminate MRI in those at risk of HCC, resolving scans with HCC, showing APHE only and no washout, and reducing the time to diagnosis. CEUS is increasingly involved in ablative therapy for HCC and secondary surveillance post treatment.

Although the adoption of CEUS and CEUS LI-RADS is progressing in the United States, it is currently still limited. Therefore, the majority of institutions may not be able to implement the practice as described earlier.

Conclusion

CEUS for noncardiac application was introduced in the United States much later than in other countries related to delayed Food and Drug Administration approval until April 2016. As a result, the recognition as well as availability of CEUS is still very low. In spite of this situation, CEUS offers many advantages including high sensitivity to microbubble contrast, real-time imaging capability, and the advantages of a purely intravascular contrast agent as well as a high safety profile with no nephrotoxicity.

Differences of CEUS from CT/MRI are largely beneficial to CEUS, especially for ICC diagnosis. CEUS will show washout in malignancy whereas CT/MRI may show progressive or sustained enhancement due to interstitial contrast leakage causing pseudoenhancement.

CEUS LI-RADS 5 allows the noninvasive diagnosis of HCC, which can be treated without biopsy and for differentiation from other lesions that may require biopsy for diagnosis, such as LR-3, LR-4, and LR-M. The data supporting CEUS are accurate and should be included as an imaging modality for HCC diagnosis.

References

1. Heimbach J, et al. AASLD guidelines for the treatment of hepatocellular carcinoma. Hepatology. 2017;
2. European Association for the Study of the Liver. Electronic address, e.e.e. and L. European Association for the Study of the, EASL clinical practice guidelines: management of hepatocellular carcinoma. J Hepatol. 2018;69(1):182–236.
3. Piscaglia F, et al. The safety of Sonovue in abdominal applications: retrospective analysis of 23188 investigations. Ultrasound Med Biol. 2006;32(9):1369–75.
4. Darge K, et al. Safety of contrast-enhanced ultrasound in children for non-cardiac applications: a review by the Society for Pediatric Radiology (SPR) and the International Contrast Ultrasound Society (ICUS). Pediatr Radiol. 2013;43(9):1063–73.
5. Kono Y, et al. Clinical impact of CEUS on non-characterizable observations and observations with intermediate probability of malignancy on CT/MRI in patients at risk for HCC. Abdom Radiol (NY). 2024;49(8):2639–49.
6. Burns PN, Wilson SR. Focal liver masses: enhancement patterns on contrast-enhanced images–concordance of US scans with CT scans and MR images. Radiology. 2007;242(1):162–74.
7. Wilson SR, et al. Enhancement patterns of focal liver masses: discordance between contrast-enhanced sonography and contrast-enhanced CT and MRI. AJR Am J Roentgenol. 2007;189(1):W7–W12.
8. Omata M, et al. Asia-Pacific clinical practice guidelines on the management of hepatocellular carcinoma: a 2017 update. Hepatol Int. 2017;11(4):317–70.
9. Kudo M, et al. JSH consensus-based clinical practice guidelines for the management of hepatocellular carcinoma: 2014 update by the liver cancer study group of Japan. Liver Cancer. 2014;3(3–4):458–68.
10. Korean Liver Cancer Study, G. and K. National Cancer Center. 2014 Korean Liver Cancer Study Group-National Cancer Center Korea practice guideline for the management of hepatocellular carcinoma. Korean J Radiol. 2015;16(3):465–522.
11. Burak KW, Sherman M. Hepatocellular carcinoma: consensus, controversies and future directions. A report from the Canadian Association for the study of the liver hepatocellular carcinoma meeting. Can J Gastroenterol Hepatol. 2015;29(4):178–84.
12. Italian Association for the Study of the Liver, et al. Position paper of the Italian Association for the Study of the Liver (AISF): the multidisciplinary clinical approach to hepatocellular carcinoma. Dig Liver Dis. 2013;45(9):712–23.
13. Claudon M, et al. Guidelines and good clinical practice recommendations for Contrast Enhanced Ultrasound (CEUS) in the liver – update 2012: a WFUMB-EFSUMB initiative in cooperation with representatives of AFSUMB, AIUM, ASUM, FLAUS and ICUS. Ultrasound Med Biol. 2013;39(2):187–210.
14. Bruix J, Sherman M, D. American Association for the study of liver, management of hepatocellular carcinoma: an update. Hepatology. 2011;53(3):1020–2.
15. EASL-EORTC clinical practice guidelines. Management of hepatocellular carcinoma. J Hepatol. 2012;56(4):908–43.
16. Vilana R, et al. Intrahepatic peripheral cholangiocarcinoma in cirrhosis patients may display a vascular pattern similar to hepatocellular carcinoma on contrast-enhanced ultrasound. Hepatology. 2010;51(6):2020–9.

17. Han J, et al. The degree of contrast washout on contrast-enhanced ultrasound in distinguishing intrahepatic cholangiocarcinoma from hepatocellular carcinoma. Ultrasound Med Biol. 2015;41(12):3088–95.
18. Li R, et al. Detailed analysis of temporal features on contrast enhanced ultrasound may help differentiate intrahepatic cholangiocarcinoma from hepatocellular carcinoma in cirrhosis. PLoS One. 2014;9(5):e98612.
19. Wildner D, et al. CEUS in hepatocellular carcinoma and intrahepatic cholangiocellular carcinoma in 320 patients – early or late washout matters: a subanalysis of the DEGUM multicenter trial. Ultraschall Med. 2015;36(2):132–9.
20. Seitz K, et al. Contrast-Enhanced Ultrasound (CEUS) for the characterization of focal liver lesions – prospective comparison in clinical practice: CEUS vs. CT (DEGUM multicenter trial). Parts of this manuscript were presented at the ultrasound Dreilandertreffen 2008, Davos. Ultraschall Med. 2009;30(4):383–9.
21. Guang Y, et al. Diagnosis value of focal liver lesions with SonoVue(R)-enhanced ultrasound compared with contrast-enhanced computed tomography and contrast-enhanced MRI: a meta-analysis. J Cancer Res Clin Oncol. 2011;137(11):1595–605.
22. Liu GJ, et al. Contrast-enhanced ultrasound for the characterization of hepatocellular carcinoma and intrahepatic cholangiocarcinoma. Liver Cancer. 2015;4(4):241–52.
23. de Sio I, et al. Optimized contrast-enhanced ultrasonography for characterization of focal liver lesions in cirrhosis: a single-center retrospective study. United European Gastroenterol J. 2014;2(4):279–87.
24. Huang JY, et al. Diagnostic accuracy of CEUS LI-RADS for the characterization of liver nodules 20 mm or smaller in patients at risk for hepatocellular carcinoma. Radiology. 2019:191086.
25. Terzi E, et al. Contrast ultrasound LI-RADS LR-5 identifies hepatocellular carcinoma in cirrhosis in a multicenter restropective study of 1,006 nodules. J Hepatol. 2018;68(3):485–92.
26. Ling W, et al. The preliminary application of liver imaging reporting and data system (LI-RADS) with contrast-enhanced ultrasound (CEUS) on small hepatic nodules (</= 2cm). J Cancer. 2018;9(16):2946–52.
27. Makoyeva A, et al. Use of CEUS LI-RADS for the accurate diagnosis of nodules in patients at risk for hepatocellular carcinoma: a validation study. Radiol Imaging Cancer. 2020;2(2):e190014.
28. van der Pol CB, et al. Accuracy of the liver imaging reporting and data system in computed tomography and magnetic resonance image analysis of hepatocellular carcinoma or overall malignancy-a systematic review. Gastroenterology. 2019;156(4):976–86.
29. Lyshchik A, et al. Clinical validation of contrast-enhanced ultrasound liver imaging reporting and data system in a prospective multinational study in North America and Europe. Hepatology. 2023;79(2):380–91.
30. Marrero JA, et al. Diagnosis, staging, and management of hepatocellular carcinoma: 2018 practice guidance by the american association for the study of liver diseases. Hepatology. 2018;68(2):723–50.
31. Singal AG, et al. AASLD practice guidance on prevention, diagnosis, and treatment of hepatocellular carcinoma. Hepatology. 2023;78(6):1922–65.
32. Hu J, et al. Resolution of indeterminate MRI with CEUS in patients at high risk for hepatocellular carcinoma. Abdom Radiol (NY). 2020;45(1):123–33.

Part IV

Prognosis

Should Barcelona Clinic Liver Cancer Staging Be Used as the Preferred Staging System for Hepatocellular Carcinoma?

16

Alessandro Vitale and Michele Finotti

Main Points

1. The prognosis for patients with HCC depends on tumor burden, liver dysfunction, and performance status – factors that are incorporated in the BCLC staging system.
2. The BCLC system is evidence based and remains the most widely validated staging system globally; however, its limitations including limited prognostic ability and the rigidity of its treatment algorithm have been increasingly recognized.
3. In an era of precision medicine, customized prognostic and treatment scores based on the patient stage rather than the disease stage have been claimed.

Introduction

In cancers management, staging systems play an important role, allowing an accurate prognostic stratification of the tumour, guide therapeutic options according to the stage and lead to a common stratification of the tumour to design clinical trials. Conventional staging systems consider only the morphologic features of the tumour, while prognostic scores take into account all the main features related to the final prognosis. Staging system and prognostic score must be easy to use, reproducible and transportable to different populations in order to be recommended and used on a large scale [1].

The in situ tumour spread and the patient's organs involvement are generally adequate to stage most solid tumours.

Unlike other tumours, hepatocellular carcinoma (HCC) usually rises in the context of liver cirrhosis, making its management unique and complex respect other malignancies. The prognosis of patient affected by HCC is related not only by the tumour behaviour, but also by the progression of the liver dysfunction.

In patients with HCC the prognosis is affected by three main factors [2]:

1. The stage of the tumour.
2. Liver function.
3. General health status of the patient.

Number and size of lesions, vascular invasion and metastatic spread in the body usually define tumour status.

Liver function is evaluated through biochemical markers (albumin, bilirubin, prothrombin time) and signs and symptoms of liver decompensation (ascites, encephalopathy, portal hypertension, impaired renal function, hyponatremia). Prognostic scores have been developed based on these variables, such as MELD score [3] (Model for End Stage Liver disease score), Child Pugh score [4] (CPS) or ALBI score [5] (albumin-bilirubin score).

The general health of the patient is usually evaluated through the ECOG Performance Status [6] or the Karnofsky index [7].

Staging systems and prognostic scores for HCC are based on these three main features, with the final intent to weight each factor to achieve an excellent prognostic ability.

Furthermore, an ideal staging system for HCC patients should address three main functions: to determine the prognosis for individual patients; to represent a common scale for treatment selection for individual patients (to avoid under and overtreatment); to represent a common scale for the comparison of outcomes among treatment methods and institutions and for RCT design [8].

The Barcelona Clinic Liver Cancer score (BCLC) [9] is one of the most used HCC staging systems, due to its ease of use and the therapeutic indications for each stage.

A. Vitale (✉) · M. Finotti
Department of Surgical Oncological and Gastroenterological Sciences, Padua University, Padua, Italy
e-mail: michele.finotti@studenti.unipd.it

N. S. Reau, A. G. Singal (eds.), *Controversies in Hepatocellular Carcinoma*, https://doi.org/10.1007/978-3-031-90225-3_16

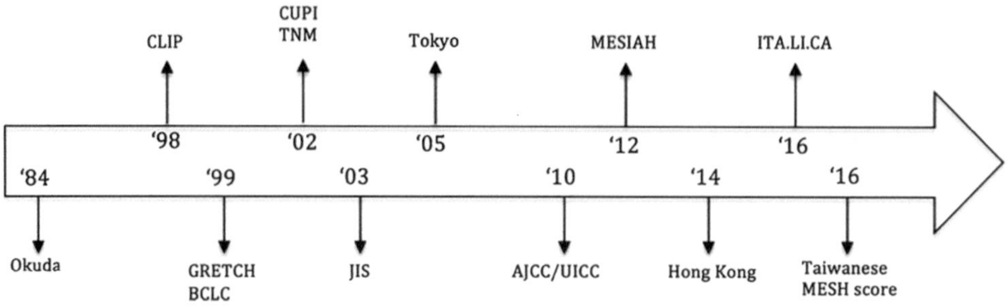

Fig. 16.1 HCC staging systems during the last decades

The topic of this chapter is to understand if the BCLC system is the superior staging system to establish patient prognosis and to help treatment decision in patients affected by HCC.

Is the BCLC Classification the Superior Staging System for HCC Patients to Determine Patient Prognosis?

During the last three decades several staging systems or prognostic scores, both from the east and the eastern world, have been proposed to estimate the prognosis of HCC patients (see Fig. 16.1).

Some general consideration must be pointed out before proceeding with the description and the comparison of existing staging systems.

First of all, selection of the clinical and tumour variables and the consequent validity of a system are influenced by the population where the stage system has been developed. In particular, Easter models are mainly based on young HBV patients with mild underlying cirrhosis; Western models are based on older population with more compromised liver function. Staging systems developed in Western and Easter regions are very difficult to compare, and different systems for each region and each tumour stage have been proposed. Furthermore, liver function and especially the tumour status can be described and defined with different methods and variables.

In addition, staging systems developed in the past can be nowadays obsolete due to diagnostic and therapeutic progression.

All these variability in the HCC stage definition led to the development of a large number of different scoring systems, using different variables based on different methodologies.

We can classify these large numbers of systems/scores into three main categories based on the methodology by which they have been created:

A. Data-based scores, from a real cohort population.
B. Evidence-based staging systems, derived from literature review.
C. Combined prognostic systems, based on literature evidence and weighted in a real population.

A. *Data-Based Scores*

Data-based scores system are developed from a real-life cohort population using objective and reproducible variables. These systems are developed using rigorous statistical methodology and the HCC patient survival is accurately predicted. However, because they are developed from a specific population, their application to the general population is not always feasible and, the most important limitation, they are not able to define tumour stages for treatment selection.

We described the most important data-based scores in detail (see Tables 16.1 and 16.2).

The staging system proposed by Okuda et al. [10] in 1984 is the first score that combines the anatomical extension of the tumour ($\leq$ or > 50% involvement of the liver) to the liver dysfunction (expressed by albumin, bilirubin, presence of ascites). This system has been widely used worldwide for over two decades. However, with the progression of diagnostic and therapeutic techniques during the last years and its inability to differentiate HCC according to the size of the lesions, Okuda system has been progressively abandoned.

Cancer of the Liver Italian Program (CLIP) [11] was developed through a retrospective cohort study of Italian patients. It was externally validated and it has been considered an excellent score to predict the prognosis, but some limits have been raised. In particular, the absence of clinical status of the patient, and the inability to identify the early stages, which are susceptible to curative treatments such as percutaneous or surgical therapies are the main limitations.

The GRETCH score [12] (Groupe d'Etude et de Traitement du Carcinome Hépatocellulaire), based on clinical and functional parameters, identifies three risk groups with different 1-year survival. However, in this staging system tumour extension is not evaluated.

The CUPI [16] (Chinese University Prognostic Index for hepatocellular carcinoma) derived from a mono-centric and mono-ethnic cohort of Chinese patients, affected principally HBV. The population is very specific and its generalizability to other populations is uncertain.

The JIS [13] (Japan Integrated Staging Score) combines TNM and Child-Pugh classifications. It is widely used in Japan but it lacks external validation in Western countries.

Table 16.1 Principal data-based scores system for HCC

		Score			
Stage System	Variable	0	1	2	3
Okuda [10]	*Tumour size*	≤ 50% of the liver	> 50% of the liver		
	Albumin (g/dL)	≥3	<3		
	Bilirubin (mg/dL)	<3	≥3		
	Ascites	No	Yes		
CLIP [11]	*Tumour morphology*	Uninodular and extension ≤50%	Multinodular and extension ≤50%	Massive or extension >50%	
	CPS	A	B	C	
	aFP (ng/mL)	<400	≥400	–	
	PVT	No	Yes	–	
French (GRETCH) [12]	*Karnofsky index*	≥50%	–	–	<50%
	Bilirubin (mg/dL)	<50	–	–	≥50
	Alkaline phosphatase	< 2 × ULN	–	≥ 2 × ULN	–
	aFP (ng/mL)	<35	–	≥ 35	–
	PVT	No	Yes	–	–
JIS [13]	*CPS*	A	B	C	
	TNM stage	I	II	III	IV
Tokyo [14]	*Albumin (g/dL)*	> 3.5	2.8–3.5	< 2.5	
	Bilirubin (mg/dL)	<1	1–2	>2	
	Tumour size (cm)	<2	2–5	>5	
	Numbers of nodules	≤3	–	>3	
Taiwanese MESH [15]	*Tumour size*	Within MC	Beyond MC		
	Vascular invasion or metastasis	No	Yes		
	CPS	5	≥6		
	Performance status	0–1	≥2		
	aFP (ng/mL)	<20	≥20		
	Alkaline phosphatase (IU/L)	<200	≥200		

aFP alpha fetoprotein, *CPS* Child Pugh Score, *MC* Milan Criteria, *PVT* Portal Vein Thrombosis

Table 16.2 Principal HCC stage systems and their associated variables

Staging system	Year	Pts number	Liver function	Performance status	aFP	HCC number	Size	Vascular invasion	Metastasis	Other
Okuda [10]	1984	600	Ascites Albumine Bilirubin	No	No	No	+	No	No	/
CLIP [11]	1998	435	CPS	No	Yes	Yes	Yes	Yes	No	/
French(GRETCH) [12]	1999	761	Bilirubin	Karnofsky	Yes	No	No	Yes	No	Alk-P
CUPI [16]	2002	926	Ascites Bilirubin	Symptoms	Yes	Yes	Yes	Yes	Yes	Alk-P
JIS [13]	2003	Review	CPS	No	No	Yes	Yes	Yes	Yes	/
Tokyo [14]	2005	403	Albumin Bilirubin	No	No	Yes	Yes	No	No	/
TIS [17]	2010	2030	CPS	No	Yes	TTV	TTV	No	No	/
MESIAH [18]	2012	477	MELD Albumin	No	Yes	Yes	Yes	Yes	Yes	Age
Taiwanese MESH [15]	2016	3182	CPS	ECOG	Yes	Yes	Yes	Yes	Yes	Alk-P

aFP alpha fetoprotein, *Alk-P* phosphatase alkaline, *CPS* Child Pugh Score, *MC* Milan Criteria, *PVT* Portal Vein Thrombosis

The Tokyo score [14] was derived from a cohort of Japanese patients with early and treated HCC (below 5 cm). This score can be applied in patients susceptible to radical therapies, while its power to define the prognosis decrease in advanced HCC.

The TIS score [17] (Taipei Integrated Scoring System) was developed from 2030 HCC patients retrospectively analysed. TIS is a point-scoring system combining AFP levels, CPS and the total tumour volume. It has not been validated in the Western country and no correlation has been described between stage and treatment options.

The MESIAH score [18] was created from a cohort of American JIS13 patients, validated in an Asiatic JIS13 population. The patient's prognosis affected by HCC is predicted through objective parameters, providing a possible supplement to the BCLC classification. The major limits of this stage system are the complexity of the mathematics formula and the lack of a corresponding therapeutic indication for each stage.

The MESH score [15] (Model to Estimate Survival for HCC), proposed by Liu et al., is based on 3182 patients prospectively enrolled. This score (from 0 to 1) combined the Milan Criteria, vascular invasion, CPS, the performance status and biomarkers (aFP and alkaline phosphatase). However, it does not propose a concrete treatment recommendations and it is not currently endorsed in European guidelines.

B. *Evidence-Based Staging Systems*

Evidence-based staging systems are commonly created by a group of experts who establish and define different prognostic stages based on the evidence of the scientific literature. These systems are useful to link HCC stages to subsequent treatment options. However, they are not based on a strong statistical methodology and they often lack prognostic power. TNM and BCLC are the main evidence-based score systems.

The main example of evidence-based staging system is the TNM staging system, approved during the American Joint Committee on Cancer (AJCC) [19]. As for other cancers, the TNM is mainly based on tumour pathological features, not taking into account other important variables in patients affected by cirrhosis such as liver function and patient general conditions. Therefore, this staging system is not useful to assess the prognosis of the general HCC population.

Barcelona Clinic Liver Cancer staging classification [9] (BCLC), firstly proposed in 1999, was the first staging system that integrates the assessment of liver function, tumour extension and clinical features. Each stage was associated with an indication of the first-line treatment.

The BCLC classification can be considered an evidence-based system, since it was generated by analysing the results of randomized controlled studies testing a given treatment modality versus placebo in patients with comparable tumour characteristics and liver function.

The BCLC staging system has been endorsed by the American Association for the Study of Liver Diseases (AASLD), the American Gastroenterology Association (AGA), the European Association for the Study of Liver (EASL) and the European Organization for Research and Treatment of Cancer (EORTC) [20, 21].

It classifies patients into five subgroups, from 0 to D, each group associated with a specific therapy and prognosis. In the years, the BCLC flow chart has been frequently modified. In Fig. 16.2, we proposed a recent BCLC flow chart version (Fig. 16.2) [22]. In the proposed algorithm, a compensated disease (e.g. no ascites) with a CPS A is considered a preserved liver function.

Stage 0 (very early stage disease) includes asymptomatic patients with preserved liver function, with a single nodule ≤2 cm and no tumour invasion into surrounding tissues.

Stage A (early disease) is classified as a solitary HCC of any size, or a maximum of three nodules, each ≤3 cm in size, in asymptomatic patients with Child-Pugh A or B. To note, in stage A the single HCC nodule lacks dimensional definition.

In stage 0 and A therapies with curative intent are available, such as surgical resection, liver transplantation, or thermal ablation. These treatments are associated with potential long-term curative effect and a 50–70% 5-year survival [23].

Stage B (intermediate disease) includes a very heterogeneous population. In this category, HCC can range from large single lesions to multi nodular tumour, with no macrovascular invasion or extrahepatic spread. In this group the liver function is still preserved (CPS A-B) with patient with good performance status (ECOG PS 0). In this stage, TACE is the proposed treatment of choice [24].

Stage C (advanced stage) is defined in the presence of macrovascular invasion or extrahepatic spread and an ECOG PS ≥ 1. In this stage, Sorafenib is the standard of care, but recently other systemic agents have been proposed, in particular Lenvatinib and Regorafenib [22, 25, 26].

Patients with an end-stage liver disease (stage D) for advanced liver failure, tumour growth with vascular involvement, extrahepatic disease, or physical impairment (PS > 2), cannot benefit from any specific cancer therapy and could only receive the best available supportive care (BSC).

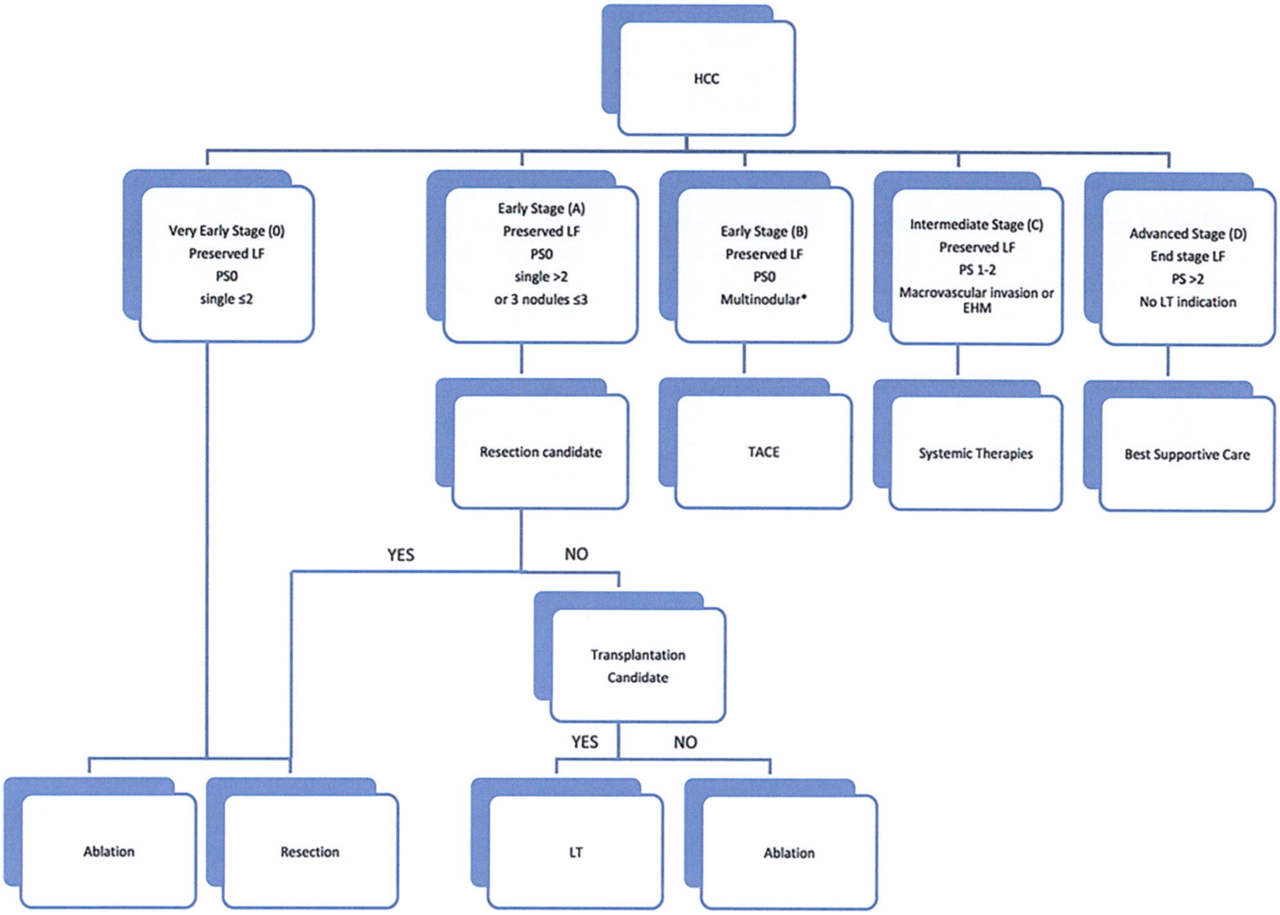

Fig. 16.2 BCLC staging system and therapeutic algorithm [22]. *Multinodular: >3 nodules, or ≥ 2 nodules if any >3 cm; *EHM* ExtraHepatic Metastases, *LF* Liver Function, *LT* Liver Transplantation, *PS* Performance Status, *TACE* Chemio Embolization

Limitations of the BCLC

In the BCLC system the most important prognostic variables for HCC patients are taken into account (i.e. tumour characteristics, liver function and patient general conditions). However, this classification has some critical limitations, mainly because it is not derived from a real population of HCC patients. Limitations are related to the prognostic capability but also to its treatment options.

BCLC Limitations to Predict Prognosis

BCLC has been developed to link prognostic variables to therapeutic options. This aim led to some limitations related to its prognostic power, compared to pure prognostic scores.

The first issue concerns the dimensional threshold of a single lesion to be included in the early stage (Stage A).

A cut-off in tumour diameter for patients with single tumour is not proposed, even if it is established in the literature that tumour diameter has a relevant prognostic impact in patient affected by single HCC. In addition, the lack of a clear cut off could represent a limit when comparing data from different studies.

A dimensional limit of 5 cm introduced by the Milan criteria [27] was proposed as a possible cut off, categorizing a solitary tumour of ≤5 cm as Early Stage A, and lesions >5 cm as Intermediate Stage B. The authors of the BCLC left the tumour resectability evaluation to the surgeon's expertise.

The second issue is related to the wide heterogeneity of the intermediate-stage disease (BCLC stage B). This stage includes from oligonodular HCC with small tumours to large multinodular HCC involving a large part of liver parenchyma.

According to BCLC system, a patient with preserved liver function (i.e. Child–Pugh class A5), two nodules of 32 and 16 mm in diameter, and on the other side, a patient with persistent ascites, jaundice (Child–Pugh class B) and multifocal large tumours, are both classified in the intermediate stage B. The prognosis and the treatment options of these two patients differ greatly.

Other important issue is the lack of distinction between intra-hepatic or segmentary vascular invasion and extra-hepatic invasion of portal trunk or main trunk of inferior vena cava.

Furthermore, a minimal deterioration of patient's performance status is considered a sufficient criterion to classify patients in the advanced stage: patients with neoplastic invasion of major vessels and patients with impaired liver function or compromised PS are both classified in group C [28, 29]. However, the firsts can benefit only from palliative chemotherapy (5-year survival rate 5%), while the seconds can receive liver transplantation with associated 5- year OS of 80% [30, 31].

BCLC Limitations to Choice Therapeutic Options

Advancements in HCC management over the last decades have led many authors to consider BCLC classification and its treatment allocation too restrictive and obsolete. The BCLC classification has been created on the principle of selecting candidates who will achieve the best results with a specific treatment, rather than selecting the treatment that provides the best result in that particular patient. In the BCLC score the therapeutic option is strictly linked to the stage of the disease. For example, transarterial chemoembolization (TACE) is the currently recommended standard of care for intermediate stage B, but it may represent an "undertreatment" for some patients and an "overtreatment" for others. A treatment-oriented sub-classification of the intermediate BCLC stage B has been recently proposed by International group of experts: Bolondi et al. developed a sub-classification of intermediate HCC in four substages, B1 to B4, according to tumour burden and liver function [32]. However, the rigidity in treatment allocation according to each stage represents the main issue of BCLC classification, especially for intermediate group B.

This rigidity in treatment allocation according to each stage represents the main issue of BCLC classification, restricting the treatment options. Recently, the concept of "stage hierarchy" and "stage migration" has been introduced to describe this limitation [33]. The rigidity of the BLCL score led many hepatobiliary centres not to strictly follow its therapeutic indications.

Two Italian studies looked at the adherence to HCC guidelines, showing that the majority of patients with HCC were not treated according to the BCLC indications, especially for those with early-stage HCC [34, 35].

Roayaie et al., in a multiregional cohort study including 8656 patients HCC, found that two-thirds of patients who underwent liver resection were not ideal candidates based on BCLC criteria [36].

Pecorelli et al. reported similar discrepancy between the clinical practice and the BCLC recommended strategy: only 50% of the patients receive the recommended therapy [37].

Similarly, Kim KM et al. reported that only half of the patients followed the BCLC-recommended treatment [41]. To note, HCC patients in which BCLC indications have not been followed, showed better outcomes [38].

The optimal treatment selection is crucial to improve patient survival rates. Determining the most appropriate treatment for HCC at any stage requires a multidisciplinary approach to tailor the most adequate treatment to each candidate.

C. *Combined Staging Systems*

Combined staging systems are developed on evidence-based variables that are weighted in a real population, allowing simultaneously and in a balanced manner a good prognosis evaluation and treatment assignment. These features theoretically make these staging systems more effective than the scores previously described. HKLC and ITA.LI.CA are the two most important combined scores.

Hong Kong Liver Cancer Staging system [39] was developed in 2014, on a cohort of patients with HCC predominantly secondary to HBV. In this score PS, CPS, tumour status (based on Milan Criteria), intra- and extra-hepatic vascular invasion or metastasis were defined as initial criteria based on literature evidence. These variables were subsequently weighted in a real population, by assigning a relative coefficient to each variable.

The HKLC, compared to BCLC classification, is associated with a better ability to characterize the BCLC intermediate and advanced stage patients, who can benefit from more aggressive treatments than those recommended by the BCLC system. The limit of this system is the lack of solid external validation in a non-Asian population.

Very few studies compared BCLC and HKLC scores and HKLC score was not associated with a better prognostic value than BCLC in Western populations [18, 38–40].

The Italian Liver Cancer (ITA.LI.CA) score [41], created in 2016 through a multicentre retrospective analysis and validated in a Taiwanese cohort, is a prognostic model able to predict patient outcomes affected by HCC. The score is based on tumour, liver function and patient-related variables (See Table 16.3). This stage system is derived from the BCLC score, but with a better definition of the tumour status and cut off dimension (especially in the early stage), based on the literature evidence.

In particular, for a better prognostic definition, the intermediate stage was split into three subgroups. Furthermore,

Table 16.3 ITA.LI.CA staging system with corresponding therapeutic options

Tumour stage	Diameter (cm)	<2	≤3	≤5	3–5	>5	≤5	>5	>5	Any	Any	Any
	Number of nodules	1	2–3	1	2–3	1	>3	2–3	>3	Any	Any	Any
	Hepatic (H) vascular invasion (VI) and/or metastases	No	No	No	No	No	No	No	No	Intra HVI	Extra HVI or metastases	Any
Functional score		CPS ≤9 and PST 0 Or CPS ≤7 and PST 1–2										CPS 8–9 and PST 1–2, or CPS >9, or PST >2
Stage		0	A		B1		B2		B3		C	D
Therapy	Expected median survival (months)											
	Best supportive care	31	22		18		17		10		9	3
	Systemic therapy	36	30		24		22		16		14	/
	Intraarterial therapy	55	45		35		33		23		/	/
	Ablation	80	65		50		48		33		/	/
	Liver resection	101	83		64		62		36		/	/
	Liver transplantation	120	112		91		90		/		/	/

intra- and extrahepatic HCC vascular invasion were identified as separate entities: HCC with intrahepatic vascular invasion is liable of therapeutic options with radical intent.

Patient status was evaluated with the CPS and the ECOG PS. Additionally, AFP was added as another important variable of the score.

Each variable showed a different impact in determining the final score: in order to better weight their prognostic influence, different points were attributed to each sub-variable. Finally, based on these results, the ITA.LI.CA integrated prognostic score has been created: the lowest score (ITA.LI.CA score = 0) of the model corresponded to the best prognosis, and the highest score (ITA.LI.CA score = 13) was associated with the worst prognosis. For each stage, the therapeutic options are reported weighting the prognosis impact.

Recently, ITA.LI.CA score was externally validated in a multicentre cohort study [42]. ITA.LI.CA has been found to perform better than other scores even in restaging patients with hepatocellular carcinoma before additional treatment decisions [43]. In conclusion, ITA.LI.CA showed to have a great ability to predict the prognosis in HCC patients. Moreover, this innovative stage system proposed therapeutic options for each stage based on the so-called "treatment hierarchy", opposed to the standard BCLC stage hierarchy concept [33, 43].

Stage Systems Comparison

The performance of a prognostic system is defined by three characteristics: homogeneity, discriminatory ability and monotonicity of gradients [44].

A system is homogeneous when differences in survival between patients of the same stage are small. Homogeneity of staging systems is usually described using likelihood ratio (LR) in a COX regression model: the higher is the LR, the lower is the difference among patients classified into the same group [45].

The discriminatory power is the ability of the system to produce great differences in survival among patients in different stages. It is usually calculated using the X^2 linear trend test (LT) and the Akaike information criteria (AIC): the higher is the LT and the lower is the AIC, the higher is the discriminatory ability of the system.

When monotonicity of gradients is fulfilled, the survival of patients in early stage is longer than survival of patients in more advance stages within the same system. The C-index is another parameter that is frequently used to determine the performance of a model: the larger the C-index, the more accurate the prognostic prediction is [46].

Many comparative studies have been proposed to define which staging system is the best in predicting survival of patients with HCC. In the study of Marrero JA et al. [47], the BCLC staging system compared with other six prognostic systems showed the best independent predictive power for survival. The superiority of BCLC staging system was supported by external validations, such as in Korean [48] and Italian populations [49]. However, in the last years inherent limitations of the BCLC score have been raised.

In 2016, Liu et al. compared 11 staging (but not the ITA.LI.CA system) between 2002 and 2013 in a large prospective database 3182 HCC patients [50]. The ability of the BCLC score to predict the HCC patient prognosis was analysed through the homogeneity and corrected Akaike information criterion (AICc), showing low AICs for the BLCL score, while the CLIP staging system resulted to be the best prognostic model in all patients and in patients with different viral aetiology and treatment strategy. To note, this study showed that a data-based score (the CLIP score) performs better than an evidence-based system.

In the study of Farinati et al., ITA.LI.CA prognostic power was also compared with that of other integrated systems BCLC, HKLC, MESIAH, CLIP, and JIS. The ITA.LI.CA score showed the best discriminatory ability and monotonicity of gradients among the most common HCC staging systems in all three study cohorts (training, internal validation and external validation). In particular, the C index of the ITA.LI.CA score in the internal and external validation cohorts was 0.71 and 0.78, respectively. The ITA.LI.CA score's prognostic ability was superior to that of BCLC stage, CLIP score, JIS stage, MESIAH score and HKLC stage. By using the likelihood ratio test to compare different survival models, the prognostic performance of the ITA. LI.CA system always resulted in significantly better discrimination ability ($p < 0.001$) than the other systems in all three study groups. The superiority of the ITA.LI.CA score was also confirmed after stratification of the analysis for time period. The ITA.LI.CA prognostic system shows a strong ability to predict individual survival in European and Asian populations [35, 41].

Conclusion

The prognosis and the therapeutic options for patients affected by HCC were mainly based on the BCLC staging system. Other staging systems have been proposed, but BCLC score has been widely adopted by many centres thanks to its simplicity of use and treatment allocation proposed for each stage. Moreover, BCLC staging system was one of the first staging systems to be designed on a literature evidence basis providing fertile ground for the subsequent design of randomized controlled trials. However, in the last years, BCLC showed its limits and has been extensively criticized, in particular due to its limited prognostic ability and its rigidity of treatment algorithm. For these reasons, in the clinical setting, the use of the BCLC score has been progressively reduced in favour of a more personalization of the treatments. In the setting of precision medicine era, customized prognostic and treatment scores based on the patient stage rather than the disease stage have been claimed. Far from being the ideal staging system, the recently proposed ITA.LI.CA based on the concept of therapeutic hierarchy offers more treatment options for each tumoral stage allowing a personalized approach to the patient.

References

1. Amin MB, Greene FL, Edge SB, et al. The eighth edition AJCC cancer staging manual: continuing to build a bridge from a population-based to a more "personalized" approach to cancer staging. CA Cancer J Clin. 2017;67(2):93–9.
2. Bosetti C, Levi F, Boffetta P, et al. Trends in mortality from hepatocellular carcinoma in Europe, 1980-2004. Hepatology. 2008;48(1):137–45.
3. Kamath PS, Kim WR, Advanced Liver Disease Study G. The model for end-stage liver disease (MELD). Hepatology. 2007;45(3):797–805.
4. Pugh RN, Murray-Lyon IM, Dawson JL, et al. Transection of the oesophagus for bleeding oesophageal varices. Br J Surg. 1973;60(8):646–9.
5. Johnson PJ, Berhane S, Kagebayashi C, et al. Assessment of liver function in patients with hepatocellular carcinoma: a new evidence-based approach-the ALBI grade. J Clin Oncol. 2015;33(6):550–8.
6. Oken MM, Creech RH, Tormey DC, et al. Toxicity and response criteria of the Eastern Cooperative Oncology Group. Am J Clin Oncol. 1982;5(6):649–55.
7. Ohya T, Kikuchi S. Clinical evaluation of chemotherapeutic agents in the treatment of primary liver cancer. Gan To Kagaku Ryoho. 1982;9(9):1623–7.
8. Kudo M, Izumi N, Kokudo N, et al. Management of hepatocellular carcinoma in Japan: consensus-based clinical practice guidelines proposed by the Japan Society of Hepatology (JSH) 2010 updated version. Dig Dis. 2011;29(3):339–64.
9. Llovet JM, Bru C, Bruix J. Prognosis of hepatocellular carcinoma: the BCLC staging classification. Semin Liver Dis. 1999;19(3):329–38.
10. Okuda K, Ohtsuki T, Obata H, et al. Natural history of hepatocellular carcinoma and prognosis in relation to treatment. Study of 850 patients. Cancer. 1985;56(4):918–28.
11. A new prognostic system for hepatocellular carcinoma: a retrospective study of 435 patients: the Cancer of the Liver Italian Program (CLIP) investigators. Hepatology. 1998;28(3):751–5.
12. Chevret S, Trinchet JC, Mathieu D, et al. A new prognostic classification for predicting survival in patients with hepatocellular carcinoma. Groupe d'Etude et de Traitement du Carcinome Hepatocellulaire. J Hepatol. 1999;31(1):133–41.
13. Kudo M, Chung H, Osaki Y. Prognostic staging system for hepatocellular carcinoma (CLIP score): its value and limitations, and a proposal for a new staging system, the Japan Integrated Staging Score (JIS score). J Gastroenterol. 2003;38(3):207–15.
14. Tateishi R, Yoshida H, Shiina S, et al. Proposal of a new prognostic model for hepatocellular carcinoma: an analysis of 403 patients. Gut. 2005;54(3):419–25.
15. Liu PH, Hsu CY, Hsia CY, et al. Proposal and validation of a new model to estimate survival for hepatocellular carcinoma patients. Eur J Cancer. 2016;63:25–33.
16. Leung TW, Tang AM, Zee B, et al. Construction of the Chinese University prognostic index for hepatocellular carcinoma and comparison with the TNM staging system, the Okuda staging system, and the cancer of the liver Italian program staging system: a study based on 926 patients. Cancer. 2002;94(6):1760–9.
17. Hsu CY, Huang YH, Hsia CY, et al. A new prognostic model for hepatocellular carcinoma based on total tumor volume: the Taipei Integrated Scoring System. J Hepatol. 2010;53(1):108–17.
18. Yang JD, Kim WR, Park KW, et al. Model to estimate survival in ambulatory patients with hepatocellular carcinoma. Hepatology. 2012;56(2):614–21.
19. Edge SB, Compton CC. The American Joint Committee on cancer: the 7th edition of the AJCC cancer staging manual and the future of TNM. Ann Surg Oncol. 2010;17(6):1471–4.
20. Bruix J, Sherman M. American association for the study of liver D. Management of hepatocellular carcinoma: an update. Hepatology. 2011;53(3):1020–2.
21. European Association For The Study Of The Liver, European Organisation For R, Treatment Of C. EASL-EORTC clinical practice guidelines: management of hepatocellular carcinoma. J Hepatol. 2012;56(4):908–43.
22. Villanueva A. Hepatocellular carcinoma. N Engl J Med. 2019;380(15):1450–62.
23. Forner A, Llovet JM, Bruix J. Hepatocellular carcinoma. Lancet. 2012;379(9822):1245–55.

24. Llovet JM, Real MI, Montana X, et al. Arterial embolisation or chemoembolisation versus symptomatic treatment in patients with unresectable hepatocellular carcinoma: a randomised controlled trial. Lancet. 2002;359(9319):1734–9.
25. Ben Mousa A. Sorafenib in the treatment of advanced hepatocellular carcinoma. Saudi J Gastroenterol. 2008;14(1):40–2.
26. Bruix J, da Fonseca LG, Reig M. Insights into the success and failure of systemic therapy for hepatocellular carcinoma. Nat Rev Gastroenterol Hepatol. 2019;16(10):617–30.
27. Mazzaferro V, Regalia E, Doci R, et al. Liver transplantation for the treatment of small hepatocellular carcinomas in patients with cirrhosis. N Engl J Med. 1996;334(11):693–9.
28. Hsu CY, Lee YH, Hsia CY, et al. Performance status in patients with hepatocellular carcinoma: determinants, prognostic impact, and ability to improve the Barcelona clinic liver cancer system. Hepatology. 2013;57(1):112–9.
29. Hsu CY, Liu PH, Hsia CY, et al. Nomogram of the Barcelona clinic liver cancer system for individual prognostic prediction in hepatocellular carcinoma. Liver Int. 2016;36(10):1498–506.
30. Vitale A, Morales RR, Zanus G, et al. Barcelona clinic liver cancer staging and transplant survival benefit for patients with hepatocellular carcinoma: a multicentre, cohort study. Lancet Oncol. 2011;12(7):654–62.
31. Finotti M, Vitale A, Volk M, et al. A 2020 update on liver transplant for hepatocellular carcinoma. Expert Rev Gastroenterol Hepatol. 2020;
32. Bolondi L, Burroughs A, Dufour JF, et al. Heterogeneity of patients with intermediate (BCLC B) hepatocellular carcinoma: proposal for a subclassification to facilitate treatment decisions. Semin Liver Dis. 2012;32(4):348–59.
33. Vitale A, Trevisani F, Farinati F, et al. Treatment of hepatocellular carcinoma in the precision medicine era: from treatment stage migration to therapeutic hierarchy. Hepatology. 2020;72(6):2206–18.
34. Borzio M, Fornari F, De Sio I, et al. Adherence to American Association for the study of liver diseases guidelines for the management of hepatocellular carcinoma: results of an Italian field practice multicenter study. Future Oncol. 2013;9(2):283–94.
35. Farinati F, Sergio A, Baldan A, et al. Early and very early hepatocellular carcinoma: when and how much do staging and choice of treatment really matter? A multi-center study. BMC Cancer. 2009;9:33.
36. Roayaie S, Jibara G, Tabrizian P, et al. The role of hepatic resection in the treatment of hepatocellular cancer. Hepatology. 2015;62(2):440–51.
37. Pecorelli A, Lenzi B, Gramenzi A, et al. Curative therapies are superior to standard of care (transarterial chemoembolization) for intermediate stage hepatocellular carcinoma. Liver Int. 2017;37(3):423–33.
38. Kim KM, Sinn DH, Jung SH, et al. The recommended treatment algorithms of the BCLC and HKLC staging systems: does following these always improve survival rates for HCC patients? Liver Int. 2016;36(10):1490–7.
39. Yau T, Tang VY, Yao TJ, et al. Development of Hong Kong liver cancer staging system with treatment stratification for patients with hepatocellular carcinoma. Gastroenterology. 2014;146(7):1691–700 e3.
40. Adhoute X, Penaranda G, Bronowicki JP, et al. Usefulness of the HKLC vs. the BCLC staging system in a European HCC cohort. J Hepatol. 2015;62(2):492–3.
41. Farinati F, Vitale A, Spolverato G, et al. Development and validation of a new prognostic system for patients with hepatocellular carcinoma. PLoS Med. 2016;13(4):e1002006.
42. Borzio M, Dionigi E, Rossini A, et al. External validation of the ITA.LI.CA prognostic system for patients with hepatocellular carcinoma: a multicenter cohort study. Hepatology. 2018;67(6):2215–25.
43. Vitale A, Farinati F, Noaro G, et al. Restaging patients with hepatocellular carcinoma before additional treatment decisions: a multicenter cohort study. Hepatology. 2018;68(4):1232–44.
44. Jemal A, Bray F, Center MM, et al. Global cancer statistics. CA Cancer J Clin. 2011;61(2):69–90.
45. Siegel RL, Miller KD, Jemal A. Cancer statistics, 2016. CA Cancer J Clin. 2016;66(1):7–30.
46. Maida M, Orlando E, Camma C, et al. Staging systems of hepatocellular carcinoma: a review of literature. World J Gastroenterol. 2014;20(15):4141–50.
47. Marrero JA, Fontana RJ, Barrat A, et al. Prognosis of hepatocellular carcinoma: comparison of 7 staging systems in an American cohort. Hepatology. 2005;41(4):707–16.
48. Kim BK, Kim SU, Park JY, et al. Applicability of BCLC stage for prognostic stratification in comparison with other staging systems: single centre experience from long-term clinical outcomes of 1717 treatment-naive patients with hepatocellular carcinoma. Liver Int. 2012;32(7):1120–7.
49. Cillo U, Bassanello M, Vitale A, et al. The critical issue of hepatocellular carcinoma prognostic classification: which is the best tool available? J Hepatol. 2004;40(1):124–31.
50. Liu PH, Hsu CY, Hsia CY, et al. Prognosis of hepatocellular carcinoma: assessment of eleven staging systems. J Hepatol. 2016;64(3):601–8.

Part V

Treatment

Should Resection Be the Preferred Therapeutic Option for Patients Within the Milan Criteria and Compensated Cirrhosis Without Portal Hypertension?

17

Myron Schwartz and Vennis Lourdusamy

Resection is the treatment of choice for patients with a single hepatocellular carcinoma (HCC) larger than 2 cm, Child-Pugh score (CPS) class A liver function, and no portal hypertension.

Main Points

1. Clinically significant portal hypertension is suggested by the presence of ascites, splenomegaly, esophageal varices, and/or a platelet count of 100,000 or less.
2. Portal pressure measurements can demonstrate portal hypertension in cases where clinical clues are inconsistent.
3. A gradient of 10 mm Hg between the wedged and free hepatic venous pressure is important.
4. Although overall survival (OS) is comparable between resection and liver transplantation (LT), historically HCC recurrence is less common after LT; total hepatectomy dramatically reduces both early (since the liver is the most common site of HCC metastasis) and late (since the cirrhotic, tumorigenic liver is removed) recurrence.
5. Resection requires lifelong surveillance and appropriate treatment of recurrence to achieve optimal survival rates.

Curative Options for Early Hepatocellular Carcinoma

The European Association for the Study of the Liver recommends that patients at high risk for HCC who would benefit from early HCC detection undergo surveillance imaging every 6 months [1]. The Barcelona Clinic Liver Cancer (BCLC) staging is the most widely accepted staging system for HCC, incorporating the degree of liver dysfunction, tumor characteristics (number, size, vascular invasion, metastasis), and patient performance status. BCLC stages 0 and A are considered early and fall within the Milan criteria (MC) based on the size of the nodules (MC = single tumor ≤ 5 cm, ≤ 3 tumors each ≤ 3 cm). BCLC stage 0 (very early stage) patients are those with a single tumor smaller than 2 cm, and stage A are those with a single tumor nodule, or 2–3 tumors each less than or equal to 3 cm. Patients within BCLC stage 0 and A should be considered for curative therapy, which can be achieved through LT, resection, or thermal ablation. Transarterial radiation using Y90 has evoked interest as a curative treatment for early HCC [2]; while not currently incorporated into guidelines, it may be considered in patients who are not readily treated with resection or thermal ablation. Considerations when choosing treatment for patients with early HCC include liver function, presence of portal hypertension, and the outcome and survival benefits of one treatment over another.

Factors Guiding the Choice of Treatment for Early Hepatocellular Carcinoma

Liver function is commonly assessed by CPS, which is calculated based on albumin, bilirubin, international normalized ratio, presence of ascites, and hepatic encephalopathy. Patients with CPS class A (score < 6) have preserved liver function. The presence or absence of portal hypertension should be determined in these patients. Clinically significant portal hypertension is suggested by the presence of ascites, splenomegaly, esophageal varices, and/or a platelet count of 100,000 or less. In cases where the clinical findings are inconsistent, portal pressure may be determined, albeit indirectly, by measuring the wedged hepatic venous pressure; a gradient greater than 10 mm Hg between the wedged and the

M. Schwartz (✉)
Icahn School of Medicine at Mount Sinai, New York, NY, USA
e-mail: myron.schwartz@mountsinai.org

V. Lourdusamy
Bayhealth Medical Group, Dover, DE, USA

N. S. Reau, A. G. Singal (eds.), *Controversies in Hepatocellular Carcinoma*, https://doi.org/10.1007/978-3-031-90225-3_17

free hepatic venous pressure is generally considered to indicate considerable portal hypertension. In our experience measuring portal pressure in this way is necessary in only a small minority (< 5%) of cases. Patients with CPS class A liver function and without evidence of portal hypertension are likely to tolerate resection without significant risk of decompensation. Patients who have clinically significant portal hypertension and/or CPS class B/C cirrhosis derive less benefit from resection and are better served with LT. In the real-world scenario there are always some exceptions and these guidelines are not strictly followed, for example, in a patient with a superficial HCC that is not readily amenable to ablation but that can be resected laparoscopically with minimal loss of functioning parenchyma.

Brief Overview of Surgical Resection

Resection is the treatment of choice for patients with single a HCC larger than 2 cm, CPS class A liver function, and no portal hypertension. Preservation of liver parenchyma to the extent possible is key to safe resection; in cirrhotic patients even with apparently normal liver function and no portal hypertension, removal of large amounts of functioning liver as, for example, in a right hepatectomy, is dangerous. In cases where major resection is needed, preliminary portal vein embolization is typically employed to induce hypertrophy of the contralateral lobe. Laparoscopic resection should be considered when technically feasible based on center experience, as it is associated with less postoperative morbidity and decreased length of hospital stay (albeit with similar survival rates) compared to open resection. A more detailed discussion of resection techniques is beyond the scope of this chapter.

Comparative Outcomes of Resection and Liver Transplantation for Hepatocellular Carcinoma

OS, disease-free survival, and HCC recurrence are the most commonly reported outcomes to assess treatment benefit. Many retrospective studies concur with the inference that OS is comparable between LT and resection performed for ideal candidates, with a 5-year OS averaging around 50–70% as discussed later. In a meta-analysis of studies comparing resection and LT in patients with HCC within the MC, [3] in an intention-to-treat analysis, there was no statistically significant difference in the 5-year OS between those who underwent LT (59.3%) vs resection (47.9%). Intention-to-treat analysis is important in the assessment of the outcomes, especially with the organ shortages that limit LT, and is discussed later in the chapter. In a retrospective analysis by Wong et al., [4] although OS appeared to be better in the transplant-eligible patients who underwent LT when compared to surgical resection, there was no statistically significant difference in the long-term survival based on the intention-to-treat analysis. More recently, Krenzien and colleagues [5] reported similar postprocedure OS for LT and resection in patients within the MC (5-year OS 73% vs 61%, $P = 0.07$). This could be explained by continued improvement in surgical expertise and optimal selection of patients for major resection, including optimizing the future remnant liver volume and improved locoregional management of recurrent lesions. On comparing the effect of etiology of HCC on the outcomes, we reported in 2014, before the introduction of effective direct-acting antiviral agent therapy for hepatitis C virus (HCV), that patients with HCC secondary to HCV who underwent either LT or resection had statistically significantly worse outcomes, both OS and disease-free survival, compared to those with hepatitis B virus (HBV). In the current era, the ability to reliably cure HCV will no doubt improve the results both of LT and resection, with decreased late HCC recurrence after resection and better preservation of liver function after both resection and LT [6, 7].

Although OS is comparable between resection and LT, HCC recurrence is clearly less common after LT; total hepatectomy dramatically reduces both early (since the liver is the most common site of HCC metastasis) and late (since the cirrhotic, tumorigenic liver is removed) recurrence [8]. HCC recurrence after resection, prompted either by undetected micrometastasis or de novo lesions arising in the background of cirrhosis, is as high as 70% at 5 years postresection [9]. Risk factors for postresection HCC recurrence include the presence of microvascular invasion (mVI), high tumor grade, and positive surgical margin [10]. HCC recurrence, however, does not necessarily correlate with poor outcome. Most recurrences after resection are in the liver, and with close surveillance many cases are managed effectively with repeat resection, ablation, or salvage LT. Management of HCC recurrence has become progressively more efficacious in recent years as a result of advances in locoregional therapies and with the involvement of multidisciplinary team decisions [11]. That HCC recurrence is not directly linked with poor OS is corroborated in a number of recent studies [12, 13]. As mentioned earlier, historically poor outcomes of resection for HCV-related HCC are bound to improve with the ability to reliably cure HCV and will likely approach the greater than 70% 5-year survival seen in patients with HBV-related HCC, for whom effective antiviral therapy has long been available. Hence LT for HCC should not be proposed just based on the high recurrence rate after resection. Undertaking resection implies a commitment on the part of the patient and the physician to lifelong surveillance and appropriate treatment of recurrence to achieve optimal survival rates.

Patients with Portal Hypertension and/or Multiple Tumors Within Milan Criteria

It is well established that the presence of multiple tumors and/or clinically significant portal hypertension leads to worse outcome after resection [14, 15]. In Asia, where HBV has long been the primary etiology and LT has not been widely available, resection has commonly been performed in this setting, but even papers that argue for it document inferior results [16]. In a prior study comparing resection for HCC between our center and the Eastern Hepatobiliary Surgery Hospital in Shanghai, the Chinese group had statistically significantly worse OS and shorter transthyretin; the difference disappeared after propensity matching for portal hypertension and number of nodules [17]. In the aforementioned study from Japan, the OS at 5 years after resection for HCC was 71% in patients without portal hypertension when compared to 56% in patients with portal hypertension ($P = 0.008$); similarly, OS was statistically significantly worse in patients with multiple (58%) vs single tumors (68%, $P = 0.035$) [16]. While these results might be considered acceptable in a setting where LT is not feasible, they cannot be justified when the transplant alternative is available. With improved results of nonsurgical locoregional therapies, it is questionable whether even in cases where for one reason or another LT cannot be performed resection is justified.

Approach

From the prior discussion, it is clear that patients with a solitary tumor and without portal hypertension are the optimal candidates for resection. When that tumor measures between 2 and 5 cm, these patients are also eligible for LT waiting list priority (eligibility criteria for transplant priority in the United States specify that the patient is not a candidate for resection; many centers, however, consider resectability a philosophical rather than a technical issue, deeming all patients with cirrhosis unresectable). In the aggregate, the retrospective data comparing OS after LT and resection in patients who are optimal resection candidates do not appear to show a statistically significant difference; while HCC recurrence is more common after resection, with close surveillance and efficient management recurrence does necessarily translate to poor survival [13]. An additional consideration with LT for HCC is the (in)availability of donor organs, resulting in a variable waiting time based on geography during which time progression of tumor or liver disease may render the patient no longer eligible for LT. Intention-to-treat analysis must therefore be applied when weighing LT vs resection [15]. A recently published study [18] found tumor progression after locoregional therapy and duration of waitlist registration to be statistically significantly associated with LT failure, in terms of both waiting list dropout and post-LT HCC recurrence. Living-donor LT (LDLT), for patients who have a suitable donor, is a way to circumvent the waiting list and avoid the risks of tumor progression and complications of tumor treatment attendant to waiting. Results of LDLT for HCC are comparable to those achieved with deceased-donor LT [19]. LDLT offers a potential advantage to patients for whom LT is the treatment of choice, but for optimal resection candidates it should not alter the decision to proceed with resection.

The concept of transplant benefit introduced and championed by the Padua group has gained considerable traction over recent years as a basis for transplant-related decision-making. Rather than focusing on post-LT outcome, the transplant benefit paradigm considers the incremental improvement in outcome with LT as compared with the best alternative to be the most relevant parameter. Transplant benefit is relevant both in deciding for LT vs resection in an individual patient, and as a principle to guide organ allocation [20]. In an analysis of more than 1000 patients from China and Italy, LT provided benefit over resection in terms of 5-year survival only in patients with a Model for End-Stage Liver Disease (MELD) score greater than or equal to 10 (the other important variable was mVI, with benefit favoring LT only in patients without mVI, but information about mVI is not available preoperatively) [21]. When viewed over a 10-year time horizon, the relative benefit of LT vs resection increases in younger patients, but this advantage disappears after age 50 years [22].

It is important to consider the issue of LT vs resection from the standpoint of organ allocation for LT. The current system is based on a balance between the principles of justice and utility. Giving priority to the sickest patients, as actualized in the use of the MELD score to establish priority, is an example of justice at work; use of the MC to prioritize patients likely to have good post-LT outcomes exemplifies utility. Implicit in this approach is an assumption that there is no good alternative to LT; this is not the case, however, with HCC, for which, in appropriate cases, resection offers a reasonable if not comparable chance for survival. From a transplant benefit perspective, the use of organs to transplant patients with preserved liver function and well-treated tumors for whom death is far from imminent can in no way be justified when patients with natural MELD scores in the 30 s for whom death within months is ensured are waiting. This is reflected in the recent US allocation policy changes incorporating a mandatory 6-month wait before priority is granted, and limiting priority to a median MELD score of 3 without the progressive increase in priority that was previously granted to HCC patients. The listing and transplantation of patients with compensated cirrhosis and well-treated early HCC, many of whom are approaching or are even beyond age 70 years, is an important contributor to the current donor organ shortage; from this perspective, it seems clear that optimal candidates for resection should be resected rather than pushed to LT.

While LT does indeed eliminate cirrhosis and dramatically reduce HCC recurrence, it by no means restores a person to "normal"; having a liver transplant is a disease unto itself, with an array of short- and long-term complications and substantially decreased survival as compared with a healthy population. In a Scandinavian study [23] that analyzed standardized mortality rates of patients beginning 1 year after LT (thereby excluding early transplant-related mortality) compared with the general population, post-LT patients who survived 1 year subsequently faced a 2.4-fold increased overall mortality risk and a 5.8-fold increased risk of premature mortality (death < age 75). This translates into a 21% survival gap at 10 years after LT and a 36% gap at 20 years.

Conclusion

Cirrhotic patients with CPS class A cirrhosis and no evidence of portal hypertension who have a solitary HCC that is technically resectable should preferably undergo resection (or, for tumors $\leq$ 2 cm that are favorably situated, thermal ablation). LT should be considered for patients within the MC who have multiple tumors and/or more advanced cirrhosis with liver dysfunction and/or portal hypertension. This approach maximizes the benefit provided by LT by limiting its application to HCC patients for whom it can provide a substantially better result than the available alternatives, and in the process offers a better chance to receive a life-saving transplant for patients with end-stage liver disease for whom there is no other option.

References

1. European Association for the Study of the Liver. EASL clinical practice guidelines: management of hepatocellular carcinoma. J Hepatol. 2018;69(1):182–236. https://doi.org/10.1016/j.jhep.2018.03.019.
2. Biederman DM, Titano JJ, Korff RA, et al. Radiation segmentectomy versus selective chemoembolization in the treatment of early-stage hepatocellular carcinoma. J Vasc Interv Radiol. 2018;29(1):30–37.e2. https://doi.org/10.1016/j.jvir.2017.08.026.
3. Menahem B, Lubrano J, Duvoux C, et al. Liver transplantation versus liver resection for hepatocellular carcinoma in intention to treat: an attempt to perform an ideal meta-analysis. Liver Transpl. 2017;23(6):836–44. https://doi.org/10.1002/lt.24758.
4. Wong RJ, Wantuck J, Valenzuela A, et al. Primary surgical resection versus liver transplantation for transplant-eligible hepatocellular carcinoma patients. Dig Dis Sci. 2014;59(1):183–91. https://doi.org/10.1007/s10620-013-2947-8.
5. Krenzien F, Schmelzle M, Struecker B, et al. Liver transplantation and liver resection for cirrhotic patients with hepatocellular carcinoma: comparison of long-term survivals. J Gastrointest Surg. 2018;22(5):840–8. https://doi.org/10.1007/s11605-018-3690-4.
6. Ikeda K, Kawamura Y, Kobayashi M, et al. Direct-acting antivirals decreased tumor recurrence after initial treatment of hepatitis C virus-related hepatocellular carcinoma. Dig Dis Sci. 2017;62(10):2932–42. https://doi.org/10.1007/s10620-017-4739-z.
7. Galati G, Muley M, Viganò M, et al. Occurrence of hepatocellular carcinoma after direct-acting antiviral therapy for hepatitis C virus infection: literature review and risk analysis. Expert Opin Drug Saf. 2019;18(7):603–10. https://doi.org/10.1080/14740338.2019.1617272.
8. Mazzaferro V, Regalia E, Doci R, et al. Liver transplantation for the treatment of small hepatocellular carcinomas in patients with cirrhosis. N Engl J Med. 1996;334(11):693–9. https://doi.org/10.1056/NEJM199603143341104.
9. Llovet JM, Zucman-Rossi J, Pikarsky E, et al. Hepatocellular carcinoma. Nat Rev Dis Primers. 2016;2:16018. https://doi.org/10.1038/nrdp.2016.18.
10. Schwartz M, Roayaie S, Konstadoulakis M. Strategies for the management of hepatocellular carcinoma. Nat Clin Pract Oncol. 2007;4(7):424–32. https://doi.org/10.1038/ncponc0844.
11. Llovet JM, Montal R, Sia D, Finn RS. Molecular therapies and precision medicine for hepatocellular carcinoma. Nat Rev Clin Oncol. 2018;15(10):599–616. https://doi.org/10.1038/s41571-018-0073-4.
12. Cucchetti A, Zanello M, Cescon M, et al. Improved diagnostic imaging and interventional therapies prolong survival after resection for hepatocellular carcinoma in cirrhosis: The University of Bologna experience over 10 years. Ann Surg Oncol. 2011;18(6):1630–7. https://doi.org/10.1245/s10434-010-1463-8.
13. Pinna AD, Yang T, Mazzaferro V, et al. Liver transplantation and hepatic resection can achieve cure for hepatocellular carcinoma. Ann Surg. 2018;268(5):868–75. https://doi.org/10.1097/SLA.0000000000002889.
14. Forner A, Bruix J. East meets the west – portal pressure predicts outcome of surgical resection for hepatocellular carcinoma. Nat Clin Pract Gastroenterol Hepatol. 2009;6(1):14–5. https://doi.org/10.1038/ncpgasthep1300.
15. Llovet JM, Fuster J, Bruix J. Intention-to-treat analysis of surgical treatment for early hepatocellular carcinoma: resection versus transplantation. Hepatology. 1999;30(6):1434–40. https://doi.org/10.1002/hep.510300629.
16. Ishizawa T, Hasegawa K, Aoki T, et al. Neither multiple tumors nor portal hypertension are surgical contraindications for hepatocellular carcinoma. Gastroenterology. 2008;134(7):1908–16. https://doi.org/10.1053/j.gastro.2008.02.091.
17. Yang T, Tabrizian P, Zhang H, et al. Comparison of patterns and outcomes of liver resection for hepatocellular carcinoma: east vs west. Clin Gastroenterol Hepatol. 2017;15(12):1972–4. https://doi.org/10.1016/j.cgh.2017.06.025.
18. Lai Q, Vitale A, Iesari S, European Hepatocellular Cancer Liver Transplant Study Group, et al. The intention-to-treat effect of bridging treatments in the setting of Milan criteria–in patients waiting for liver transplantation. Liver Transpl. 2019;25(7):1023–33. https://doi.org/10.1002/lt.25492.
19. Kulik LM, Fisher RA, Rodrigo DR, A2ALL Study Group, et al. Outcomes of living and deceased donor liver transplant recipients with hepatocellular carcinoma: results of the A2ALL cohort. Am J Transplant. 2012;12(11):2997–3007. https://doi.org/10.1111/j.1600-6143.2012.04272.x.

20. Vitale A, Lai Q. Selection of patients with hepatocellular cancer: a difficult balancing between equity, utility, and benefit. Transl Gastroenterol Hepatol. 2017;2:75. https://doi.org/10.21037/tgh.2017.09.04.
21. Vitale A, Huo TL, Cucchetti A, et al. Survival benefit of liver transplantation versus resection for hepatocellular carcinoma: impact of MELD score. Ann Surg Oncol. 2015;22(6):1901–7. https://doi.org/10.1245/s10434-014-4099-2.
22. Cillo U, Vitale A, Volk ML, et al. The survival benefit of liver transplantation in hepatocellular carcinoma patients. Dig Liver Dis. 2010;42(9):642–9. https://doi.org/10.1016/j.dld.2010.02.010.
23. Åberg F, Gissler M, Karlsen TH, et al. Differences in long-term survival among liver transplant recipients and the general population: a population-based Nordic study. Hepatology. 2015;61(2):668–77. https://doi.org/10.1002/hep.27538.

Should Liver Transplantation Be the Preferred Therapeutic Option in Patients with Compensated Cirrhosis and Portal Hypertension?

18

Miles W. Grunvald and Edie Y. Chan

Patients with a hepatic venous pressure gradient (HVPG) greater than 10 mm Hg should be excluded from surgical resection and be considered for transplantation.

Main Points

1. Although expanding surgical indications for resection could reduce the demand for hepatic grafts, sequelae of portal hypertension (pHTN)—ascites, variceal disease, shunting, and encephalopathy—are exacerbated by resection of liver parenchyma.
2. The Barcelona Liver Clinic (BCLC) staging and treatment strategy (widely adopted in Western countries) recommends against hepatic resection in patients with elevated portal pressures.
3. Prospective studies confirm that the rates of 3- and 5-year mortality are increased in patients with clinically relevant pHTN after resection of hepatocellular carcinoma (HCC).

The treatment of hepatocellular carcinoma (HCC) has many different approaches depending on disease severity and patient factors. Traditionally patients with pHTN and compensated cirrhosis have been offered liver transplantation (LT) preferentially over resection. This practice pattern is the result of many years of research, trials, and intense debate. Therefore, the majority of this chapter will review the literature that has outlined the controversy surrounding the use of resection in patients with pHTN.

M. W. Grunvald (✉)
Colon and Rectal Surgery Fellow, University of Chicago Medical Center, Chicago, IL, USA
e-mail: miles.grunvald@Uchicagomedicine.org

E. Y. Chan
Department of Surgery, Rush University Medical Center, Chicago, IL, USA
e-mail: Edie_Y_Chan@rush.edu

The decision to offer resection in the face of pHTN must be weighed against the risk of LT and other adjunctive therapies such as radiofrequency ablation or transcatheter arterial chemoembolization. Surgical decision-making in regard to pHTN has evolved over the past 30 years as new technologies, treatment strategies, and evidence have emerged. At times, data have been conflicting and management has differed between Eastern and Western surgical communities.

Liver Resection in Portal Hypertension

Since the introduction of orthotopic LT (OLT), surgeons have been working to identify and specify the indications for resection and OLT in the management of HCC. Expanding surgical indications for resection could reduce the demand for hepatic grafts, saving organs for individuals with more advance disease and reducing wait times on organ transplantation lists.

pHTN has long been considered a negative prognostic indicator for liver resection. The sequelae of pHTN—ascites, variceal disease, shunting, and encephalopathy—are exacerbated by resection of liver parenchyma. This physiologic insult of resection can lead to increased morbidity and mortality. Currently, the BCLC staging and treatment strategy (widely adopted in Western countries) recommends against hepatic resection in patients with elevated portal pressures [1, 2].

Potential Benefits of Liver Resection in Portal Hypertension

The crux of the debate to perform resection in patients with pHTN lies in the potential benefits a resection could offer. Resection has the theoretical advantage of delaying or preventing the need for receiving and maintaining a hepatic graft. This benefit is realized during the resection of smaller

N. S. Reau, A. G. Singal (eds.), *Controversies in Hepatocellular Carcinoma*, https://doi.org/10.1007/978-3-031-90225-3_18

tumors before they are upgraded to the early or intermediate stage in the BCLC model. OLT can come at immense cost to the patient because it is a much more radical operation and requires lifelong immunosuppression (fraught with infectious and other complications). Further, treating a patient with resection who would otherwise need transplantation decreases the demand for donated hepatic grafts, an exceedingly scarce resource.

Potential Hazards of Liver Resection in Portal Hypertension

Patients with pHTN are at greater risk of developing decompensation secondary to liver resection. The guiding principle behind this observation is that resection does not decrease the flow of blood to the liver but does decrease the outflow to the hepatic veins. Thus, individuals with compensated pHTN before resection can develop the sequelae of decompensated pHTN after resection. These complications can be life-threatening, life-limiting, and decrease the patient's quality of life. OLT may be necessary to salvage the most severe cases.

Resection is further associated with higher rates of tumor recurrence. While limited resections can remove tumor, paraneoplastic tissue often remains in situ. The milieu of this tissue, if it does not already contain microscopic neoplasm, is prone to generating new HCC [3, 4]. Evaluation of liver stiffness with magnetic resonance imaging or transient elastography can be a surrogate marker of liver cirrhosis and pHTN [5]. Increasing liver stiffness correlates with increasing risk of HCC and higher recurrence rates of HCC in those treated with resection [6]. Transplantation does not eliminate the risk of recurrence but does reduce the risk in the properly selected patient.

History of the Debate

Starting in the late 1980s and early 1990s, there were several studies that aimed to definitively identify patient characteristics that would lend themselves to resection or OLT. Iwatsuki et al. [7] were some of the first to publish comparisons of their transplanted patients to their resected patient. Five-year survival rates were universally higher in patients with cirrhosis and HCC who received transplantation. The authors conclude that "the treatment of choice for HCC confined to the liver is hepatic resection when anatomically and functionally feasible. Liver transplantation is the treatment of choice for HCC confined to the liver when the hepatic functions are poor."

This conclusion did not take into account more granular characteristics of the study's participants. The patient population had varying degrees of cirrhosis with a wide range of effects on hepatic synthetic function and portal physiology. Studies assessing more exact parameters of hepatic health would need to be conducted before complete dismissal of hepatic resection in the cirrhotic patient with HCC.

Research went on to look at many variables including Child-Pugh score (CPS), bilirubin, albumin, prior variceal bleeding and, importantly, the presence of elevated portal pressure. Bruix and colleagues [8] published one of the first series of patients who had hepatic venous wedge pressures measured for the calculation of HVPGs before treatment of HCC. Their study concluded that 50% of CPS class A cirrhotic patients would have some signs of hepatic decompensation following resection. Seventy-five percent of patients who were CPS class A and had a HVPG greater than 10 mm Hg developed lasting, decompensated cirrhosis.

At times, data from Japan have conflicted with that from Western Europe and the United States. A case review published in 2008 by Ishizawa et al. demonstrated strong survival rates for patients with stigmata of pHTN (>50% 5-year survival rate). The improvement in survival of these patients was attributed to advances in the treatment (and prevention) of the sequelae of pHTN including endoscopic interventions for varices and advanced surgical techniques such as splenectomy and devascularization of the abdominal esophagus at time of resection. Further, the authors suggested that the use of indocyanine green retention rate at 15 min (ICGR15) provided a more accurate picture of liver function and reserve, which improved patient selection (although this study did not directly compare the use of ICGR15 to portal pressure measurement and bilirubin). Imamura et al. [9] published data on 8 years of hepatectomy experience and concluded "the presence of portal hypertension was not a contributing factor to the overall postoperative complications in patients with HCC. Investigating its relation to each kind of complication separately, portal hypertension is a risk factor for postoperative ascitic complications." However, like many other Japanese studies, ICGR15 was used to assess liver function rather than direct measurement of portal pressure. The largest studied registry of liver cancer patients has been conducted by the Liver Cancer Study Group of Japan. Periodic surveillance of their primary liver cancer patients has led to robust data regarding their management of various forms of liver cancer. The 15th follow-up survey in 2004 demonstrated some success in the management of HCC with resection in the setting of cirrhosis. A 45% 5-year survival rate was appreciated among this population, but unfortunately the published survey is not granular enough to identify the patients who had pHTN within this cirrhotic group [10].

Since the mid-1990s there have been many retrospective studies and several prospective studies assessing the role of liver resection in pHTN [1, 8, 11–14]. Their results have varied in degree of hazard but generally have concurred that pHTN is a negative prognostic factor for resection.

The Current State of Resection in the Setting of Portal Hypertension

There have been only four prospective trials to date that have investigated survival of resection in HCC patients with pHTN: Boleslawski et al. (France), [11] Bruix et al. (Spain), [8] Llop et al. (Spain), [12] and Llovet et at (Spain) [15]. All these studies corroborate that the rates of 3- and 5-year mortality are increased in patients with clinically relevant pHTN. Specifically, 3-year mortality in patients with pHTN ranges from 39% to 71% and 5-year mortality ranges from 25% to 56% [16].

One meta-analysis conducted in 2015 by Berzigotti and colleagues [16] assessing these prospective studies along with several other retrospective studies reached the same conclusion: pHTN always increases the odds of mortality with resection at 3 and 5 years. The 5-year mortality pooled odds ratio was 1.75; however, the authors noted that studies using HVPG for the assessment of portal hypertension (vs the use of surrogate markers such as varices, splenomegaly, thrombocytopenia, etc.) had a pooled odds ratio of 8.00.

Progression of Hepatocellular Carcinoma While Waiting for Liver Transplantation

To assess the efficacy and economy of liver resection in patients with pHTN, one must compare the outcomes to those of the alternative therapies. The alternative treatment, OLT, is burdened with scarcity of donor grafts and the necessity of being listed for one of these grafts. Thus, intention-to-treat analyses may provide a more realistic comparison between transplant and resection.

Llovet et al. [15] published the first intention-to-treat analysis, demonstrating that resection plays an important role in the management of HCC. They maintain that patients with elevated bilirubin and HVPG should be excluded from resection because it may shorten survival (HCC without surgery was found to have a 50% 5-year survival). However, the authors note that "the main drawback of OLT are the dropouts caused by the long waiting time. By now, its rate varies among ABO blood groups, but as the waiting time increases, all groups will face the same problem."

Currently, 25% of all liver transplants in the United States are indicated for the treatment of HCC. This number is expected to rise as hepatitis C virus infection reaches its peak and nonalcoholic cirrhosis continues to rise in prevalence. Historically, patients with HCC were known to have higher dropout rates from transplant lists because of tumors becoming unresectable while waiting for transplantation [15, 17]. In an attempt to reduce the disparity of dropout between HCC and non-HCC patients, the United States formed policies in 2002 to give priority to those with HCC [18]. Over the next decade the aforementioned policy seemed to overcorrect and created a nonequipoise in the other direction. These inequities prompted the addition of some restrictions for HCC patients, including a 6-month waiting period. Between 2005 and 2009, the probability of waitlist dropout at 24 months was 11.6%; this increased to 19.5% between 2010 and 2014. Dropout constitutes a considerable concern for patients seeking treatment for HCC [17].

Listing practices change regularly, and surgeons should attempt to balance the risks associated with waiting for transplantation with the risks of complications of resection. This calculation is made even more complex by the mounting body of research advocating for the use of locoregional therapies before transplantation: "ablate and wait" [19]. Some have proposed using resection as a bridge to transplantation; however, this practice is currently regarded as controversial and has not been widely adopted. LT for recurrence after resection or for pathology that is highly unfavorable has been studied and remains controversial [20–22].

Neoadjuvant Therapies

Concurrent to the research on resection vs OLT in HCC has been the introduction of many minimally invasive treatment modalities. Transcatheter radioembolization, hepatic artery embolization, and transarterial chemoembolization (TACE) have all been identified as having a potential benefit in the management of HCC. In particular, TACE has shown to increase the resectability of HCC. TACE can help "shrink" the tumor to sizes that allow for resection. These treatment modalities are not for every patient. Management of HCC, with its plethora of therapies, requires a diverse and committed team [23].

While the aforementioned therapies achieve their goal by reducing tumor size and burden, other therapies have sought to increase the size of the future liver remnant. Principally, portal vein embolization has been identified as a way to increase the amount of liver tissue (and synthetic capacity) following resection. Portal vein embolization has demonstrated efficacy in HCC with CPS class A cirrhosis. Caution should be taken in the patient with pHTN or increased portal pressures because embolization of parts of the portal system will exacerbate the pressures [24].

Conclusion

Hepatic resection in the setting of CPS class A cirrhosis with pHTN has tempted surgeons for at least the last 30 years. Outcomes for these patients do seem to be improving and at times have demonstrated greater than 50% 5-year survival. However, the vast majority of studies demonstrate statistically inferior 3- and 5-year survival rates with increases in decompensated liver disease. This has been particularly true in studies that have used direct measurement portal pressures. At this time patients with HVPG greater than 10 mm Hg should be excluded from surgical resection and be considered for transplantation. Continued exploration of neoadjuvant interventions may increase the safety and broaden the indications for resection in this group of patients.

Acknowledgments The authors would like to acknowledge the Rush University Medical Center Department of Transplant Surgery for their support and guidance.

References

1. Llovet JM, Fuster J, Bruix J, Barcelona-Clínic Liver Cancer Group. The Barcelona approach: diagnosis, staging, and treatment of hepatocellular carcinoma. Liver Transpl. 2004;10(2 Suppl 1):S115–20. https://doi.org/10.1002/lt.20034.
2. Bruix J, Sherman M, American Association for the Study of Liver Diseases. Management of hepatocellular carcinoma: an update. Hepatology. 2011;53(3):1020–2. https://doi.org/10.1002/hep.24199.
3. Vargas V, Castells L, Balsells J, et al. Hepatic resection or orthotopic liver transplant in cirrhotic patients with small hepatocellular carcinoma. Transplant Proc. 1995;27(1):1243–4.
4. Tabrizian P, Jibara G, Shrager B, Schwartz M, Roayaie S. Recurrence of hepatocellular cancer after resection: patterns, treatments, and prognosis. Ann Surg. 2015;261(5):947–55. https://doi.org/10.1097/SLA.0000000000000710.
5. Singh S, Fujii LL, Murad MH, et al. Liver stiffness is associated with risk of decompensation, liver cancer, and death in patients with chronic liver diseases: a systematic review and meta-analysis. Clin Gastroenterol Hepatol 2013;11(12):1573–84.e1–2; quiz e88–9. https://doi.org/10.1016/j.cgh.2013.07.034.
6. Jung KS, Kim SU, Choi GH, et al. Prediction of recurrence after curative resection of hepatocellular carcinoma using liver stiffness measurement (FibroScan). Ann Surg Oncol. 2012;19(13):4278–86. https://doi.org/10.1245/s10434-012-2422-3.
7. Iwatsuki S, Starzl TE, Sheahan DG, et al. Hepatic resection versus transplantation for hepatocellular carcinoma. Ann Surg. 1991;214(3):221–9. https://doi.org/10.1097/00000658-199109000-00005.
8. Bruix J, Castells A, Bosch J, et al. Surgical resection of hepatocellular carcinoma in cirrhotic patients: prognostic value of preoperative portal pressure. Gastroenterology. 1996;111(4):1018–22. https://doi.org/10.1016/S0016-5085(96)70070-7.
9. Imamura H, Seyama Y, Kokudo N, et al. One thousand fifty-six hepatectomies without mortality in 8 years. Arch Surg. 2003;138(11):1198–206. https://doi.org/10.1001/archsurg.138.11.1198.
10. Ikai I, Itai Y, Okita K, et al. Report of the 15th follow-up survey of primary liver cancer. Hepatol Res. 2004;28(1):21–9. https://doi.org/10.1016/j.hepres.2003.08.002.
11. Boleslawski E, Petrovai G, Truant S, et al. Hepatic venous pressure gradient in the assessment of portal hypertension before liver resection in patients with cirrhosis. Br J Surg. 2012;99(6):855–63. https://doi.org/10.1002/bjs.8753.
12. Llop E, Berzigotti A, Reig M, et al. Assessment of portal hypertension by transient elastography in patients with compensated cirrhosis and potentially resectable liver tumors. J Hepatol. 2012;56(1):103–8. https://doi.org/10.1016/j.jhep.2011.06.027.
13. Liu J, Zhang H, Xia Y, et al. Impact of clinically significant portal hypertension on outcomes after partial hepatectomy for hepatocellular carcinoma: a systematic review and meta-analysis. HPB (Oxford). 2019;21(1):1–13. https://doi.org/10.1016/j.hpb.2018.07.005.
14. Takemura N, Aoki T, Hasegawa K, et al. Hepatectomy for hepatocellular carcinoma after perioperative management of portal hypertension. Br J Surg. 2019;106(8):1066–74. https://doi.org/10.1002/bjs.11153.
15. Llovet JM, Fuster J, Bruix J. Intention-to-treat analysis of surgical treatment for early hepatocellular carcinoma: resection versus transplantation. Hepatology. 1999;30(6):1434–40. https://doi.org/10.1002/hep.510300629.
16. Berzigotti A, Reig M, Abraldes JG, Bosch J, Bruix J. Portal hypertension and the outcome of surgery for hepatocellular carcinoma in compensated cirrhosis: a systematic review and meta-analysis. Hepatology. 2015;61(2):526–36. https://doi.org/10.1002/hep.27431.
17. Mehta N, Dodge JL, Hirose R, Roberts JP, Yao FY. Increasing liver transplantation wait-list dropout for hepatocellular carcinoma with widening geographical disparities: implications for organ allocation. Liver Transpl. 2018;24(10):1346–56. https://doi.org/10.1002/lt.25317.
18. Freeman RB Jr, Wiesner RH, Harper A, UNOS/OPTN Liver Disease Severity Score, UNOS/OPTN Liver and Intestine, and UNOS/OPTN Pediatric Transplantation Committees, et al. The new liver allocation system: moving toward evidence-based transplantation policy. Liver Transpl. 2002;8(9):851–8. https://doi.org/10.1053/jlts.2002.35927.
19. Roberts JP, Venook A, Kerlan R, Yao F. Hepatocellular carcinoma: ablate and wait versus rapid transplantation. Liver Transpl. 2010;16(8):925–9. https://doi.org/10.1002/lt.22103.
20. Schwartz ME. Liver transplantation for hepatocellular carcinoma: the best treatment, but for which patient? Hepatology. 1996;24(6):1539–41. https://doi.org/10.1002/hep.510240640.
21. Truty MJ, Vauthey JN. Surgical resection of high-risk hepatocellular carcinoma: patient selection, preoperative considerations, and operative technique. Ann Surg Oncol. 2010;17(5):1219–25. https://doi.org/10.1245/s10434-010-0976-5.
22. Belghiti J, Cortes A, Abdalla EK, et al. Resection prior to liver transplantation for hepatocellular carcinoma. Ann Surg. 2003;238(6):885–92.; discussion 892–893. https://doi.org/10.1097/01.sla.0000098621.74851.65.
23. Akateh C, Black SM, Conteh L, et al. Neoadjuvant and adjuvant treatment strategies for hepatocellular carcinoma. World J Gastroenterol. 2019;25(28):3704–21. https://doi.org/10.3748/wjg.v25.i28.3704.
24. Shindoh J, Tzeng CWD, Vauthey JN. Portal vein embolization for hepatocellular carcinoma. Liver Cancer. 2012;1(3–4):159–67. https://doi.org/10.1159/000343829.

19 Should Radiation-Based Therapy Be Considered in All Patients with Hepatocellular Carcinoma?

Alexander A. Harris, Daniel T. Chang, William Small Jr, and Tarita O. Thomas

From palliation to definitive treatment, radiation therapy (RT) should be considered in all patients with hepatocellular carcinoma (HCC).

Main Points

1. Radiation improves quality of life in patients with advanced or metastatic HCC.
2. Stereotactic body radiation therapy (SBRT) is a form of external beam radiotherapy (EBRT) that allows for the noninvasive delivery of tumoricidal doses of radiation in highly conformal manner over a small number of fractions.
3. SBRT in HCC shows local control (LC) rates of 87–92%, comparable to other liver-directed therapies (LDTs).
4. EBRT in combination with other LDTs or systemic therapy has promising results in those with advanced, unresectable HCC.
5. EBRT is effective in treating patients with tumor thrombus or as a bridge to transplant.

A. A. Harris
Radiation Oncology Consultants Ltd., Ascension Alexian Brothers Medical Center, Elk Grove, IL, USA

D. T. Chang
Department of Radiation Oncology, Stanford University, Stanford, CA, USA
e-mail: dtchangd@med.umich.edu

W. Small Jr
Department of Radiation Oncology, Stritch School of Medicine, Loyola University, Maywood, IL, USA
e-mail: wmsmall@lumc.edu

T. O. Thomas (✉)
Department of Radiation Oncology, Feinberg School of Medicine, Northwestern University, Chicago, IL, USA
e-mail: tarita.thomas@nm.org

The gold standard in the management of HCC is surgical resection or liver transplantation (LT). However, only 15–30% of patients with HCC are candidates for surgical resection because of tumor extent or underlying liver dysfunction [1]. The other 70–85% of patients can receive nonsurgical liver-directed therapies (LDT) including, but not limited to, radiofrequency ablation (RFA), transarterial chemoembolization (TACE), transarterial radioembolization (TARE), and EBRT.

EBRT in the management of HCC has evolved over the years and is now included as one of the recommended treatment modalities for unresectable or medically inoperable HCC in the National Comprehensive Cancer Center Network guidelines [2]. SBRT, also known as *stereotactic ablative body radiation*, is a specialized way of delivering high, conformal doses of EBRT in 5 or fewer treatments. SBRT offers the ability to spare large portions of the normal liver and adjacent gastrointestinal (GI) structures while simultaneously delivering ablative radiation doses to the tumor in a noninvasive manner.

This chapter will discuss the evolution of RT in managing HCC, outline and compare clinical data of EBRT and other LDTs, identify optimal candidates and special cases to be treated with SBRT, highlight novel radiation techniques used to improve SBRT delivery, and recognize the multiple ongoing clinical trials of SBRT in HCC. The goal of this chapter is to showcase why external beam radiation-based therapy should be considered an appropriate option in patients with HCC.

Evolution of Radiation Therapy in Hepatocellular Carcinoma: From Palliation to Ablation

Traditionally, the liver had been a challenging organ to radiate because of difficulty in localizing tumors, respiratory motion, and other set-up challenges. These challenges

N. S. Reau, A. G. Singal (eds.), *Controversies in Hepatocellular Carcinoma*, https://doi.org/10.1007/978-3-031-90225-3_19

necessitated radiation oncologists to treat large volumes of normal liver to ensure they adequately cover the tumor, or in many cases, treating the entire liver, which, coupled with the relatively high radiosensitivity of the liver, leads to a high risk of treatment-related toxicity, known as *radiation-induced liver disease* (RILD). RILD is a clinical syndrome occurring from venous occlusion secondary to radiation treatment and results in anicteric hepatomegaly, ascites, and elevated liver enzymes (particularly serum alkaline phosphatase) occurring 2 weeks to 4 months after RT. RILD can occur after any type of radiation-based LDT, including EBRT and TARE. The risk of RILD initially limited RT in HCC to a palliative role [3, 4].

Palliation of Advanced or Metastatic Hepatocellular Carcinoma Is an Important Role of Radiation in the Management of HCC

Radiation can be used to reduce mass effect, decrease GI bleeding, and prevent tumor rupture when delivered to large, symptomatic HCC tumors. Palliative radiation delivered to lymph node metastases, bone metastases, and brain metastases have also been shown to be effective at reducing symptoms, with pain relief rates as high as 84–97%, as well as potentially improving survival [5–11]. One National Cancer Database analysis of more than 3000 patients treated with palliative therapy from 1998 to 2011 for unresectable HCC showed that 25% of patients that received some form of palliative RT had an improved overall survival (OS) over those that received palliative surgical resection alone, palliative chemotherapy alone, or pain control alone (hazard ratio [HR] = 0.65; CI, 0.5–0.83) [12]. These types of palliative radiation treatments can often be performed in a single radiation treatment, providing a quick, effective, and affordable way to improve patients' quality of life.

Technological advances in the delivery of radiation allow for radiation oncologists to deliver radiation to the liver more safely. These techniques, including three-dimensional conformal radiation therapy (3DCRT) and intensity-modulated radiation therapy (IMRT) allow for dose escalation by delivering a more conformal radiation dose and preventing excessive liver damage. These higher doses led to greater LC, which in turn improved survival of patients treated with RT for HCC. Seong et al. [13] reported a tumor-specific response rate of 67% after a mean dose of 52 Gy at 1.8 Gy per fraction for patients who failed TACE and were treated with conformal radiotherapy. Dawson and colleagues [14] observed that patients who received radiation doses of more than 70 Gy at 1.5 Gy per fraction had better median survival (>16.4 months) than those who received lower doses (11.6 months).

It is important to note that the biological effects from radiation are not only a function of total dose delivered but also the dose per fraction and the tissue being irradiated. These factors are used to calculate a biologically effective dose (BED). The BED allows for expressing the biological effect that a specific radiation dose fractionation will have on specific tissue. BED is calculated using the following equation; $\mathrm{BED} = nd\left[1+\frac{d}{\alpha/\beta}\right]$, where n is the number of fractions, d is dose per fraction, and α/β is a radiobiological relationship of the rate of cell killing as a function of dose and the tissue being irradiated [15]. Calculating the BED also allows for different radiation fractionation regimens to be compared on the basis of anticipated biological effect. For example, the delivery of 50 Gy in 25 fractions, at 2 Gy per fraction, results in a BED of 60 Gy. However, if the 50 Gy was delivered in 5 fractions of 10 Gy per fraction, the BED is substantially higher at 100 Gy. In general, increasing the dose per fraction increases the tumoricidal potential but also increased the risk of late-treatment toxicity effects [16]. Therefore, escalation of dose per fraction can be very useful to improve LC, but radiation oncologists were again limited from further dose escalation out of concern over RILD.

Practitioners at the University of Michigan helped address this issue by creating an individualized method of assessing the risk of RILD for patients with hepatic malignancy treated with EBRT using a normal tissue complication probability (NTCP) model [17, 18]. The NTCP model is used to calculate the probability of developing a specific normal tissue toxicity as a function of the volume of tissue irradiated. Using this model, a protocol was created that allows for the delivery of a maximum dose with no more than 10% risk of RILD [19]. Practitioners at Princess Margaret Hospital built on this protocol to develop an isotoxic RT schedule that allows patients with HCC unsuitable for standard therapies to be treated in 6 fractions over 2 weeks using SBRT. The dose per fraction was determined based on the effective volume of normal liver irradiated (V_{eff}), which enables comparison of the dose received by a given volume of tissue for different RT plans. When smaller volumes of normal liver were irradiated ($V_{eff} < 25\%$), higher doses (54 Gy) were delivered safely to HCCs, with excellent LC. When a larger amount of normal liver was to be irradiated (V_{eff} 25–80%), lower doses (24–54 Gy) were delivered safely, although LC was reduced [20]. These findings, coupled with advances in radiation treatment planning, motion management, and image-guided radiation therapy, made it possible to deliver higher tumoricidal doses of radiation to parts of the liver safely with conformal radiation via advanced radiation techniques, including 3DCRT, IMRT, SBRT, or proton beam therapy [21–24].

Clinical Radiation Data in the Definitive Setting

As radiation technology has advanced over the years, so have the methods of treating HCC with EBRT. Because of these changes, the response rates and LC rates of patients treated with EBRT have improved over time. As radiation oncologists moved from treating HCC with 3DCRT to IMRT to SBRT, there has been a steady increase in LC and OS [25–34]. Table 19.1 outlines a number of studies that highlight EBRT's definitive treatment of HCC across various radiation modalities.

SBRT is a relatively new modality in the definitive treatment of HCC. As a result, many of the mature data regarding the efficacy of SBRT in HCC are retrospective. These retrospective studies showed considerable promise, with 3-year LC rates of 91–92% and 1-year OS rates of 87–95% [29, 30]. Recently, prospective clinical trials using SBRT in the definitive setting have demonstrated similarly high 3-year LC rates of 91–96% and 1-year OS rates of 75–93% [31–34].

Comparing Outcomes Between Stereotactic Body Radiation Therapy and Other Liver-Directed Therapies

There are many choices of LDTs that can be used to treat patients with HCC: resection, RFA, TACE, TARE, and SBRT. The choice of therapy should be based on patient- and cancer-specific factors such as tumor size, number of lesions, location, liver function, comorbidities, institutional expertise, and patient preference. Unfortunately, there are few phase 3 data comparing SBRT and other LDTs to date to help elucidate the optimal treatment for specific patients. The vast majority of the available data comes from single-institutional, retrospective studies.

In comparing SBRT with resection, Su et al. [35] performed a propensity-matched retrospective analysis in patients with Child-Pugh score (CPS) class A with 1–2 small HCCs without prior LDTs. SBRT was administered to patients in this study when they were deemed medically inoperable because of their comorbidities and/or tumor loca-

Table 19.1 Summary of studies of definitive external beam radiotherapy in the treatment of hepatocellular carcinoma across various radiation modalities

Study	Study design	No.	Patient liver function	Median tumor size	Radiation modality	Dose and fractions	LC or RR	OS	Comments
Kim, 2006 [25]	Retrospective	70	CP-A: 82% CP-B: 12%	7.5 cm	3D-CRT	44–54 Gy in 22-27 fx	RR: 54.3%	43% at 1 y	84% failed prior LDT; 59% with PVT
Mornex, 2006 [26]	Prospective	27	CP-A: 59% CP-B: 41%	3.2 cm	3D-CRT	66 Gy in 33 fx	RR: 92%	NR	All ineligible for other LDT
McIntosh, 2009 [27]	Retrospective	20	CP-A: 55% CP-B: 45%	9 cm	IMRT	50 Gy in 20 fx	RR: 66%	75% at 1 y	55% failed prior TACE; all received capecitabine
Chi, 2010 [28]	Prospective	23	CP-A: 65% CP-B: 35%	NR	IMRT	52.5 Gy in 15 fx	RR. 74%	70% at 1 y	All received concurrent sunitinib
Sanuki, 2014 [29]	Retrospective	42	CP-B: 90% CP-B: 10%	2.1	SBRT	35–40 Gy in 5 fx	LC: 97% at 2 y	81% at 2 y	93% of patients with complete response
Jung, 2013 [30]	Retrospective	92	CP-A: 74% CP-B: 26%	2.5 cm	SBRT	30-60 Gy in 3-5 fx	LC: 92% at 3 y	87% at 1 y	All but 1 failed prior LDT
Kang, 2012 [31]	Prospective	47	CP-A: 87% CP-B: 13%	2.9 cm	SBRT	42-60 Gy in 3 fx	LC: 95% at 2 y	95% at 2 y	Salvage after incomplete TACE
Bujold, 2013 [32]	Prospective	31	CP-A: 100%	7.2 cm	SBRT	36 Gy in 6 fx	87%	87%	61% with multiple lesions; 55% with PVT
Lasley, 2015 [33]	Prospective	59	CP-A: 64% CP-B: 36%	Volume: 33.6 cm^3	SBRT	CP-A: 48 in 5 fx CP-B: 40 in 5 fx	CP-A: 91% CP-B: 82%	CP-A: 94% CP-B: 57%	20% with PVT 15% prior LDT
Takeda, 2016 [34]	Prospective	90	CP-A: 91% CP-B: 9%	2.3 cm	SBRT	35-40 Gy in 5 fx	LC: 96.3% at 3 y	66.7% at 3 y	42% failed prior LDT

3D-CRT 3-dimensional conformal radiation therapy, *CP* Child-Pugh score class, *fx* fractions, *Gy* Gray, *IMRT* intensity-modulated radiation therapy, *LC* local control, *LDT* liver-directed therapy, *NR* not reported, *OS* overall survival, *PVT* portal venous thrombosis, *RR* response rate, *SBRT* stereotactic body radiation therapy, *TACE* transarterial chemoembolization

tion. The authors found no differences in 5-year OS (74.3% vs 69.2%, $P = 0.405$) or in intrahepatic progression-free survival (PFS) (43.9% vs 35.9%, $P = 0.945$) between those treated with SBRT and those treated with surgical resection.

In comparing SBRT and RFA, Wahl and colleagues [36] performed a single-institution, retrospective study of patients with inoperable HCC and found 1- and 2-year freedom from local progression rates of 97.4% and 83.8% for those treated with SBRT and 83.6% and 80.2% for those treated with RFA. RFA treatment was associated with a higher rate of local progression (HR = 2.63; $P = 0.016$), which was driven primarily by a higher rate of local progression in tumors greater than or equal to 2 cm treated with RFA as compared with those treated with SBRT (HR = 3.35; $P = 0.025$). The OS at 1 and 2 years after treatment was 70% and 53% after RFA and 74% and 46% after SBRT, with no statistically significant difference. In a different study, Shiozawa et al. [37] reviewed 76 patients with a solitary HCC smaller than or equal to 5 cm treated with RFA or SBRT. The 1-year LC rates were 97.4% and 97.1% in those treated with RFA and SBRT, respectively, and the 1-year cumulative survival rates were 100% and 95.2% with no statistically significant difference in either rate between the 2 groups.

In comparing SBRT and TACE, Sapir et al. [38] propensity score–matched patients with 1 to 2 previously untreated HCCs. They found 1- and 2-year LC rates of 96.5% and 91.3% for patients treated with SBRT compared to 47.1% and 22.9% for patients treated with TACE (HR = 15.7; $P < 0.001$). The SBRT-treated patients also exhibited higher rates of freedom from in-liver progression, with 1 and 2 years with rates of 56.5% and 26.9% for SBRT and 35.9% and 10.7% for TACE. Patients treated with TACE also required more additional LDT (68% vs 34%, $P < 0.001$) and systemic therapy (16.7% vs 6.4%, $P = 0.01$) as opposed to those treated with SBRT. Of the patients who went on to undergo LT, pathologic complete response was seen in 58% of lesions after SBRT and in 24% lesions after TACE ($P = 0.054$). Furthermore, Bush and colleagues [39] prospectively compared TACE with EBRT radiation in a randomized study. The radiation was delivered with proton beam radiation at a dose of 70.2 Gy in 15 fractions. The interim analysis showed a trend toward improved 2-year LC (88% vs 45%, $P = 0.06$) and PFS (48% vs 31%, $P = 0.06$) in the radiation arm over the TACE arm.

Other studies have examined the addition of SBRT to TACE as compared to TACE alone and have shown improved outcomes with the combination therapy [40, 41]. Similar studies have shown the combination of TACE with conventionally fractionated RT (1.8–2.0 Gy per fraction) via 3DCRT also improves outcomes over TACE alone [42, 43]. A meta-analysis of 25 trials, which compiled RT given as whole-liver, standard fractionation, as well as SBRT, confirmed these findings by showing a statistically significant improvement in median OS in patients treated with RT and TACE (22.7 months) compared with TACE alone (13.5 months) [44]. In a slightly different type of study, Hardy-Abeloos et al. [45] examined the efficacy and safety of SBRT after TARE and compared them to SBRT after TACE. They found no statistically significant difference in the treatment-related toxicity, LC rates, or in OS between the groups treated with SBRT after TARE and those treated with SBRT after TACE. At this time, there are limited published data comparing SBRT and TARE in treating HCC in the definitive setting.

Combining Systemic Therapies with Radiation

Recent developments in systemic therapy have led to additional therapies having an important role in managing advanced, unresectable HCC. Sorafenib is a systemic agent that has been shown to be effective in managing advanced HCC. It received Food and Drug Administration approval after a randomized phase 3 trial showed improvement in OS vs placebo alone in patients with advanced HCC and CPS class A cirrhosis [46]. More recently there have been additional systemic therapies approved for locally advanced or metastatic HCC.

Recent studies aimed to compare sorafenib in combination with other LDT in unresectable or locally advanced patients. Sorafenib has been combined with TACE in multiple prospective randomized trials with mixed results. The SPACE trial and the TACE-2 trial compared TACE with or without sorafenib in patients with unresectable HCC and found no difference in PFS [47, 48]. However, the TACTICS trial did show an improvement in PFS with the combination of TACE and sorafenib vs TACE alone [49]. A retrospective, single-institution study looked into the combination of Y-90 with sorafenib in locally advanced HCC and found promising initial results with PFS of 6.6 months and OS of 19.52 months [50].

There are limited data examining the combination of sorafenib and EBRT at this time. In a phase 2 study, Chen et al. [51] treated patients with advanced, unresectable HCC unfit for TACE with standard fractionation-conformal RT with concurrent and sequential sorafenib. The tolerance and toxicities of treatment were acceptable. The authors reported complete or partial response rates of 55% with a 2-year infield PFS of 39% with few patients experiencing grade 2 and grade 3 hepatic toxicity, respectively. A single-institution phase 1 trial at Princess Margaret Hospital tested the safety of the combination of neoadjuvant, concurrent, and adjuvant sorafenib delivered with SBRT. They found that clinically significant toxicity was dependent on the irradiated volume and dose of sorafenib, with the combination of treatment being safe and effective in those with low V_{eff} ($V_{eff} < 30\%$) [52]. An ongoing phase 3, multi-institutional trial, RTOG 1112, looks to shed further light on the combination of SBRT and sorafenib. In this trial, patients with Barcelona Clinic Liver Cancer stage intermediate or advanced HCC who are unsuitable for surgical resection or other LDTs are to be randomly assigned to either sorafenib alone or sorafenib with SBRT (NCT01730937).

Who Is Eligible for Stereotactic Body Radiation Therapy?

Careful patient selection is needed when treating patients with HCC with SBRT. Like all other LDTs, the treatment of choice should be made by the patient and a multidisciplinary oncology team after considering all LDT options. SBRT provides a number of advantages over other LDTs. Because SBRT is noninvasive, it can be delivered to patients unable to medically tolerate invasive procedures. Given the noninvasive nature of EBRT, patients are at lower risk of developing posttreatment pain, fever, or postembolization syndrome after EBRT than those treated with transarterial therapies. One prospective, randomized trial showed that patients treated with EBRT had fewer hospitalization days within 30 days of treatment than those treated with TACE (0.73 days vs 4.6 days per patient; $P < 0.001$) [39].

EBRT can target lesions in difficult to access anatomic regions or high-risk surgical regions like near delicate vasculature or biliary structures. Lesions in these high-risk surgical areas, including the central portal area, adjacent to great vessels, or near the biliary tree, have been treated safely and effectively with SBRT [53]. In addition, lesions in areas difficult to access endovascularly, such as peripheral lesions or lesions near the liver dome, can easily be treated with SBRT, which is not dependent on these same anatomic restrictions.

One of the key considerations radiation oncologists rely on in deciding appropriate liver SBRT candidates is the amount of functional liver that would remain after treatment. Data from partial hepatectomy studies were used to derive a threshold of at least 700 cc of normal functioning liver spared from receiving ablative doses of radiation during SBRT [54]. For the same reasons, many trials and practitioners limit treatment to patients with adequate residual liver function, with treatment being limited to those with CPS class A or early B disease. More recently, studies have examined the effect of the albumin-bilirubin score on RILD and survival in patients with HCC treated with SBRT and found lower rates of RILD with longer survival in patients with lower albumin-bilirubin grade [55–57]. The criteria used to identify patients who are optimal candidates for LDT via SBRT at Loyola University Medical Center are listed in Table 19.2.

Table 19.2 Institutional eligibility criteria for stereotactic body radiation therapy for hepatocellular carcinoma

Tumor size, cm	≤ 5
No. of tumors	≤ 5 lesions
Liver volume	> 700 cc residual, functional liver after treatment
Liver function	Child-Pugh score class A, B7, or B8

Special Cases Treated with External Beam Radiotherapy

Patients with benign or tumor-related portal venous thrombosis (PVT) or inferior vena cava thrombosis in the setting of HCC present unique challenges for many other LDTs. At the time of HCC diagnosis, 6.5% of patients have demonstrable PVT, which is likely underreported as one autopsy series noted an incidence as high as 44%. These lesions can be treated with EBRT alone or with a combination of EBRT and other LDTs. In a series of patients who had HCC with PVT or inferior vena cava thrombosis treated with SBRT, a median OS of 13 months was reported with an overall objective response rate of greater than 75% after SBRT. Toxicity was minimal, with only a single reported grade 3 toxicity event (elevated bilirubin). The combination of EBRT and TACE may provide better response rates than either one of the treatments alone or other treatment modalities [58, 59]. In a study of 77 patients with HCC and PVT treated with TACE alone or TACE with SBRT, there was a statistically significant improvement in median OS with the addition of SBRT to TACE over TACE alone, 10.9 months vs 4.1 months [59]. In a prospective, randomized trial, Yoon et al. [60] compared the combination of TACE and EBRT with sorafenib in 90 patients with liver-confined HCC with macroscopic vascular invasion. They found a higher rate of radiographic response (33.3% vs 2.2%, $P < 0.001$), longer median time to progression (31.0 vs 11.7 weeks, $P < 0.001$), and a longer OS (55.0 vs 43.0 weeks, $P = 0.04$) in the patients treated with EBRT and TAC than in those treated with sorafenib.

In addition, EBRT can be used effectively as a noninvasive bridging therapy. Four studies using hypofractionated EBRT or SBRT showed in-field tumor control rates of 83–100% while awaiting definitive therapy via LT [61–64]. In comparing the efficacy of SBRT to TACE or RFA as bridging therapies, Sapisochin and colleagues [65] found no differences in the rates of toxicity, rates of patients proceeding to transplantation, or OS after transplantation between the study arms. Similarly, Mohamed et al. [66] found no difference in the radiographic response to bridging therapies before transplantation between SBRT, TACE, RFA, and TARE. However, they did find a statistically significant difference in the rates of grade 3 toxicity after bridging therapy between SBRT, TARE, TACE, and RFA (0%, 0%, 11%, 22%; $P = 0.046$), favoring bridging therapy via SBRT or TARE.

Radiation Techniques

The first step in treating a patient with HCC with EBRT or SBRT is radiation planning. This process involves a computed tomography (CT) scan performed in the treatment position. Techniques are used to optimize target visibility, limit tumor motion, and maximize reproducibility. Fusion of previously obtained diagnostic scans with the simulation scan further aids in target visualization. The treating radiation oncologist uses these images to outline the target, identify the organs at risk (OAR), prescribe the dose and fractionation, and assign dose limits for neighboring OARs. The treatment plan then moves to a team of medical physicists and dosimetrists who, aided by sophisticated computer software, balance multiple normal-tissue constraints and priorities to create a radiation plan that optimally balances adequate tumor coverage with minimal dose to OARS. Before the delivery of EBRT, the radiation oncologist reviews the dose distribution and the dose-volume-histogram to ensure adequate tumor coverage while minimizing the risk of toxicity to OARS. An example of a CT-based SBRT treatment plan for HCC and the corresponding dose-volume-histogram is shown in Fig. 19.1.

These complex plans allow for very conformal doses paired with very high doses (120% prescribed dose) to the center of the tumor, high doses to the periphery of the tumor (100% prescription dose), and very rapid dose fall-off limiting dose to adjacent OARS (often 0–10% prescription dose). Because of this, image guidance is one of the cornerstones of SBRT, as small changes in patient position can lead to target misses or damage to OARS.

There have been many novel ways to further reduce the risk of near misses or OAR damage while delivering SBRT. Adaptive radiation therapy (ART) is a technological advancement that allows for on-the-fly plan adjustment at the time of treatment [67]. With ART, if a patient is on the treatment table preparing for an SBRT treatment and a change is noted on the image guidance, a new plan can be generated rapidly to account for the changes and ensure a safe delivery dose with a modified plan that can be identified at the time of each individual treatment. SBRT precision in treating HCC is even further amplified with the combination of ART with a magnetic resonance imaging (MRI)-based linear accelerator [68].

MRI imaging has substantially improved soft-tissue visualization and provides considerable benefits in localizing and characterizing HCCs. Before 2017, high-quality, real-time images of tumors with onboard linear accelerator imaging were not possible because of the technical limitations of MRI and linear accelerator compatibility. However, in 2017, an MRI-based linear accelerator obtained Food and Drug Administration approval for use in treating patients. This new technology provided substantial improvements in real-time target tracking and adaptive treatment of soft tissue and gastrointestinal organs using MRI. Researchers at the Henry Ford Health System outlined some of their early experience using MRI-based linear accelerator in treating liver lesions

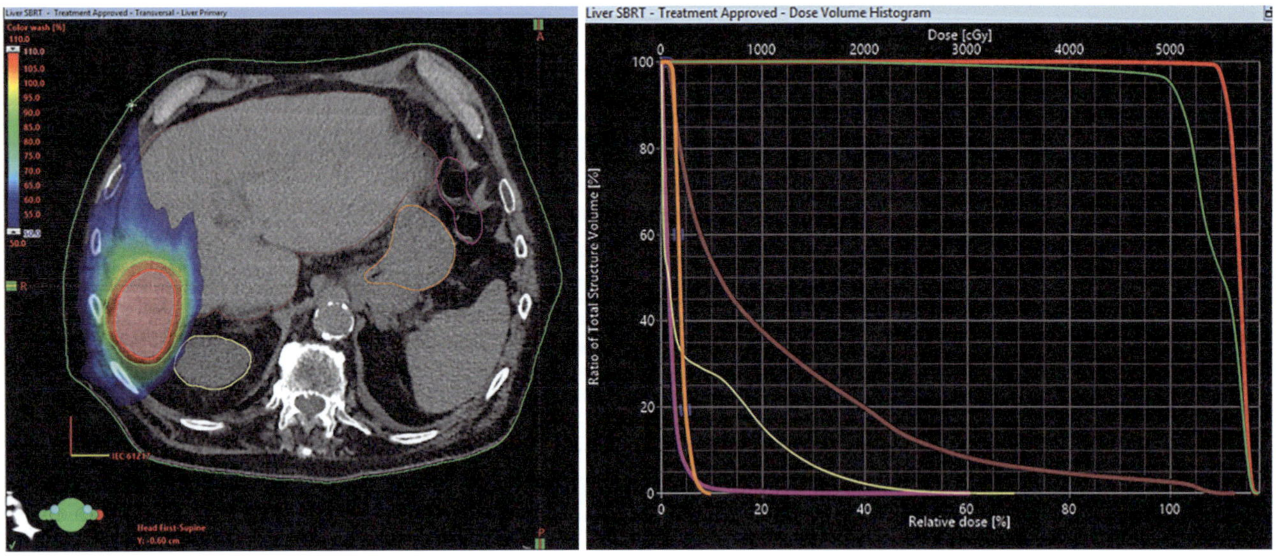

Fig. 19.1 CT-guided SBRT plan with corresponding dose-volume histogram (DVH). Image on the left is the CT-guided radiation plan, with targets and OARS contoured: tumor (red), planned treatment volume (green), liver less tumor (brown), stomach (orange), kidneys (yellow), and bowel (magenta). The plan displays the dose distribution with the central, red colors identifying high dose (110% prescription) and the outer, blue colors identifying low dose (50% prescription). The image on the right is the DVH that corresponds to the plan on the left

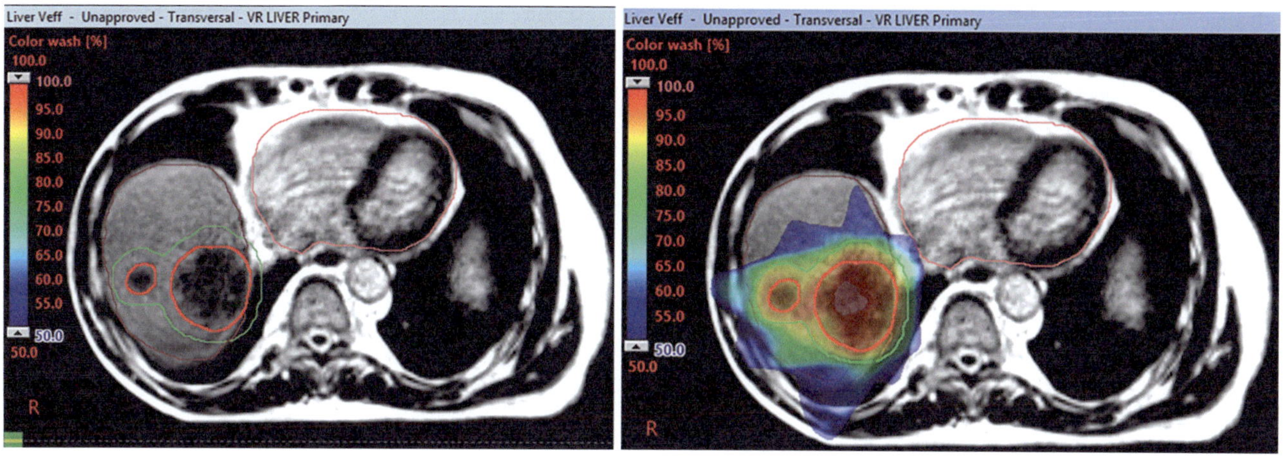

Fig. 19.2 MRI-guided SBRT plan for HCC. Image on the left is the contouring of targets and OARs done via an MRI-guided plan: tumor (red), planned treatment volume (green), liver less tumor (brown), and heart (pink). The plan displays the dose distribution with the central, red colors identifying high dose (100% prescription) and the outer, blue colors identifying low dose (50% prescription)

and compared MRI-based and CT-based plans on select patients [69]. Owing to the low visualization and large tumor motion, the CT-based target volume measured 40.8 cm^3, whereas, thanks to tumor tracking and superior soft-tissue resolution, the MRI-based target volume was substantially smaller, with a volume of 25.14 cm^3 [70]. Rosenberg et al. [71] described the results of a multi-institutional experience treating both HCC and secondary liver cancers with MRI-linear accelerators, and found very low rates of grade 3 GI toxicity (7.7%) while treating large tumor volumes (median 98.2 cm^3; range, 13–2024). An example of an MRI-guided SBRT plan with its corresponding target and OAR delineation is shown in Fig. 19.2.

Another important advancement in RT technology was the development of proton beam irradiation. Proton RT is a type of RT that uses charged particles (protons) instead of photons. The chief benefit of proton beam RT is the ability to deposit the dose to a specific depth and subsequently limit the exit dose. This provides different ways to limit radiation exposure to adjacent critical structures and well as normal liver. Fukumitsu and colleagues [72] reported on their prospectively collected experience of patients with either CPS class A or B liver disease and potentially resectable single HCCs treated with proton beam with doses of 66 Gy in 10 fractions. LC rates were 94.5% and 87.8% at 3 and 5 years after treatment. Hong et al. [73] found similarly high rates of LC in HCC patients treated with proton beam radiation. After delivering a median dose of 58 Gy in 15 fractions with proton beam radiation, they reported a 2-year LC rate of 94.8% and a 2-year OS rate of 63.2% in their HCC cohort. There have been a number of other studies that have shown similarly high rates of tumor control with low rates of toxicity in patients treated with proton beam therapy to their HCC [74–79].

Ongoing Comparative Trials

As stated previously, the majority of the data examining radiation and SBRT in the setting of HCC are retrospective or small prospective studies. Fortunately, there are a number of ongoing clinical trials designed to add additional, level-one data to help guide us to the best way to use RT in the setting of HCC.

There is an ongoing, single-institutional, phase 3 randomized trial from China that is randomly assigning patients with small HCC (<5 cm) between RFA and SBRT as first-line therapy (NCT03898921). RTOG 1112 is a phase 3, randomized trial comparing patients with Barcelona Clinic Liver Cancer stage intermediate or advanced HCC that are unsuitable for surgical resection or other LDTs to be randomly assigned to either sorafenib alone or sorafenib with SBRT (NCT01730937). The CLEAR trial at the University of Pennsylvania, Loyola University, and Oregon Health is a pilot study evaluating the combination of SBRT and TACE vs TACE alone in patients with a CPS score A or B HCC awaiting a liver transplant (NCT01918683). An international study based out of Stanford is attempting to directly compare SBRT and TACE in patients with residual or recurrent HCC after initial TACE (NCT02762266). As more and more clinical trials are developed to compare and/or combine different therapies, oncologists will be better positioned to tailor treatments to patients with HCC.

Conclusion

Radiation in HCC management was originally reserved for palliation, where it continues to improve the quality of life in patients with advanced or metastatic HCC. However, advances in radiation techniques and understanding of liver toxicity have allowed for dose escalation and thus improved LC. SBRT is a form of EBRT that allows for the noninvasive delivery of tumoricidal doses of radiation in a highly conformal manner over a small number of fractions. Prospective trials using SBRT in HCC show LC rates of 87–92%, which are comparable to other LDTs. In addition, trials suggest that EBRT in combination with other LDTs or systemic therapy have promising results. EBRT is effective in treating patients with tumor thrombus or as a bridge to transplant. Further advances in radiation techniques and technologies, including adaptive radiation planning, MRI-guided linear accelerators, and proton beam radiation, are revolutionizing the way radiation oncologists can treat HCC. Current and future ongoing clinical trials will provide additional evidence for the optimal role RT will play in the management of HCC. From palliation to definitive treatment, RT should be considered in all patients with HCC.

References

1. Bruix J, Sherman M, Llovet JM, EASL Panel of Experts on HCC. Clinical management of hepatocellular carcinoma. Conclusions of the Barcelona–2000 EASL conference. European Association for the study of the liver. J Hepatol. 2001;35(3):421–30. https://doi.org/10.1016/s0168-8278(01)00130-1.
2. Benson AB, D'Angelica MI, Abbott DE, et al. NCCN guidelines insights: hepatobiliary cancers, version 1.2017. J Natl Compr Cancer Netw. 2017;15(5):563–73. https://doi.org/10.6004/jnccn.2017.0059.
3. Lawrence TS, Robertson JM, Anscher MS, Jirtle RL, Ensminger WD, Fajardo LF. Hepatic toxicity resulting from cancer treatment. Int J Radiat Oncol Biol Phys. 1995;31(5):1237–48. https://doi.org/10.1016/0360-3016(94)00418-K.
4. Dawson LA, Normolle D, Balter JM, McGinn CJ, Lawrence TS, Ten Haken RK. Analysis of radiation-induced liver disease using the Lyman NTCP model. Int J Radiat Oncol Biol Phys. 2002;53(4):810–21. https://doi.org/10.1016/s0360-3016(02)02846-8.
5. Zeng ZC, Tang ZY, Fan J, et al. Consideration of role of radiotherapy for lymph node metastases in patients with HCC: retrospective analysis for prognostic factors from 125 patients. Int J Radiat Oncol Biol Phys. 2005;63(4):1067–76. https://doi.org/10.1016/j.ijrobp.2005.03.058.
6. Seong J, Koom WS, Park HC. Radiotherapy for painful bone metastases from hepatocellular carcinoma. Liver Int. 2005;25(2):261–5. https://doi.org/10.1111/j.1478-3231.2005.01094.x.
7. Kaizu T, Karasawa K, Tanaka Y, et al. Radiotherapy for osseous metastases from hepatocellular carcinoma: a retrospective study of 57 patients. Am J Gastroenterol. 1998;93(11):2167–71. https://doi.org/10.1111/j.1572-0241.1998.00614.x.
8. Chang L, Chen YL, Kao MC. Intracranial metastasis of hepatocellular carcinoma: review of 45 cases. Surg Neurol. 2004;62(2):172–7. https://doi.org/10.1016/j.surneu.2003.10.002.
9. Tunc B, Filik L, Tezer-Filik I, Sahin B. Brain metastasis of hepatocellular carcinoma: a case report and review of the literature. World J Gastroenterol. 2004;10(11):1688–9. https://doi.org/10.3748/wjg.v10.i11.1688.
10. Hawkins MA, Dawson LA. Radiation therapy for hepatocellular carcinoma: from palliation to cure. Cancer. 2006;106(8):1653–63. https://doi.org/10.1002/cncr.21811.
11. He J, Shi S, Ye L, et al. A randomized trial of conventional fraction versus hypofraction radiotherapy for bone metastases from hepatocellular carcinoma. J Cancer. 2019;10(17):4031–7. https://doi.org/10.7150/jca.28674.
12. Hammad AY, Robbins JR, Turaga KK, Christians KK, Gamblin TC, Johnston FM. Palliative interventions for hepatocellular carcinoma patients: analysis of the National Cancer Database. Ann Palliat Med. 2017;6(1):26–35. https://doi.org/10.21037/apm.2016.11.02.
13. Seong J, Park HC, Han KH, et al. Local radiotherapy for unresectable hepatocellular carcinoma patients who failed with transcatheter arterial chemoembolization. Int J Radiat Oncol Biol Phys. 2000;47(5):1331–5. https://doi.org/10.1016/s0360-3016(00)00519-8.
14. Dawson LA, McGinn CJ, Normolle D, et al. Escalated focal liver radiation and concurrent hepatic artery fluorodeoxyuridine for unresectable intrahepatic malignancies. J Clin Oncol. 2000;18(11):2210–8. https://doi.org/10.1200/JCO.2000.18.11.2210.
15. Hall EJ, Giaccia AJ. Radiobiology for the radiologist. 7th ed. Wolters Kluwer Lippincott Williams & Wilkins; 2012.
16. Williams MV, Denekamp J, Fowler JF. A review of alpha/beta ratios for experimental tumors: implications for clinical studies of altered fractionation. Int J Radiat Oncol Biol Phys. 1985;11(1):87–96. https://doi.org/10.1016/0360-3016(85)90366-9.
17. Dawson LA, Normolle D, Balter JM, McGinn CJ, Lawrence TS, Ten Haken RK. Analysis of radiation-induced liver disease using the Lyman NTCP model. Int J Radiat Oncol Biol Phys. 2002;53:810–21. [PubMed] [Google Scholar]SAME AS 4
18. McGinn CJ, Ten Haken RK, Ensminger WD, Walker S, Wang S, Lawrence TS. Treatment of intrahepatic cancers with radiation doses based on a normal tissue complication probability model. J Clin Oncol. 1998;16(6):2246–52. https://doi.org/10.1200/JCO.1998.16.6.2246.
19. Ten Haken RK, Martel MK, Kessler ML, et al. Use of Veff and iso-NTCP in the implementation of dose escalation protocols. Int J Radiat Oncol Biol Phys. 1993;27(3):689–95. https://doi.org/10.1016/0360-3016(93)90398-f.
20. Dawson LA, Eccles C, Craig T. Individualized image guided iso-NTCP based liver cancer SBRT. Acta Oncol. 2006;45(7):856–64. https://doi.org/10.1080/02841860600936369.
21. Klein J, Dawson LA. Hepatocellular carcinoma radiation therapy: review of evidence and future opportunities [published correction appears in Int J Radiat Oncol Biol Phys. 2014;88(2):461-462]. Int J Radiat Oncol Biol Phys. 2013;87(1):22–32. https://doi.org/10.1016/j.ijrobp.2012.08.043.
22. Chen CP. Role of radiotherapy in the treatment of hepatocellular carcinoma. J Clin Transl Hepatol. 2019;7(2):183–90. https://doi.org/10.14218/JCTH.2018.00060.
23. Sanuki N, Takeda A, Kunieda E. Role of stereotactic body radiation therapy for hepatocellular carcinoma. World J Gastroenterol. 2014;20(12):3100–11. https://doi.org/10.3748/wjg.v20.i12.3100.
24. Yoo GS, Yu JI, Park HC. Proton therapy for hepatocellular carcinoma: current knowledges and future perspectives. World J Gastroenterol. 2018;24(28):3090–100. https://doi.org/10.3748/wjg.v24.i28.3090.
25. Kim TH, Kim DY, Park JW, et al. Three-dimensional conformal radiotherapy of unresectable hepatocellular carcinoma patients for whom transcatheter arterial chemoembolization was ineffective or unsuitable. Am J Clin Oncol Cancer Clin Trials. 2006;29(6):568–75. https://doi.org/10.1097/01.coc.0000239147.60196.11.

26. Mornex F, Girard N, Beziat C, et al. Feasibility and efficacy of high-dose three-dimensional-conformal radiotherapy in cirrhotic patients with small-size hepatocellular carcinoma non-eligible for curative therapies – mature results of the French phase II RTF-1 trial. Int J Radiat Oncol Biol Phys. 2006;66(4):1152–8. https://doi.org/10.1016/j.ijrobp.2006.06.015.
27. McIntosh A, Hagspiel KD, Al-Osaimi AM, et al. Accelerated treatment using intensity-modulated radiation therapy plus concurrent capecitabine for unresectable hepatocellular carcinoma. Cancer. 2009;115(21):5117–25. https://doi.org/10.1002/cncr.24552.
28. Chi KH, Liao CS, Chang CC, et al. Angiogenic blockade and radiotherapy in hepatocellular carcinoma. Int J Radiat Oncol Biol Phys. 2010;78(1):188–93. https://doi.org/10.1016/j.ijrobp.2009.07.1725.
29. Sanuki N, Takeda A, Oku Y, et al. Stereotactic body radiotherapy for small hepatocellular carcinoma: a retrospective outcome analysis in 185 patients. Acta Oncol. 2014;53(3):399–404. https://doi.org/10.3109/0284186X.2013.820342.
30. Jung J, Yoon SM, Kim SY, et al. Radiation-induced liver disease after stereotactic body radiotherapy for small hepatocellular carcinoma: clinical and dose-volumetric parameters. Radiat Oncol. 2013;8:249. https://doi.org/10.1186/1748-717X-8-249.
31. Kang JK, Kim MS, Cho CK, et al. Stereotactic body radiation therapy for inoperable hepatocellular carcinoma as a local salvage treatment after incomplete transarterial chemoembolization. Cancer. 2012;118(21):5424–31. https://doi.org/10.1002/cncr.27533.
32. Cárdenes HR, Price TR, Perkins SM, et al. Phase I feasibility trial of stereotactic body radiation therapy for primary hepatocellular carcinoma. Clin Transl Oncol. 2010;12(3):218–25. https://doi.org/10.1007/s12094-010-0492-x.
33. Lasley FD, Mannina EM, Johnson CS, et al. Treatment variables related to liver toxicity in patients with hepatocellular carcinoma, child-Pugh class a and B enrolled in a phase 1-2 trial of stereotactic body radiation therapy. Pract Radiat Oncol. 2015;5(5):e443–9. https://doi.org/10.1016/j.prro.2015.02.007.
34. Takeda A, Sanuki N, Tsurugai Y, et al. Phase 2 study of stereotactic body radiotherapy and optional transarterial chemoembolization for solitary hepatocellular carcinoma not amenable to resection and radiofrequency ablation. Cancer. 2016;122(13):2041–9. https://doi.org/10.1002/cncr.30008.
35. Su TS, Liang P, Liang J, et al. Long-term survival analysis of stereotactic ablative radiotherapy versus liver resection for small hepatocellular carcinoma. Int J Radiat Oncol Biol Phys. 2017;98(3):639–46. https://doi.org/10.1016/j.ijrobp.2017.02.095.
36. Wahl DR, Stenmark MH, Tao Y, et al. Outcomes after stereotactic body radiotherapy or radiofrequency ablation for hepatocellular carcinoma. J Clin Oncol. 2016;34(5):452–9. https://doi.org/10.1200/JCO.2015.61.4925.
37. Shiozawa K, Watanabe M, Ikehara T, et al. Comparison of percutaneous radiofrequency ablation and CyberKnife for initial solitary hepatocellular carcinoma: a pilot study. World J Gastroenterol. 2015;21:13490–9. https://doi.org/10.3748/wjg.v21.i48.13490.
38. Sapir E, Tao Y, Schipper MJ, et al. Stereotactic body radiation therapy as an alternative to transarterial chemoembolization for hepatocellular carcinoma. Int J Radiat Oncol Biol Phys. 2018;100(1):122–30. https://doi.org/10.1016/j.ijrobp.2017.09.001.
39. Bush DA, Smith JC, Slater JD, et al. Randomized clinical trial comparing proton beam radiation therapy with transarterial chemoembolization for hepatocellular carcinoma: results of an interim analysis. Int J Radiat Oncol Biol Phys. 2016;95(1):477–82. https://doi.org/10.1016/j.ijrobp.2016.02.027.
40. Zhang Z, Yang X, Wena M, Wan J. Evaluation of TACE combined with gamma-knife radiotherapy for primary hepatocellular carcinoma. J Interventional Radiol. 2012;7(21):596–9. https://doi.org/10.3969/j.issn.1008-794X.2012.07.017.
41. Honda Y, Kimura T, Aikata H, et al. Stereotactic body radiation therapy combined with transcatheter arterial chemoembolization for small hepatocellular carcinoma. J Gastroenterol Hepatol. 2013;28(3):530–6. https://doi.org/10.1111/jgh.12087.
42. Li XL, Guo WX, Hong XD, et al. Efficacy of the treatment of transarterial chemoembolization combined with radiotherapy for hepatocellular carcinoma with portal vein tumor thrombus: a propensity score analysis. Hepatol Res. 2016;46(11):1088–98. https://doi.org/10.1111/hepr.12657.
43. Guo WJ, Yu EX, Liu LM, et al. Comparison between chemoembolization combined with radiotherapy and chemoembolization alone for large hepatocellular carcinoma. World J Gastroenterol. 2003;9(8):1697–701. https://doi.org/10.3748/wjg.v9.i8.1697.
44. Huo YR, Eslick GD. Transcatheter arterial chemoembolization plus radiotherapy compared with chemoembolization alone for hepatocellular carcinoma: a systematic review and meta-analysis. JAMA Oncol. 2015;1(6):756–65. https://doi.org/10.1001/jamaoncol.2015.2189.
45. Hardy-Abeloos C, Lazarev S, Ru M, et al. Safety and efficacy of liver stereotactic body radiation therapy for hepatocellular carcinoma after segmental transarterial radioembolization. Int J Radiat Oncol Biol Phys. 2019;105(5):968–76. https://doi.org/10.1016/j.ijrobp.2019.09.006.
46. Spinzi G, Paggi S. Sorafenib in advanced hepatocellular carcinoma. N Engl J Med. 2008;359(23):2497–8.; author reply 2498–2499. https://doi.org/10.1056/nejmc081780.
47. Lencioni R, Llovet JM, Han G, et al. Sorafenib or placebo plus TACE with doxorubicin-eluting beads for intermediate stage HCC: the SPACE trial. J Hepatol. 2016;64(5):1090–8. https://doi.org/10.1016/j.jhep.2016.01.012.
48. Meyer T, Fox R, Ma YT, et al. Sorafenib in combination with transarterial chemoembolisation in patients with unresectable hepatocellular carcinoma (TACE 2): a randomised placebo-controlled, double-blind, phase 3 trial. Lancet Gastroenterol Hepatol. 2017;2(8):565–75. https://doi.org/10.1016/S2468-1253(17)30156-5.
49. Kudo M, Ueshima K, Ikeda M, TACTICS Study Group, et al. Randomised, multicentre prospective trial of transarterial chemoembolisation (TACE) plus sorafenib as compared with TACE alone in patients with hepatocellular carcinoma: TACTICS trial. Gut. 2020;69(8):1492–501. https://doi.org/10.1136/gutjnl-2019-318934.
50. Mahvash A, Murthy R, Odisio BC, et al. Yttrium-90 resin microspheres as an adjunct to sorafenib in patients with unresectable hepatocellular carcinoma. J Hepatocell Carcinoma. 2016;3:1–7. https://doi.org/10.2147/JHC.S62261.
51. Chen SW, Lin LC, Kuo YC, Liang JA, Kuo CC, Chiou JF. Phase 2 study of combined sorafenib and radiation therapy in patients with advanced hepatocellular carcinoma. Int J Radiat Oncol Biol Phys. 2014;88(5):1041–7. https://doi.org/10.1016/j.ijrobp.2014.01.017.
52. Brade AM, Ng S, Brierley J, et al. Phase 1 trial of sorafenib and stereotactic body radiation therapy for hepatocellular carcinoma. Int J Radiat Oncol Biol Phys. 2016;94(3):580–7. https://doi.org/10.1016/j.ijrobp.2015.11.048.
53. Eriguchi T, Takeda A, Sanuki N, et al. Acceptable toxicity after stereotactic body radiation therapy for liver tumors adjacent to the central biliary system. Int J Radiat Oncol Biol Phys. 2013;85(4):1006–11. https://doi.org/10.1016/j.ijrobp.2012.09.012.
54. Guglielmi A, Ruzzenente A, Conci S, Valdegamberi A, Iacono C. How much remnant is enough in liver resection? Dig Surg. 2012;29(1):6–17. https://doi.org/10.1159/000335713.
55. Murray LJ, Sykes J, Brierley J, et al. Baseline albumin-bilirubin (ALBI) score in Western patients with hepatocellular carcinoma treated with stereotactic body radiation therapy (SBRT). Int J Radiat Oncol Biol Phys. 2018;101(4):900–9. https://doi.org/10.1016/j.ijrobp.2018.04.011.
56. Su TS, Yang HM, Zhou Y, et al. Albumin-bilirubin (ALBI) versus Child-Turcotte-Pugh (CTP) in prognosis of HCC after stereotactic

body radiation therapy. Radiat Oncol. 2019;14(1):50. https://doi.org/10.1186/s13014-019-1251-y.
57. Schaub SK, Bowen SR, Chapman TR, Nyflot MJ, Apisarnthanarax S. ALBI-RT: a novel radiation therapy specific hepatotoxicity prediction model for hepatocellular carcinoma patients. J Clin Oncol. 2018;36(4 Suppl):444. https://doi.org/10.1200/JCO.2018.36.4_suppl.444.
58. Xi M, Zhang L, Zhao L, et al. Effectiveness of stereotactic body radiotherapy for hepatocellular carcinoma with portal vein and/or inferior vena cava tumor thrombosis. PLoS One. 2013;8(5):e63864. https://doi.org/10.1371/journal.pone.0063864.
59. Li XL, Guo WX, Hong XD, et al. Efficacy of the treatment of transarterial chemoembolization combined with radiotherapy for hepatocellular carcinoma with portal vein tumor thrombus: a propensity score analysis. Hepatol Res. 2016; [SAME AS 42]
60. Yoon SM, Ryoo BY, Lee SJ, et al. Efficacy and safety of transarterial chemoembolization plus external beam radiotherapy vs sorafenib in hepatocellular carcinoma with macroscopic vascular invasion: a randomized clinical trial. JAMA Oncol. 2018;4(5):661–9. https://doi.org/10.1001/jamaoncol.2017.5847.
61. Katz AW, Chawla S, Qu Z, Kashyap R, Milano MT, Hezel AF. Stereotactic hypofractionated radiation therapy as a bridge to transplantation for hepatocellular carcinoma: clinical outcome and pathologic correlation. Int J Radiat Oncol Biol Phys. 2012;83(3):895–900. https://doi.org/10.1016/j.ijrobp.2011.08.032.
62. Al Hamad AA, Hassanain M, Michel RP, Metrakos P, Roberge D. Stereotactic radiotherapy of the liver: a bridge to transplantation stereotactic radiotherapy of the liver: a bridge to transplantation. Technol Cancer Res Treat. 2009;8(6):401–5. https://doi.org/10.1177/153303460900800601.
63. O'Connor JK, Trotter J, Davis GL, Dempster J, Klintmalm GB, Goldstein RM. Long-term outcomes of stereotactic body radiation therapy in the treatment of hepatocellular cancer as a bridge to transplantation. Liver Transpl. 2012;18(8):949–54. https://doi.org/10.1002/lt.23439.
64. Sandroussi C, Dawson LA, Lee M, et al. Radiotherapy as a bridge to liver transplantation for hepatocellular carcinoma. Transpl Int. 2010;23(3):299–306. https://doi.org/10.1111/j.1432-2277.2009.00980.
65. Sapisochin G, Barry A, Doherty M, et al. Stereotactic body radiotherapy vs. TACE or RFA as a bridge to transplant in patients with hepatocellular carcinoma. An intention-to-treat analysis. J Hepatol. 2017;67(1):92–9. https://doi.org/10.1016/j.jhep.2017.02.022.
66. Mohamed M, Katz AW, Tejani MA, et al. Comparison of outcomes between SBRT, yttrium-90 radioembolization, transarterial chemoembolization, and radiofrequency ablation as bridge to transplant for hepatocellular carcinoma. Adv Radiat Oncol. 2015;1(1):35–42. https://doi.org/10.1016/j.adro.2015.12.003.
67. Acharya S, Fischer-Valuck BW, Kashani R, et al. Online magnetic resonance image guided adaptive radiation therapy: first clinical applications. Int J Radiat Oncol Biol Phys. 2016;94(2):394–403. https://doi.org/10.1016/j.ijrobp.2015.10.015.
68. Henke L, Kashani R, Robinson C, et al. Phase I trial of stereotactic MR-guided online adaptive radiation therapy (SMART) for the treatment of oligometastatic or unresectable primary malignancies of the abdomen. Radiother Oncol. 2018;126(3):519–26. https://doi.org/10.1016/j.radonc.2017.11.032.
69. Wen N, Kim J, Doemer A, et al. Evaluation of a magnetic resonance guided linear accelerator for stereotactic radiosurgery treatment. Radiother Oncol. 2018;127(3):460–6. https://doi.org/10.1016/j.radonc.2018.04.034.
70. Feldman AM, Modh A, Glide-Hurst C, Chetty IJ, Movsas B. Real-time magnetic resonance-guided liver stereotactic body radiation therapy: an institutional report using a magnetic resonance–Linac system. Cureus. 2019;11(9):e5774. https://doi.org/10.7759/cureus.5774.
71. Rosenberg SA, Henke LE, Shaverdian N, et al. A multi-institutional experience of MR-guided liver stereotactic body radiation therapy. Adv Radiat Oncol. 2019;4(1):142–9. https://doi.org/10.1016/j.adro.2018.08.005.
72. Fukumitsu N, Sugahara S, Nakayama H, et al. A prospective study of hypofractionated proton beam therapy for patients with hepatocellular carcinoma. Int J Radiat Oncol Biol Phys. 2009;74(3):831–6. https://doi.org/10.1016/j.ijrobp.2008.10.073.
73. Hong TS, Wo JY, Yeap BY, et al. Multi-institutional phase II study of high-dose hypofractionated proton beam therapy in patients with localized, unresectable hepatocellular carcinoma and intrahepatic cholangiocarcinoma. J Clin Oncol. 2016;34(5):460–8. https://doi.org/10.1200/JCO.2015.64.2710.
74. Chiba T, Tokuuye K, Matsuzaki Y, et al. Proton beam therapy for hepatocellular carcinoma: a retrospective review of 162 patients. Clin Cancer Res. 2005;11(10):3799–805. https://doi.org/10.1158/1078-0432.CCR-04-1350.
75. Bush DA, Hillebrand DJ, Slater JM, Slater JD. High-dose proton beam radiotherapy of hepatocellular carcinoma: preliminary results of a phase II trial. Gastroenterology. 2004;127(5 Suppl 1):S189–93. https://doi.org/10.1053/j.gastro.2004.09.033.
76. Kawashima M, Furuse J, Nishio T, et al. Phase II study of radiotherapy employing proton beam for hepatocellular carcinoma. J Clin Oncol. 2005;23(9):1839–46. https://doi.org/10.1200/JCO.2005.00.620.
77. Mizumoto M, Tokuuye K, Sugahara S, et al. Proton beam therapy for hepatocellular carcinoma adjacent to the porta hepatis. Int J Radiat Oncol Biol Phys. 2008;71(2):462–7. https://doi.org/10.1016/j.ijrobp.2007.09.056.
78. Hata M, Tokuuye K, Sugahara S, et al. Proton beam therapy for hepatocellular carcinoma patients with severe cirrhosis. Strahlenther Onkol. 2006;182(12):713–20. https://doi.org/10.1007/s00066-006-1564-2.
79. Sugahara S, Oshiro Y, Nakayama H, et al. Proton beam therapy for large hepatocellular carcinoma. Int J Radiat Oncol Biol Phys. 2010;76(2):460–6. https://doi.org/10.1016/j.ijrobp.2009.02.030.

20 Should Transarterial Radioembolization Be the First-Line Therapy Compared to Transarterial Chemoembolization for Intermediate-Stage Hepatocellular Carcinoma?

Alexandra Morris, Stefan Brancel, and Scott Schwartz

Transarterial radioembolization (TARE) has proven to be safe and effective in intermediate-stage hepatocellular carcinoma (HCC) with multiple advantages over transarterial chemoembolization (TACE) but with no proven survival benefit over TACE.

Main Points

1. There is no proven survival benefit between TACE and TARE, especially when confined to Barcelona Clinic Liver Cancer (BCLC) system stage B intermediate disease.
2. Studies have demonstrated increased time to progression with TARE over TACE.
3. TARE is also the preferred treatment method for patients with HCC and concurrent portal vein thrombosis (PVT).
4. TARE delivered as a radiation lobectomy is useful for bridging to resection in patients with HCC and small future liver remnants (FLRs).
5. Studies indicate TARE is more cost-effective than TACE, and prospective studies demonstrate improved quality of life (QoL) over TACE.

Hepatic tumors obtain 80% of their blood supply from the hepatic artery while normal hepatic parenchyma receives 75% of its blood flow from the portal vein. This allows for transarterial delivery of therapy selectively toward hepatic tumors. While this knowledge was initially used mainly for embolization and delivery of chemotherapeutic agents with TACE, TARE became an alternative locoregional option. Over time, TACE became the standard of care for intermediate-stage HCC and retrospective analyses have demonstrated that more patients with HCC are treated with TACE than with any other HCC therapy [1]. BCLC stage B, or intermediate-stage HCC, is defined as asymptomatic disease, with tumors larger than 3 cm or multinodular disease, and without vascular invasion. Mounting literature [2] now suggests that TARE may be the superior transarterial treatment option for select patient populations. The goal of this chapter will be to discuss why TARE should be considered the first-line therapy compared to TACE for intermediate-stage HCC.

History of Transarterial Chemoembolization

TACE has long been an accepted treatment modality for HCC, both as a bridging therapy to liver transplantation (LT) and as a treatment for patients who are not amenable to surgery. In the 1980s, multiple randomized controlled trials (RCTs) introduced the concept of chemoembolization, specifically with mitomycin C or adriamycin to embolize and treat hepatic tumors and demonstrated an objective tumor response rate in 75% of cases and a 1-year survival that was comparable to surgery [3, 4]. In the early 2000s, TACE was adopted as a locoregional therapy after demonstrating survival benefits in multiple studies. A meta-analysis comparing 7 RCTs showed that TACE offers a statistically significant survival advantage at 2 years over supportive treatment alone ($P = 0.017$) [5]. Approximately 60% of patients treated with TACE show statistically significantly reduced tumor growth, with 35% showing at least a 25% reduction in size and the remaining 25% showing evidence of tumor necrosis. Prior papers had expressed concern for liver failure post TACE procedures, yet multiple new studies have demonstrated that TACE is safe and effective, with treatment-related mortality estimated to be below 5% [6]. TACE, if applied correctly,

A. Morris (✉) · S. Brancel · S. Schwartz
Henry Ford Hospital, Detroit, MI, USA
e-mail: alexandram@rad.hfh.edu; stefanb@rad.hfh.edu; scotts@rad.hfh.edu

N. S. Reau, A. G. Singal (eds.), *Controversies in Hepatocellular Carcinoma*, https://doi.org/10.1007/978-3-031-90225-3_20

can lead to survival benefits without adversely affecting hepatic functional reserve.

As an advancement to further refine TACE, nonresorbable drug-eluting beads (DEBs) were developed. Theoretically, DEBs gradually release the chemotherapeutic agent after injection and increase the duration of ischemia, thus enhancing the effects of drug delivery to the tumor, while at the same time reducing systemic availability [7]. A meta-analysis of 4 RCTs supported this claim and clinically showed a better objective tumor response rate in DEB-TACE vs conventional TACE (cTACE) ($P = 0.04$); however, 3-year survival and incidence of adverse effects was similar in both groups [8]. The PRECISION V study showed that patients with Child-Pugh score class B, bilobar disease, and recurrent disease showed a statistically significant increase in objective response after DEB-TACE vs cTACE ($P = 0.038$). Moreover, this study demonstrated a statistically significantly decreased liver toxicity in DEB-TACE ($P < 0.001$) and a statistically significantly lower rate of doxorubicin-related side effects ($P = 0.0001$) [9]. In one study of 71 patients, DEB-TACE had an increased survival from diagnosis compared to conventional TACE (610 days for DEB-TACE vs 284 days for cTACE, $P = 0.03$) [10]. Conversely, a subsequent study of 177 patients showed equal tumor response and frequency of adverse events for TACE vs DEB-TACE; however, less post-procedural abdominal pain was recorded in the DEB-TACE group [11]. Many of the studies evaluating the effectiveness of TACE are difficult to compare because there is no standardized regimen including type of embolic particle, type of chemotherapy, type of emulsifying agent, and number of treatment sessions. In summary, cTACE and DEB-TACE are widely accepted treatment modalities for unresectable HCC; however, the degree to which it is superior to other transarterial therapies remains uncertain.

History of Transarterial Radioembolization

Despite the wide use of TACE, TARE is an alternative that likely improves outcomes for specific patient populations. It was first described as early as the mid-1960s by Ariel [12] and Simon et al. [13] who delivered yttrium-90 (Y90) into the celiac artery to treat hepatic tumors. This showed tumor response; however, it led to poor outcomes secondary to toxicities. In 1973, TARE was first used for colon cancer metastatic to the liver, which showed an objective decrease in tumor size [14]. After these early studies, in the late 1980s and early 1990s, multiple phase 1 trials with TheraSphere (BTG) (glass microspheres) and SIR-Spheres (Sirtex) (resin microspheres) were performed, which demonstrated safety and efficacy [15–17]. These pilot studies led to Food and Drug Administration (FDA) approval of TheraSpheres under a humanitarian device exemption. In 2002, SIR-Spheres were approved by the FDA for the treatment of colorectal metastasis to the liver; however, their use for HCC and other hepatic malignancies remains off-label. Since the early 2000s, Y90 has gained increasing acceptance as a locoregional therapy, and multiple studies have shown its superiority over other treatment options. An analysis of Y90 in 108 patients demonstrated an average time to progression of 10 months, median overall survival (OS) of 16.4 months, and no evidence of lung or visceral toxicity. The most common adverse event was transient fatigue syndrome [18]. A less common, but more serious, side effect is radioembolization-induced liver disease (REILD), which is still poorly defined, but is most frequently considered either symptomatic liver disease without cancer progression or liver enzyme elevation after TARE. It has a reported mortality in 0–5% of patients [19]. Risk factors for developing REILD include prior liver-directed therapy, higher radiation dose, and whole-liver radiation [19]. Other studies demonstrated that patients could develop radioembolization-induced chronic hepatoxicity up to 1 year after the initial treatment [20]. This study demonstrated that 13% of patients developed radioembolization-induced chronic hepatotoxicity, which resulted in death in 5% of patients from hepatic decompensation after radioembolization. Statistically significant risk factors for REILD cited in this study include a higher intrahepatic tumor volume and a higher number of hepatic comorbidities leading to cirrhosis. Of note, this study involved 98 patients at a single institution and 54% of the patients treated with radioembolization had neuroendocrine tumors. While there are considerable side effects associated with TARE, it remains an important treatment option for HCC patients because of its favorable side effect profile as compared to TACE [21].

Multiple studies have suggested that TARE is efficacious compared to other treatment modalities for intermediate-stage HCC. In the SIRveNIB trial, Y90 was compared to sorafenib. This study showed similar OS between the 2 treatment arms, with worse side effects associated with sorafenib, including fatigue, hematologic abnormalities, diarrhea, abdominal pain, and dermatologic reactions. There was higher tumor response rate with Y90 [22]. In the SARAH trial, 459 patients were randomly assigned to receive to sorafenib or Y90. While the median OS was not statistically significantly different between the 2 treatment groups, objective response rate was statistically significantly higher with Y90 and patients had improved QoL with decreased adverse effects [23]. These landmark trials placed Y90 as a favorable treatment option for patients with advanced, liver-confined HCC. While TACE was long considered the standard of care for transarterial therapy for intermediate-stage HCC, in 2017, Salem et al. [24] published an analysis with 1000 patients over a 15-year time period that indicated TARE should be considered as the primary treatment for intermediate stage HCC.

Barcelona-Clinic Liver Cancer Algorithm and Transarterial Chemoembolization

While there are many other staging systems that are discussed in Controversy 14, BCLC remains the most ubiquitous in Western medicine. In 1999, the BCLC staging guidelines were published with proposed stages 0, A, B, C, D and associated therapeutic recommendations for each [25]. For early disease or BCLC A, these patients are eligible for curative therapies, which include resection and LT. At the time of the initial guidelines, BCLC stage B or intermediate stage, defined as asymptomatic multifocal disease or lesions larger than 3 cm not amenable to other therapies, had little treatment options that achieved the goals of palliation, bridge to transplant, or downstaging. The recommendation at that time was for clinical trials with transarterial therapies as no other treatment had shown survival benefit. When updated BCLC guidelines were published in 2004, TACE was just gaining momentum with 2 RCTs published showing survival benefits [26]. In 2002, one study showed survival benefit in 40 patients with predominantly hepatitis-related HCC who underwent chemoembolization with lipiodol and cisplatin [27]. Another study also demonstrated this survival benefit in patients with predominantly hepatitis C-related HCC [28]. When the BCLC staging guidelines arrived in 2004, the only RCT evidence available for locoregional therapy was for TACE. It is important to note that while meta-analysis confirmed survival benefits of TACE at the time, there was only a 35% response rate on follow-up imaging when averaged across studies [29]. Since these guidelines were published in 2004 and TACE became the standard of care for BCLC stage B, there has been no inclusion of other locoregional therapies in updated recommendations as recently as 2018 [30]. This is likely because of the multiple studies that have not shown survival benefit of TARE over TACE.

Direct Comparisons of Transarterial Chemoembolization and Transarterial Radioembolization

While TACE is an effective treatment with an acceptable safety profile for intermediate-stage HCC, TARE has gained considerable interest leading to multiple RCTs and numerous prospective and retrospective studies. The PREMIERE trial is an RCT with 45 patients who were randomly assigned to receive to either TACE or TARE. These patients were mainly BCLC A patients who were ineligible for resection or ablation. The primary outcome of this trial was time to progression (TTP), which was defined as a 25% increase in arterial enhancement, PVT, index lesions requiring retreatment, or new lesions. Median TTP was statistically significantly increased in TARE patients at 26 months vs 6.8 months in cTACE patients [2]. This is important because increased TTP and improved local control could give patients more time on the transplant list. Another RCT was a European multicenter study that randomly assigned 28 patients to TACE or TARE. These patients were mixed BCLC A, B, C, but the majority were stage B with the primary outcome being QoL. The researchers observed no difference in QoL or OS between the 2 treatment arms [31]. Another single-European center RCT was performed with mainly BCLC B patients undergoing segmental TACE vs lobar TARE. This study demonstrated no difference in OS, but more progression in TACE and more liver failure in TARE [32]. Noteworthy, lobar, rather than segmental TARE, was performed, which would explain the increased liver failure seen in the Y90 treatment arm. To summarize these RCTs, a meta-analysis was performed that found no difference in disease progression in pooled studies between the 2 treatment modalities and no difference in 1- or 2-year survival [33]. These results were further confirmed by a meta-analysis of 5 studies and 553 patients that revealed no statistically significant survival differences for up to 4 years between TACE vs Y90 for unresectable HCC with similar complication profiles except for posttreatment pain, which was greater in TACE [34]. A further meta-analysis of 1652 patients comparing TARE vs TACE demonstrated that TARE increased 2-year OS rates and resulted in better objective response rates according to the mRECIST criteria with lower risk of adverse events [35]. Another study comparing Y90, TACE, radiofrequency ablation (RFA), and stereotactic body radiotherapy (SBRT) confirmed that Y90 has the highest rates of pathologic complete response and radiologic response compared to the other treatment modalities [36]. More institutions began implementing Y90 as the standard-of-care transarterial therapy after the 2017 prospective cohort study that demonstrated superiority of TARE over TACE in multiple scenarios [24]. In this paper, 1000 patients were treated with TARE with a follow-up of 15 years. This paper explained how TARE is more broadly applicable than TACE across BCLC stages. For example, patients can receive boost dosing if they have PVT (B-SIRT) [37] or ablative dosing for segmentectomy. Even in BCLC D patients, this study showed 14 patients who underwent LT after Y90 with 5-year survival post LT as high as 92%. These results demonstrated that while Y90 should be considered as a primary treatment option in patients with BCLC B, it is also an option for patients with BCLC A not amenable to other treatment options and can even be employed to treat BCLC D patients.

Improved Quality of Life and Fewer Adverse Effects with Transarterial Radioembolization

While OS has not been consistently shown to differ between the 2 treatment modalities, measures of QoL and adverse events have been shown to favor TARE over TACE. The aforementioned meta-analysis comparing Y90, TACE, RFA, and SBRT revealed that toxicities were statistically significantly higher for patients receiving TACE and RFA than for patients receiving Y90 [36]. Retrospective studies show lower rates of postembolization syndromes and postprocedure hospitalizations in patients treated with TARE [21, 38]. More analyses revealed statistically significantly lower instances of transaminitis and abdominal pain [39]. The PREMIERE trial demonstrated that Y90 had lower rates of diarrhea and hypoalbuminemia [2]. Furthermore, a prospective study demonstrated that despite having more advanced disease on average, patients treated with Y90 had improved QoL at 2 and 4 weeks post treatment compared to patients treated with TACE [40].

Transarterial Radioembolization and Portal Venous Thrombosis

It has long been recognized that intermediate-stage HCC is a very heterogeneous disease classification and as such there has always been treatment variation within this grouping. A primary example is intermediate-stage HCC with PVT, which is considered a contraindication to TACE because of the risk of acute liver failure. TACE is a macroembolic technique that can lead to arterial hypoperfusion. In cases of HCC with associated PVT, TARE offers a better alternative because of its microembolic nature. In a small group of patients, one study demonstrated that HCC patients with portal vein invasion or occlusion treated with TACE had greater than 50% 30-day mortality [4]. One study demonstrated that TARE provides similar OS in HCC patients with PVT when compared with patients without PVT. [41] Another study demonstrated that in 15 patients with HCC and PVT, TARE was well tolerated with bilirubin toxicity noted in only 5 patients, and this was linked to disease progression rather than treatment effect [42]. Thus, TARE became the preferred treatment for HCC with concurrent PVT.

Transarterial Radioembolization as a Bridge to Liver Resection or Transplantation

TARE has additional advantages in the overall management of HCC with the goal of eventual resection or transplantation. Radiation lobectomy, a high-dose lobar treatment, causes tumor necrosis while also leading to hypertrophy of the FLR for eventual planned resection. TARE is likely to increase FLR size even after failed portal vein embolization and can maximize tumor necrosis before surgery [43]. A study of 31 patients showed median FLR hypertrophy after TARE before resection of 23.3% (72% 3 months after resection) after radiation lobectomy and 9% (94% 3 months after resection) after radiation segmentectomy. The majority of patients showed more than 50% tumor necrosis with 3-year OS rate at 86% [44]. TACE has not been shown to have an effect on FLR size. Furthermore, Y90 is better than TACE for downstaging HCC patients from UNOS T3 to T2 (58% vs 31% in 86 patients), thus allowing patients to be assessed for LT [45]. Increased TTP lowers transplant list dropout rates, favoring Y90 as the preferred transarterial therapy for bridge to transplant.

Cost Analysis of Transarterial Radioembolization Versus Transarterial Chemoembolization

Evidence of increased cost-effectiveness of TARE over TACE further supports TARE as a preferred transarterial treatment option. The main difference in cost between the 2 treatments can be explained by the fact that Y90 is usually administered in fewer sessions when compared to TACE. Additionally, Y90 typically requires fewer inpatient admissions [21]. One study showed that TARE is more cost-effective than TACE for BCLC C patients, but not BCLC A or B patients [46]. Another study with 630 patients compared the cost-effectiveness of TARE vs sorafenib for intermediate- and advanced-stage HCC. For both stages, TARE led to lower health care costs and greater health improvements compared with sorafenib [47].

Counterpoint: Situations of Transarterial Chemoembolization as the Preferred Treatment Modality

While TACE and TARE are both safe and effective treatments, TACE remains the most widely used and historically recommended transarterial therapy for intermediate-stage HCC. This translates to many institutions having higher levels of experience and comfort with TACE. There are also practical limitations related to institutional review board oversight for TARE, specifically for TheraSphere because of its FDA approval under a humanitarian device exemption. Additionally, while Y90 shows promise in treating a large variety of these intermediate-stage HCC patients, contraindications to Y90 treatment exist. End-stage liver disease and HCC have been associated with increased arteriovenous shunting, which can be problematic during Y90 if the microspheres pass through the tumor and hepatic parenchyma and travel to the lungs, leading to radiation pneumonitis. This is assessed before treatment with a mapping procedure in which Tc-99 m macroaggregated albumin is injected in the intended distribution and imaged with SPECT (single-photon emission computed tomography). Depending on the degree of shunting, reduced dosages of Y90 may be required, which limit tumor response. TARE is contraindicated when lung dose is more than 30 Gy per single session, or more than 50 Gy cumulatively. In these cases, TACE is the preferred locoregional treatment option.

Future Directions of Locoregional Therapy

While many trials are promising for Y90, many institutions still favor TACE based on established guidelines and feel that further investigations need to occur with more standardization of techniques before treatment algorithms change. There is concern that many of the studies used outdated techniques and did not report or compare delivered dose. Studies have shown there is improved necrosis when the dose exceeded 190 Gy (90–100%). Garin et al. [48] evaluated the effect of tumor dose (TD) based on Tc-99 m macroaggregated albumin quantification. They showed the response rate was only 9.1% for patients with a TD of less than 205 Gy and 89.7% with a TD greater than 205 Gy. This suggests all ongoing and future trials should include and analyze TD. Furthermore, this trial recommended using a personalized dosimetric approach, which could improve the overall efficacy of Y90.

While further comparison studies are needed, future directions of research also include superselective delivery of both Y90 and chemotherapeutic agents as well as development of transarterial delivery of biologic agents, including viral vectors, natural killer cells, and low-density lipoprotein–docosahexaenoic acid (LDL-DHA) [49]. Two exciting examples of viral vectors are the use of adenovirus delivered transarterially to either produce tumor suppressor proteins, or directly causing oncolysis via viral replication. He and colleagues [50] were able to show prolonged survival of intermediate to advanced HCC with TACE with H101 as compared to just TACE in a retrospective study of 590 patients. H101 is a virus that replicates only in cells with inactivated or absent p53, in which the virus causes cell lysis. Alternatively, an engineered adenovirus (rAD-p53) produces p53 in infected cells. In a prospective study of 102 patients with unresectable HCC, participants underwent either transarterial segmental delivery of rAD-p53 or cTACE using lobaplatin. Patients who received rAD-p53 showed improved OS compared to those receiving TACE (median OS 28 vs 22 months) (hazard ratio 0.58; 95% CI, 0.35–0.96; $P = 0.035$) [51].

Conclusion

There is no proven survival benefit between TACE and TARE, especially when confined to BCLC stage B intermediate disease. Although TACE has been the standard of care in the treatment of intermediate-stage HCC, there is an accumulating body of evidence to suggest that TARE outperforms TACE. Studies have demonstrated increased TTP with TARE over TACE, which can decrease transplant list dropout rates. TARE is also the preferred treatment method for patients with HCC and concurrent PVT. Additionally, TARE delivered as a radiation lobectomy is useful for bridging to resection in patients with HCC and small FLRs. Studies also indicate TARE is more cost-effective than TACE, and prospective studies demonstrate improved QoL over TACE. While additional analyses need to be performed to further compare these transarterial therapies, TARE has proven to be safe and effective in intermediate-stage HCC with multiple advantages over TACE.

References

1. White JA, Gray SH, Li P, et al. Current guidelines for chemoembolization for hepatocellular carcinoma: room for improvement? Hepatol Commun. 2017;1(4):338–46. https://doi.org/10.1002/hep4.1046.
2. Salem R, Gordon AC, Mouli S, et al. Y90 Radioembolization significantly prolongs time to progression compared with chemoembolization in patients with hepatocellular carcinoma. Gastroenterology. 2016;151(6):1155–1163.e2. https://doi.org/10.1053/j.gastro.2016.08.029.
3. Okamoto Y, Konno A, Togawa K, Kato T, Tamakawa Y, Amano Y. Arterial chemoembolization with cisplatin microcapsules. Br J Cancer. 1986,53(3):369–75. https://doi.org/10.1038/bjc.1986.61.
4. Yamada R, Sato M, Kawabata M, Nakatsuka H, Nakamura K, Takashima S. Hepatic artery embolization in 120 patients with unresectable hepatoma. Radiology. 1983;148(2):397–401. https://doi.org/10.1148/radiology.148.2.6306721.
5. Llovet JM, Bruix J. Systematic review of randomized trials for unresectable hepatocellular carcinoma: chemoembolization improves survival. Hepatology. 2003;37(2):429–42. https://doi.org/10.1053/jhep.2003.50047.
6. Garwood ER, Fidelman N, Hoch SE, Kerlan RK Jr, Yao FY. Morbidity and mortality following transarterial liver chemoembolization in patients with hepatocellular carcinoma and synthetic hepatic dysfunction. Liver Transpl. 2013;19(2):164–73. https://doi.org/10.1002/lt.23552.
7. Varela M, Real MI, Burrel M, et al. Chemoembolization of hepatocellular carcinoma with drug eluting beads: efficacy and doxorubicin pharmacokinetics. J Hepatol. 2007;46(3):474–81. https://doi.org/10.1016/j.jhep.2006.10.020.
8. Facciorusso A, Licinio R, Muscatiello N, Di Leo A, Barone M. Transarterial chemoembolization: evidences from the literature and applications in hepatocellular carcinoma patients. World J Hepatol. 2015;7(16):2009–19. https://doi.org/10.4254/wjh.v7.i16.2009.
9. Lammer J, Malagari K, Vogl T, et al. Prospective randomized study of doxorubicin-eluting-bead embolization in the treatment of hepatocellular carcinoma: results of the PRECISION V study.

Cardiovasc Intervent Radiol. 2010;33(1):41–52. https://doi.org/10.1007/s00270-009-9711-7.
10. Dhanasekaran R, Kooby DA, Staley CA, Kauh JS, Khanna V, Kim HS. Comparison of conventional transarterial chemoembolization (TACE) and chemoembolization with doxorubicin drug eluting beads (DEB) for unresectable hepatocelluar carcinoma (HCC). J Surg Oncol. 2010;101(6):476–80. https://doi.org/10.1002/jso.21522.
11. Golfieri R, Giampalma E, Renzulli M, et al. Randomised controlled trial of doxorubicin-eluting beads vs conventional chemoembolisation for hepatocellular carcinoma. Br J Cancer. 2014;111(2):255–64. https://doi.org/10.1038/bjc.2014.199.
12. Ariel IM. Treatment of inoperable primary pancreatic and liver cancer by the intra-arterial administration of radioactive isotopes (Y90 radiating microspheres). Ann Surg. 1965;162(2):267–78. https://doi.org/10.1097/00000658-196508000-00018.
13. Simon N, Warner RR, Baron MG, Rudavsky AZ. Intra-arterial irradiation of carcinoid tumors of the liver. Am J Roentgenol Radium Therapy, Nucl Med. 1968;102(3):552–61. https://doi.org/10.2214/ajr.102.3.552.
14. Grady ED. Internal radiation therapy of hepatic cancer. Dis Colon Rectum. 1979;22(6):371–5. https://doi.org/10.1007/BF02586901.
15. Andrews JC, Walker SC, Ackermann RJ, Cotton LA, Ensminger WD, Shapiro B. Hepatic radioembolization with yttrium-90 containing glass microspheres: preliminary results and clinical follow-up. J Nucl Med. 1994;35(10):1637–44.
16. Houle S, Yip TK, Shepherd FA, et al. Hepatocellular carcinoma: pilot trial of treatment with Y-90 microspheres. Radiology. 1989;172(3):857–60. https://doi.org/10.1148/radiology.172.3.2549567.
17. Dancey JE, Shepherd FA, Paul K, et al. Treatment of nonresectable hepatocellular carcinoma with intrahepatic 90Y-microspheres. J Nucl Med. 2000;41(10):1673–81.
18. Hilgard P, Hamami M, El Fouly A, et al. Radioembolization with yttrium-90 glass microspheres in hepatocellular carcinoma: European experience on safety and long-term survival. Hepatology. 2010;52(5):1741–9. https://doi.org/10.1002/hep.23944.
19. Braat MNGJA, van Erpecum KJ, Zonnenberg BA, van den Bosch MAJ, Lam MGEH. Radioembolization-induced liver disease: a systematic review. Eur J Gastroenterol Hepatol. 2017;29(2):144–52. https://doi.org/10.1097/MEG.0000000000000772.
20. Currie BM, Hoteit MA, Ben-Josef E, Nadolski GJ, Soulen MC. Radioembolization-induced chronic hepatotoxicity: a single-center cohort analysis. J Vasc Interv Radiol. 2019;30(12):1915–23. https://doi.org/10.1016/j.jvir.2019.06.003.
21. El Fouly A, Ertle J, El Dorry A, et al. In intermediate stage hepatocellular carcinoma: radioembolization with yttrium 90 or chemoembolization? Liver Int. 2015;35(2):627–35. https://doi.org/10.1111/liv.12637.
22. Chow PKH, Gandhi M, Tan SB, et al. SIRveNIB: selective internal radiation therapy versus sorafenib in Asia-Pacific patients with hepatocellular carcinoma. J Clin Oncol. 2018;36(19):1913–21. https://doi.org/10.1200/JCO.2017.76.0892.
23. Vilgrain V, Pereira H, Assenat E, et al. Efficacy and safety of selective internal radiotherapy with yttrium-90 resin microspheres compared with sorafenib in locally advanced and inoperable hepatocellular carcinoma (SARAH): an open-label randomised controlled phase 3 trial. Lancet Oncol. 2017;18(12):1624–36. https://doi.org/10.1016/S1470-2045(17)30683-6.
24. Salem R, Gabr A, Riaz A, et al. Institutional decision to adopt Y90 as primary treatment for hepatocellular carcinoma informed by a 1,000-patient 15-year experience. Hepatology. 2018;68(4):1429–40. https://doi.org/10.1002/hep.29691.
25. Llovet JM, Brú C, Bruix J. Prognosis of hepatocellular carcinoma: the BCLC staging classification. Semin Liver Dis. 1999;19(3):329–38. https://doi.org/10.1055/s-2007-1007122.
26. Llovet JM, Fuster J, Bruix J, Barcelona-Clínic Liver Cancer Group. The Barcelona approach: diagnosis, staging, and treatment of hepatocellular carcinoma. Liver Transpl. 2004;10(2 Suppl 1):S115–20. https://doi.org/10.1002/lt.20034.
27. Lo CM, Ngan H, Tso WK, et al. Randomized controlled trial of transarterial lipiodol chemoembolization for unresectable hepatocellular carcinoma. Hepatology. 2002;35(5):1164–71. https://doi.org/10.1053/jhep.2002.33156.
28. Llovet JM, Real MI, Montaña X, et al. Arterial embolisation or chemoembolisation versus symptomatic treatment in patients with unresectable hepatocellular carcinoma: a randomised controlled trial. Lancet. 2002;359(9319):1734–9. https://doi.org/10.1016/S0140-6736(02)08649-X.
29. Lencioni R, Petruzzi P, Crocetti L. Chemoembolization of hepatocellular carcinoma. Semin Intervent Radiol. 2013;30(1):3–11. https://doi.org/10.1055/s-0033-1333648.
30. Forner A, Reig M, Bruix J. Hepatocellular carcinoma. Lancet. 2018;391(10127):1301–14. https://doi.org/10.1016/S0140-6736(18)30010-2.
31. Kolligs FT, Bilbao JI, Jakobs T, et al. Pilot randomized trial of selective internal radiation therapy vs. chemoembolization in unresectable hepatocellular carcinoma. Liver Int. 2015;35(6):1715–21. https://doi.org/10.1111/liv.12750.
32. Pitton MB, Kloeckner R, Ruckes C, et al. Randomized comparison of selective internal radiotherapy (SIRT) versus drug-eluting bead transarterial chemoembolization (DEB-TACE) for the treatment of hepatocellular carcinoma. Cardiovasc Intervent Radiol. 2015;38(2):352–60. https://doi.org/10.1007/s00270-014-1012-0.
33. Casadei Gardini A, Tamburini E, Iñarrairaegui M, Frassineti GL, Sangro B. Radioembolization versus chemoembolization for unresectable hepatocellular carcinoma: a meta-analysis of randomized trials. Onco Targets Ther. 2018;11:7315–21. https://doi.org/10.2147/OTT.S175715.
34. Lobo L, Yakoub D, Picado O, et al. Unresectable hepatocellular carcinoma: radioembolization versus chemoembolization: a systematic review and meta-analysis. Cardiovasc Intervent Radiol. 2016;39(11):1580–8. https://doi.org/10.1007/s00270-016-1426-y.
35. Yang Y, Si T. Yttrium-90 transarterial radioembolization versus conventional transarterial chemoembolization for patients with hepatocellular carcinoma: a systematic review and meta-analysis. Cancer Biol Med. 2018;15(3):299–310. https://doi.org/10.20892/j.issn.2095-3941.2017.0177.
36. Mohamed M, Katz AW, Tejani MA, et al. Comparison of outcomes between SBRT, yttrium-90 radioembolization, transarterial chemoembolization, and radiofrequency ablation as bridge to transplant for hepatocellular carcinoma. Adv Radiat Oncol. 2016;1(1):35–42. https://doi.org/10.1016/j.adro.2015.12.003.
37. Garin E, Lenoir L, Edeline J, et al. Boosted selective internal radiation therapy with 90Y-loaded glass microspheres (B-SIRT) for hepatocellular carcinoma patients: a new personalized promising concept. Eur J Nucl Med Mol Imaging. 2013;40(7):1057–68. https://doi.org/10.1007/s00259-013-2395-x.
38. Zhang Y, Li Y, Ji H, Zhao X, Lu H. Transarterial Y90 radioembolization versus chemoembolization for patients with hepatocellular carcinoma: a meta-analysis. Biosci Trends. 2015;9(5):289–98. https://doi.org/10.5582/bst.2015.01089.
39. Salem R, Lewandowski RJ, Mulcahy MF, et al. Radioembolization for hepatocellular carcinoma using yttrium-90 microspheres: a comprehensive report of long-term outcomes. Gastroenterology. 2010;138(1):52–64. https://doi.org/10.1053/j.gastro.2009.09.006.
40. Salem R, Gilbertsen M, Butt Z, et al. Increased quality of life among hepatocellular carcinoma patients treated with radioembolization, compared with chemoembolization. Clin Gastroenterol Hepatol. 2013;11(10):1358–1365.e1. https://doi.org/10.1016/j.cgh.2013.04.028.

41. Ozkan ZG, Poyanli A, Ucar A, et al. Favorable survival time provided with radioembolization in hepatocellular carcinoma patients with and without portal vein thrombosis. Cancer Biother Radiopharm. 2015;30(3):132–8. https://doi.org/10.1089/cbr.2014.1748.
42. Salem R, Lewandowski R, Roberts C, et al. Use of yttrium-90 glass microspheres (TheraSphere) for the treatment of unresectable hepatocellular carcinoma in patients with portal vein thrombosis. J Vasc Interv Radiol. 2004;15(4):335–45. https://doi.org/10.1097/01.rvi.0000123319.20705.92.
43. Teo JY, Goh BKP. Contra-lateral liver lobe hypertrophy after unilobar Y90 radioembolization: an alternative to portal vein embolization? World J Gastroenterol. 2015;21(11):3170–3. https://doi.org/10.3748/wjg.v21.i11.3170.
44. Gabr A, Abouchaleh N, Ali R, et al. Outcomes of surgical resection after radioembolization for hepatocellular carcinoma. J Vasc Interv Radiol. 2018;29(11):1502–1510.e1. https://doi.org/10.1016/j.jvir.2018.06.027.
45. Lewandowski RJ, Kulik LM, Riaz A, et al. A comparative analysis of transarterial downstaging for hepatocellular carcinoma: chemoembolization versus radioembolization. Am J Transplant. 2009;9(8):1920–8. https://doi.org/10.1111/j.1600-6143.2009.02695.x.
46. Rostambeigi N, Dekarske AS, Austin EE, Golzarian J, Cressman EN. Cost effectiveness of radioembolization compared with conventional transarterial chemoembolization for treatment of hepatocellular carcinoma. J Vasc Interv Radiol. 2014;25(7):1075–84. https://doi.org/10.1016/j.jvir.2014.04.014.
47. Rognoni C, Ciani O, Sommariva S, Tarricone R. Cost-effectiveness analysis of treatments involving radioembolization in intermediate-stage hepatocellular carcinoma. J Comp Eff Res. 2018;7(3):209–21. https://doi.org/10.2217/cer-2017-0050.
48. Garin E, Rolland Y, Pracht M, et al. High impact of macroaggregated albumin-based tumour dose on response and overall survival in hepatocellular carcinoma patients treated with ^{90}Y-loaded glass microsphere radioembolization. Liver Int. 2017;37(1):101–10. https://doi.org/10.1111/liv.13220.
49. Sutphin PD, Lamus D, Kalva SP, Li J, Corbin IR, Hoshida Y. Interventional radiologic therapies for hepatocellular carcinoma: from where we began to where we are going. In: Hoshida Y, editor. Hepatocellular carcinoma: translational precision medicine approaches. Human Press; 2019. p. 169–94.
50. He CB, Lao XM, Lin XJ. Transarterial chemoembolization combined with recombinant human adenovirus type 5 H101 prolongs overall survival of patients with intermediate to advanced hepatocellular carcinoma: a prognostic nomogram study. Chin J Cancer. 2017;36(1):59. https://doi.org/10.1186/s40880-017-0227-2.
51. Shen A, Liu S, Yu W, Deng H, Li Q. p53 gene therapy-based transarterial chemoembolization for unresectable hepatocellular carcinoma: a prospective cohort study. J Gastroenterol Hepatol. 2015;30(11):1651–6. https://doi.org/10.1111/jgh.13009.

Should All Patients Listed for Liver Transplantation Receive Bridging Therapy Before Transplant?

21

Laura Kulik

Patients who can safely undergo locoregional therapy (LRT) should be treated in an attempt to decrease waitlist dropout and gain insight into tumor biology using both tumor and alpha fetoprotein (AFP) response.

Main Points

1. LRT is frequently employed for patients on the waitlist with the intent of controlling tumor progression beyond the Milan criteria (MC) and decreasing waitlist dropout.
2. While LRT is used in the vast majority of patients awaiting liver transplantation (LT), there are no randomized controlled trials (RCTs) that have addressed the efficacy of LRT vs no therapy.
3. Several important questions remain unanswered when considering LRT as a bridge to transplant, including whether one form of LRT is more advantageous than others and the effect of combining systemic therapy with LRT.
4. Given the allocation for LT in hepatocellular carcinoma (HCC), it is unlikely many patients will be observed without receiving bridging therapy because of concerns for tumor progression and waitlist dropout.
5. The type of LRT used is dependent on several variables, including tumor size/number/location, pretreatment bilirubin level, Child-Pugh score (CPS) status, and patient choice, which should be ultimately decided in the contact of a multidisciplinary tumor board to optimize post–LT outcomes.

LT remains the best therapeutic option for select patients with early HCC, serving the dual purpose of removing the tumor and the underlying cirrhotic liver. LRT is frequently employed for patients on the waitlist with the intent of controlling tumor progression beyond the MC and decreasing waitlist dropout, which is estimated to be as high as 24% over 3 years for patients with HCC [1]. LRTs include percutaneous ablation (microwave, radiofrequency, or cryoablation), intra-arterial embolization (chemoembolization or radioembolization), and external beam radiation. Limited data exist on systemic agents as bridging therapy; in the one RCT of adding bridging sorafenib to LRT, time to tumor progression (TTP) was similar compared to placebo [2]. LRT is recommended in patients anticipated to wait more than 6 months for LT [3]. However, as of October 2015, patients with HCC meeting the MC are mandated to wait 6 months before receiving an HCC Model for End-Stage Liver Disease (MELD) upgrade [4]. More recently, the MELD upgrade for HCC has been standardized, using the median MELD at the time of transplant in each donor service area minus 3 points, without further points granted every 3 months [4]. Such changes are anticipated to lead to longer waiting times for those with HCC. With this in mind, LRT may become an even more pivotal strategy in bridging patients with HCC to LT. A summary of arguments for and against providing LRT to all patients with HCC listed for LT is presented in Table 21.1.

L. Kulik (✉)
Medicine (Gastroenterology and Hepatology), Radiology, Surgery (Organ Transplantation), Northwestern University, Chicago, IL, USA
e-mail: Laura.Kulik@nm.org

N. S. Reau, A. G. Singal (eds.), *Controversies in Hepatocellular Carcinoma*, https://doi.org/10.1007/978-3-031-90225-3_21

Table 21.1 Summary of arguments for and against providing locoregional therapy to all patients with hepatocellular carcinoma listed for liver transplantation

For	Against
Delay tumor progression	Complications of LRT
Unmask tumor biology	T1 lesions
Downstaging	Inappropriate downstaging with higher risk of recurrence post LT

Abbreviations: *LRT* locoregional therapy, *LT* liver transplantation

Benefits of Locoregional Therapy

There are conflicting data regarding the effect of LRT on post-LT outcomes. While LRT is used in the vast majority of patients awaiting LT, there are no RCTs that have addressed the efficacy of LRT vs no therapy. Such a trial is unlikely to be performed because LRT has become the standard of care. Existing studies have been limited by the fact that groups not receiving LRT generally have shorter waiting times, higher biological MELD scores, and smaller tumors relative to contemporary groups receiving LRT. These factors are associated with improved survival and therefore confound studies to date.

Some studies have found a benefit to LRT. An extensive analysis of Scientific Registry of Transplant Recipients data from 1997 to 2006 found improved 3-year post-LT survival in patients receiving LRT compared to no therapy (76% vs 71%, respectively, $P = 0.03$) [5]. More recently, a study using the European Liver Transplant Registry ($N = 4978$) found that LRT before transplant was associated with superior 5-year overall survival (OS) (hazard ratio [HR] = 0.84; 95% CI, 0.73–0.96) and HCC-specific OS (HR = 0.76; 95% CI, 0.59–0.98) compared to no therapy. The effect of LRT was greater in those with vascular invasion (VI) (HR = 0.71; 95% CI, 0.55–0.92) compared to those without VI (HR = 0.88; 95% CI, 0.75–1.05). This speaks for LRT having the ability to favorably affect aggressive tumor phenotypes. Lastly, receipt of 1 or 2 sessions of LRT was associated with improved survival, whereas 3 or more sessions had no statistically significant effect. The response to LRT was not analyzed, as neither the degree of pathologic necrosis nor tumor grade was reported [6].

Another study using the United Network for Organ Sharing (UNOS) database between 2002 and 2013 ($N = 2794$) reported statistically significantly lower post-LT HCC recurrence in those receiving bridging LRT compared to those without LRT (8.6% vs 12.4%, respectively, $P = 0.02$) [7]. The benefit of LRT persisted in the multivariable analysis, which controlled for tumor factors (size, number), patient demographics, and waitlist time. In addition, having received LRT before LT was associated with improved median OS after the development of HCC recurrence post LT (75.9 months vs. 53.1 months; $P < 0.001$). It should be noted that only 24.6% of patients in the study received LRT. This study and other registry analyses were limited, however, by missing data regarding HCC recurrence, HCC-related mortality, and explant pathology, with less than one-quarter of patients having explant pathology available.

On the other hand, a number of studies have failed to find benefit to LRT. Several studies from the 1990s and early 2000s found no difference in OS from receipt of LRT [8–10]. A small study recently revealed no difference in OS or HCC recurrence with LRT [11]. A recent systematic review reported that while the use of LRT for T2 tumors was associated with a nonstatistically significant trend toward lower waitlist dropout from tumor progression, there was no statistically significant difference in post-LT survival or HCC recurrence between groups [12]. Overall, the data suggest at least a modest benefit to LRT, but further studies are needed.

The Importance of Complete Pathologic Response to Locoregional Therapy

Beyond decreasing waitlist dropout, tumor progression on LRT can be used to identify patients at higher risk of HCC recurrence post LT, another argument for treating all patients with LRT. Some studies suggest that LRT can invoke an immune response associated with a favorable prognosis [13–15]. In contrast, tumor progression despite LRT has been shown to portend poor outcomes including higher HCC recurrence post LT even compared to patients not receiving LRT. It is therefore not clear whether LRT response is a surrogate for tumor biology, with benefit limited to those with less aggressive tumors, or whether it confers a true advantage in preventing tumor growth and invasiveness.

Patients with MELD scores of less than 15 and a single lesion smaller than 3 cm who demonstrate complete radiographic response and AFP of less than 20 ng/mL after LRT have exceedingly low rates of waitlist dropout and inferior survival benefit from LT. Such patients should receive lower prioritization for LT, allowing more appropriate allocation of organs to those with greater potential to benefit [16, 17]. An example of this would be a patient with chronic hepatitis C infection and advanced fibrosis who achieves sustained virologic response. In the era of highly effective hepatitis C virus therapies, this may become a common clinical scenario.

In a study by the US Multicenter HCC Transplant Consortium, [18] 79.3% of patients ($n = 2854$) received LRT before undergoing LT including transarterial chemoembolization (TACE) (53.4%), ablation (12.9%), TACE plus ablation (8.3%), and 4.7% LRT other than TACE or ablation. The 20.5% of patients achieving a complete pathologic response (cPR, defined as no viable tumor on explant) had superior 5-year recurrence-free survival compared to those without LRT before LT ($n = 747$) and those with LRT who did not

Table 21.2 Factors associated with complete pathologic response to locoregional therapy [19, 20]

Factor	Odds ratio for cPR
Lower AFP level	1.6 per log decrease
Lower tumor number	–
Smaller tumor size	–
Age	1.01 per y decrease
Female sex	1.23
MELD score	1.06 per point decrease
Etiology of liver disease	0.76 (HCV) 0.33 (cryptogenic)
Geographic region	0.47 (shorter wait time)
≥ 3 LRT treatments	0.48 (TACE) 0.24 (ablation)
Disrupted tumor capsule	–

Abbreviations: *AFP* alpha fetoprotein, *cPR* complete pathologic response, *HCV* hepatitis C virus, *LRT* locoregional therapy, *MELD* model for end-stage liver disease, *TACE* transarterial chemoembolization

achieve cPR. In addition, recurrence was higher among those without cPR compared to no LRT in multivariate analysis. Further analysis of the patients that received LRT showed a statistically significantly lower HCC recurrence in cPR (5.8%) compared to non-cPR (16%) at 5 years post-LT [19]. Importantly, statistically significantly improved 5-year OS was mediated by lower HCC recurrence as the non-HCC related mortality was nearly equivalent between the cPR and non-cPR groups. Factors associated with cPR from this and other studies are presented in Table 21.2. A model was constructed using multivariate competing risk regression that allowed stratification of the non-cPR group into low, moderate, and high risks of post-LT HCC recurrence (5-year HCC recurrence 6.1%, 24%, and 69%; $P < 0.0001$). The variables that conferred statistically significant HCC recurrence risk included maximal tumor size (HR 9.22, per log cm increase) macro (HR 2.78)/micro (2.30) VI, poorly (HR 2.42)/moderately (HR 1.59) differentiated grade, pre-LT AFP (HR 1.45 per log unit increase), and neutrophil to lymphocyte ratio (HR 1.34 per log unit increase). A smaller study ($N = 274$) from European transplant centers confirmed that residual tumor viability on explant greater than or equal to 2 cm was a strong predictor of post-LT HCC recurrence [21].

Recently, UNOS added AFP level as a criterion to receive a MELD HCC exception [4]. This stresses that radiographic response alone is not adequate to gauge tumor biology and decisions of transplant candidacy. To this end, OPTN has incorporated AFP into criteria for HCC MELD upgrade irrespective of meeting the MC. If AFP is greater than 1000 ng/mL at listing, there must be a decline to less than 500 ng/mL post LRT to obtain a standard MELD upgrade. This criterion is supported by previous studies linking a pre-LT increase in AFP to post-LT mortality, independent of tumor burden [22]. While the overall goal of LRT should be to achieve cPR to reduce post-LT HCC recurrence, receipt of 3 or more LRTs has been associated with a statistically significantly higher rate of recurrence [18]. This speaks for the number of LRTs being a surrogate for aggressive tumor biology.

Does Locoregional Therapy Modality Matter?

If all patients should receive LRT, does the modality influence efficacy? Several studies have attempted to address this question, and based on available evidence, no modality has emerged superior. TACE is the most commonly used LRT before LT. Newer forms of therapy, including stereotactic body radiation therapy (SBRT), radioembolization (usually using yttrium-90 [Y90]), or RFA, have either been excluded from analyses or lumped together. There is a paucity of data directly comparing LRT modalities and cPR. Such comparisons should also account for pretreatment size of the targeted lesion.

There are several studies suggesting that RFA is more likely to lead to complete tumor necrosis compared to TACE [23]. In the study using the European Liver Transplant Registry ($N = 4978$), [6] RFA was associated with the highest 5-year OS (80.9%) compared to other modalities, an effect that persisted in a subanalysis of patients with more than 5 nodules larger than 3 cm. Moreover, in multivariate analysis, RFA had the strongest influence on 5-year HCC-specific mortality (HR = 0.43; 95% CI, 0.26–0.69).

Another small study examined 60 patients who received 79 LRT sessions as a bridge to LT: SBRT (n Z = 24), TACE (n Z = 37), RFA ($n = 9$), and Y90 ($n = 9$) [24]. cPR was demonstrated in 40% of patients. Y90 achieved the highest rate of cPR (75%) compared to SBRT (28.5%), TACE (41%), and RFA (60%). The median size of target tumors was 3.4 cm (Y90), 3.0 cm (SBRT), 2.6 cm (TACE), and 2.5 cm (RFA). A single-center study that randomly assigned 45 BCLC A/B patients to receive TACE or Y90 reported a statistically significantly longer TTP in the Y90 group (TACE: 6.8 months, Y90: > 26 months; HR = 0.122; $P = 0.007$) [25]. However, OS censored to LT was comparable between TACE (17.7 months) and Y90 (18.6 months). Larger RCTs are needed, but the feasibility of such trials is doubtful. Therefore, at this time, no particular LRT modality is recommended over another.

Counterpoint: Complications of Locoregional Therapy

An argument for *not* providing LRT to all patients is that LRT may cause adverse events. These include hepatic decompensation, tract seeding, embolization, and vascular and biliary complications. Overuse of LRT in all patients awaiting LT runs the risk of subsequent hepatic decompensation resulting in urgent LT or the inability to perform LT. Therefore, patient selection is paramount, as is weighing the competing risks of tumor progression and hepatic decompensation leading to waitlist dropout.

LRT may induce ischemia, which can adversely affect future therapies including LT. In one propensity-matched analysis of 57 pairs of patients, patients receiving pre-LT TACE had higher ischemia-related complications at the time of LT, including adhesions, cholecystitis, and bile duct necrosis. However, post-LT complications were similar between groups, suggesting that the benefits outweighed these complications [26]. And a recent meta-analysis found a higher occurrence of post-LT hepatic artery complications including thrombosis, stenosis, and pseudoaneurysms in patients receiving TACE before LT [27].

Hepatic decompensation is another risk of LRT that may precipitate acute liver failure that may require urgent LT or may lead to waitlist dropout because of critical illness. In one study of 251 patients who underwent TACE, reversible hepatotoxicity occurred in 31% of patients, and irreversible injury occurred in 15% of patients. Four percent of patients died within 6 weeks of TACE and 2% had to undergo urgent LT. In another study of 614 patients, 17% of patients with ascites developed liver failure after TACE. RFA, too, may precipitate liver failure; in one small single-center study (N = 350 RFA treatments in 255 patients), there were 13 (3.7%) cases of liver failure. Patient factors predicting hepatic decompensation with LRT are listed in Table 21.3. In addition, reactivation of hepatitis B has been described following TACE, pointing to the importance of viral suppression before delivery of therapy.

Tract, peritoneal, or abdominal wall seeding are additional risks of LRT, particularly with the percutaneous approach used in ablation procedures. Case reports have described recurrent HCC with metastases in the abdominal wall and greater omentum along a prior RFA tract. The risk of seeding after any percutaneous biopsy or treatment was estimated at 2% in one review article. A more recent study from Kaiser Permanente (N = 1015) found that 4 out of 220 patients (1.8%) undergoing ablation had seeding, all of whom had both ablation and biopsy. No patients having ablation alone without biopsy had seeding.

Table 21.3 Selected factors associated with hepatic decompensation after locoregional therapy

Factor	Type of LRT
Ascites	TACE
Child-Pugh class B or C	TACE, RFA
Tumor size	RFA
MELD ≥20	TACE
Encephalopathy	TACE

Abbreviations: *LRT* locoregional therapy, *MELD* model for end-state liver disease, *RFA* radiofrequency ablation, *TACE* transarterial chemoembolization

Counterpoint: T1 Lesions and Downstaging

Two situations in which LRT is *not* indicated for all patients, but still could benefit some patients, are T1 lesions and downstaging.

Downstaging, the use of LRT to move a patient outside the MC to within the MC, improves access to LT for patients not previously eligible. UNOS selection criteria for downstaging include diameter of less than 8 cm for a single lesion, a combined diameter of less than 8 cm for multiple lesions, and no evidence of metastatic disease. One small, single-center study of 108 patients who underwent LT after downstaging reported 1- and 5-year survival rates of 95% and 80%, respectively. However, patients with AFP greater than 1000 ng/mL and CPS B or C cirrhosis all failed downstaging. Another study found that patients meeting UNOS downstaging criteria (N = 422) had similar post-LT survival compared to patients originally within the MC (N = 3276), although recurrence risk was higher in the downstaging group. Patients outside UNOS guidelines (N = 121) who were nonetheless downstaged had worse post-LT survival and higher rates of HCC recurrence. This suggests that not all patients outside the MC should undergo downstaging; those with better compensated liver disease and lower AFP are more likely to benefit. In addition, a structured protocol for downstaging is recommended. However, this area warrants further study.

The use of LRT for T1 HCC lesions remains controversial. Patients with T1 lesions who are not already listed for LT should be treated with resection or ablation. In those *already* listed for LT, American Association for the Study of Liver Diseases guidelines recommend observation over routine LRT; however, this recommendation is conditional and based on very low-quality evidence. A single-center study of patients with T1 lesions who did not undergo LRT (N = 114) found that the median time from progression to T2 was 6.9 months. In another small, single-center retrospective study (N = 87), patients with T1 tumors who received immediate LRT vs observation until tumor progression to T2 were compared. MELD score, tumor size, and AFP levels were similar between groups. There was no difference in progression outside the MC (19% for LRT vs 23% for observation, P = 0.49) or OS (55 months for LRT vs 36 months for observation, P = 0.22). Larger studies should be performed to further assess whether patients with T1 tumors could benefit from LRT. This potentially varies by geographic region and wait time to LT.

Conclusion

The interaction of LRT on post-LT outcomes is complex. The majority of studies report differences in baseline characteristics between those treated with LRT and those not, specifically shorter waiting times, smaller tumors, and higher biological MELD scores in the latter. Given recent changes to HCC exception points, future studies may be better equipped to address the true benefit of LRT during a period of "observation" that may select the most aggressive tumors before LT. On the other hand, the number of patients without LRT before LT will likely continue to decline. There are still several questions that remain unanswered: Does LRT lead to an immune response that augments tumor necrosis, prevents tumor growth, and metastasis? Is one form of LRT more immunogenic than another? Will combining systemic therapies with LRT lead to a more robust response? If so, is there a safe combination and time frame to withhold systemic agents that will need to elapse before transplant to prevent graft loss?.

In the meantime, patients who can safely undergo LRT should be treated, in an attempt to decrease waitlist dropout and gain insight into tumor biology using both tumor and AFP response. The type of LRT used depends on several variables, including tumor size/number/location, pretreatment bilirubin level, CPS status, and patient choice, which should be ultimately decided in the contact of a multidisciplinary tumor board to optimize post-LT outcomes.

References

1. Agopian VG, Harlander-Locke MP, Ruiz RM, et al. Impact of pretransplant bridging locoregional therapy for patients with hepatocellular carcinoma within Milan criteria undergoing liver transplantation: analysis of 3601 patients from the US Multicenter HCC Transplant Consortium. Ann Surg. 2017;266(3):525–35. https://doi.org/10.1097/SLA.0000000000002381.
2. DiNorcia J, Florman SS, Haydel B, et al. Pathologic response to pretransplant locoregional therapy is predictive of patient outcome after liver transplantation for hepatocellular carcinoma: analysis from the US Multicenter HCC Transplant Consortium. Ann Surg. 2020;271(4):616–24. https://doi.org/10.1097/SLA.0000000000003253.
3. Yang K, Sung PS, You YK, et al. Pathologic complete response to chemoembolization improves survival outcomes after curative surgery for hepatocellular carcinoma: predictive factors of response. HPB (Oxford). 2019;21(12):1718–26. https://doi.org/10.1016/j.hpb.2019.04.017.
4. Manzia TM, Lai Q, Iesari S, et al. Impact of remnant vital tissue after locoregional treatment and liver transplant in hepatocellular cancer patients, a multicentre cohort study. Transpl Int. 2018;31(9):988–98. https://doi.org/10.1111/tri.13153.
5. Berry K, Ioannou GN. Serum alpha-fetoprotein level independently predicts posttransplant survival in patients with hepatocellular carcinoma. Liver Transpl. 2013;19(6):634–45. https://doi.org/10.1002/lt.23652.
6. Cescon M, Cucchetti A, Ravaioli M, Pinna AD. Hepatocellular carcinoma locoregional therapies for patients in the waiting list. Impact on transplantability and recurrence rate. J Hepatol. 2013;58(3):609–18. https://doi.org/10.1016/j.jhep.2012.09.021.
7. Mohamed M, Katz AW, Tejani MA, et al. Comparison of outcomes between SBRT, yttrium-90 radioembolization, transarterial chemoembolization, and radiofrequency ablation as bridge to transplant for hepatocellular carcinoma. Adv Radiat Oncol. 2016;1(1):35–42. https://doi.org/10.1016/j.adro.2015.12.003.
8. Salem R, Gordon AC, Mouli S, et al. Y90 Radioembolization significantly prolongs time to progression compared with chemoembolization in patients with hepatocellular carcinoma. Gastroenterology. 2016;151(6):1155–63. https://doi.org/10.1053/j.gastro.2016.08.029.
9. da Silva Boteon APC, Longatto Boteon Y, Vinuela EF, et al. The impact of transarterial chemoembolization induced complications on outcomes after liver transplantation: a propensity-matched study. Clin Transpl. 2018;32(5):e13255. https://doi.org/10.1111/ctr.13255.
10. Sneiders D, Houwen T, Pengel LHM, Polak WG, Dor FJMF, Hartog H. Systematic review and meta-analysis of posttransplant hepatic artery and biliary complications in patients treated with transarterial chemoembolization before liver transplantation. Transplantation. 2018;102(1):88–96. https://doi.org/10.1097/TP.0000000000001936.
11. Garwood ER, Fidelman N, Hoch SE, Kerlan RK Jr, Yao FY. Morbidity and mortality following transarterial liver chemoembolization in patients with hepatocellular carcinoma and synthetic hepatic dysfunction. Liver Transpl. 2013;19(2):164–73. https://doi.org/10.1002/lt.23552.
12. Hsin IF, Hsu CY, Huang HC, et al. Liver failure after transarterial chemoembolization for patients with hepatocellular carcinoma and ascites: incidence, risk factors, and prognostic prediction. J Clin Gastroenterol. 2011;45(6):556–62. https://doi.org/10.1097/MCG.0b013e318210ff17.
13. Kong WT, Zhang WW, Qiu YD, et al. Major complications after radiofrequency ablation for liver tumors: analysis of 255 patients. World J Gastroenterol. 2009;15(21):2651–6. https://doi.org/10.3748/wjg.15.2651.
14. Chen TM, Huang PT, Lin LF, Tung JN. Major complications of ultrasound-guided percutaneous radiofrequency ablations for liver malignancies: single center experience. J Gastroenterol Hepatol. 2008;23(8 Pt 2):e445–50. https://doi.org/10.1111/j.1440-1746.2007.05078.x.
15. Jang JW, Kim YW, Lee SW, et al. Reactivation of hepatitis B virus in HBsAg-negative patients with hepatocellular carcinoma. PLoS One. 2015;10(3):e0122041. https://doi.org/10.1371/journal.pone.0122041.
16. Nakamura M, Hayami S, Ueno M, et al. Detection of needle tract implantation and peritoneal seeding after radiofrequency ablation using intraoperative near-infrared fluorescence system for recurrent hepatocellular carcinoma: a case report. Surg Case Rep. 2018;4(1):76. https://doi.org/10.1186/s40792-018-0485-5.
17. Stigliano R, Marelli L, Yu D, Davies N, Patch D, Burroughs AK. Seeding following percutaneous diagnostic and therapeutic approaches for hepatocellular carcinoma. What is the risk and the outcome? Seeding risk for percutaneous approach of HCC. Cancer Treat Rev. 2007;33(5):437–47. https://doi.org/10.1016/j.ctrv.2007.04.001.

18. Szpakowski JL, Drasin TE, Lyon LL. Rate of seeding with biopsies and ablations of hepatocellular carcinoma: a retrospective cohort study. Hepatol Commun. 2017;1(9):841–51. https://doi.org/10.1002/hep4.1089.
19. Mehta N, Guy J, Frenette CT, et al. Excellent outcomes of liver transplantation following down-staging of hepatocellular carcinoma to within Milan criteria: a multicenter study. Clin Gastroenterol Hepatol. 2018;16(6):955–64. https://doi.org/10.1016/j.cgh.2017.11.037.
20. Mehta N, Dodge JL, Grab JD, Yao FY. National experience on down-staging of hepatocellular carcinoma before liver transplant: influence of tumor burden, alpha-fetoprotein, and wait time. Hepatology. 2020;71(3):943–54. https://doi.org/10.1002/hep.30879.
21. Yao FY, Fidelman N. Reassessing the boundaries of liver transplantation for hepatocellular carcinoma: where do we stand with tumor down-staging? Hepatology. 2016;63(3):1014–25. https://doi.org/10.1002/hep.28139.
22. Marrero JA, Kulik LM, Sirlin CB, et al. Diagnosis, staging, and management of hepatocellular carcinoma: 2018 practice guidance by the American Association for the study of liver diseases. Hepatology. 2018;68(2):723–50. https://doi.org/10.1002/hep.29913.
23. Mehta N, Sarkar M, Dodge JL, Fidelman N, Roberts JP, Yao FY. Intention to treat outcome of T1 hepatocellular carcinoma with the "wait and not ablate" approach until meeting T2 criteria for liver transplant listing. Liver Transpl. 2016;22(2):178–87. https://doi.org/10.1002/lt.24360.
24. Tholey DM, Hornung B, Enestvedt CK, et al. Close observation versus upfront treatment in hepatocellular carcinoma: are the exception points worth the risk? BMJ Open Gastroenterol. 2017;4(1):e000157. https://doi.org/10.1136/bmjgast-2017-000157.
25. Nordness MF, Hamel S, Godfrey CM, et al. Fatal hepatic necrosis after nivolumab as a bridge to liver transplant for HCC: are checkpoint inhibitors safe for the pretransplant patient? Am J Transplant. 2020;20(3):879–83. https://doi.org/10.1111/ajt.15617.
26. Chang TT, Sawhney R, Monto A, et al. Implementation of a multidisciplinary treatment team for hepatocellular cancer at a Veterans Affairs Medical Center improves survival. HPB (Oxford). 2008;10(6):405–11. https://doi.org/10.1080/13651820802356572.
27. Yopp AC, Mansour JC, Beg MS, et al. Establishment of a multidisciplinary hepatocellular carcinoma clinic is associated with improved clinical outcome. Ann Surg Oncol. 2014;21(4):1287–95. https://doi.org/10.1245/s10434-013-3413-8.

Should Transplant Criteria Be Broadened Beyond United Network for Organ Sharing (UNOS) Downstaging Protocol to Include Larger Tumors?

22

Allison J. Kwong and Neil Mehta

Selected patients with initial tumors beyond United Network for Organ Sharing (UNOS) downstaging (DS) criteria may benefit from liver transplantation (LT), and current criteria for LT priority should be broadened to include larger tumors. We propose that patients initially outside the Milan criteria (MC) should demonstrate not only response to locoregional therapy (LRT) by imaging but also alpha fetoprotein (AFP) less than 100 ng/mL to be eligible for LT.

Main Points

1. Liver transplantation may be a viable treatment option for those with larger tumors beyond UNOS-DS criteria who are successfully downstaged to MC.
2. Comparable outcomes after LT can be achieved in patients outside MC (1 lesion < 5 cm or 2–3 lesions each < 3 cm) after successful downstaging as those always within the MC.
3. HCC recurrence is the major cause of post-LT mortality in those transplanted for HCC and typically presents more than 1 year post-transplant.
4. Patients with initially more advanced HCC will be at higher risk of waitlist dropout as a result of tumor progression and inability to downstage to within transplantable criteria.
5. With careful selection practices, certain patients with larger tumors can do well with LT and should not be excluded a priori based on tumor size or number alone.

A. J. Kwong (✉)
Stanford Transplant Outcomes Research Center, Stanford University, Stanford, CA, USA
e-mail: ajk@stanford.edu

N. Mehta
University of California, San Francisco, CA, USA
e-mail: Neil.Mehta@ucsf.edu

Liver transplantation (LT) is now an established treatment option for early-stage hepatocellular carcinoma (HCC) within the MC, defined as a single lesion smaller than 5 cm or 2 to 3 lesions each smaller than 3 cm [1]. In the United States, patients meeting these criteria receive LT priority through Model for End-Stage Liver Disease exception points, which approximate the urgency for LT compared to the waitlist population. In 2018, a total of 1571, or 20.4%, of LTs in the United States were performed for patients with HCC exception points (2018 ADR). LT is generally accepted as a treatment option for HCC within conventional MC; however, only a minority of patients with HCC qualify for these fairly restrictive LT criteria. Over the past 2 decades, there has been interest in expanding criteria to be able to offer this treatment option to a wider population of patients with HCC.

Relevant Outcomes

Patients with HCC generally exceed the standard for 1-year post-LT graft survival, as they have few other comorbidities at the time of LT. HCC recurrence, the major cause of post-LT mortality, typically presents after the 1-year mark. Consequently, longer-term outcomes including 3-year and 5-year survival may be more relevant metrics of success. In areas where patients with HCC draw from the deceased donor pool, it is generally accepted that survival outcomes should be comparable to those without HCC [2]. Most recently, 3- and 5-year survival was 81.7% and 76.6% for HCC patients, compared to 81.4% and 76.4% for non-HCC patients (2018 ADR). HCC recurrence is also of particular interest, considering the curative intent of LT for this indication.

Patients presenting with more advanced HCC may be less likely to eventually reach or qualify for LT, considering the underlying tumor biology and need for bridging LRT. Waitlist dropout, defined as the risk of death or disease progression

N. S. Reau, A. G. Singal (eds.), *Controversies in Hepatocellular Carcinoma*, https://doi.org/10.1007/978-3-031-90225-3_22

while awaiting LT, helps to guide patient and provider expectations and estimate the likelihood of LT. This metric is influenced not only by tumor burden but also AFP levels, waiting times, severity of liver disease, and the response to LRT. Intention-to-treat survival follows patients from the time of initial presentation, captures both waitlist and post-LT events, and may offer a more representative estimate of potential outcomes compared to waitlist or post-LT survival alone.

Beyond Milan

Over the past 2 decades, more liberal selection criteria have been proposed, expanding on the MC and exploring different cutoffs for combined tumor number and size. These include the University of California, San Francisco (UCSF) criteria (1 lesion up to 6.5 cm, or 2–3 lesion each < 4.5 cm, and total tumor diameter < 8 cm); "up-to-7" criteria (sum of the size of largest tumor [cm] and number of tumors not exceeding 7); Valencia criteria (3 tumors each < 5 cm, total tumor diameter ≤ 10 cm), and the extended Toronto criteria (no upper limit on size or number; excluding patients with cancer-related symptoms, extrahepatic disease, vascular invasion, or poorly differentiated tumors). Overall, outcomes between patients within MC and UCSF criteria were similar; however, inferior outcomes—in one study, less than 50% 5-year survival at 5 years—have been observed among patients exceeding UCSF criteria [3]. Based on these results and because of ongoing organ shortages in many parts of the world including the United States, further expansion of LT criteria based on tumor number and size has not been endorsed.

Downstaging

Downstaging is a pathway by which patients with HCC beyond MC may be eligible for LT, but only once the tumor burden is reduced to within MC with LRT. The downstaging approach allows for observation of tumor response and tumor behavior over time, which may in itself serve as a selection tool for patients who will do well after LT [4]. This strategy has been tested in patients initially exceeding the MC with varying cutoffs for tumor number in size, including the UCSF downstaging criteria (1 lesion up to 8 cm, 2–3 lesions each < 5 cm, and total tumor diameter < 8 cm), and criteria proposed by Ravaioli et al. (1 tumor < 6 cm, 2 tumors each < 5 cm, and 3–5 tumors each < 5 cm with total tumor diameter < 12 cm), among others [5, 6].

Outcomes for LT recipients successfully downstaged using LRT have not been statistically significantly different from those always within the MC, in terms of post-LT survival and HCC recurrence. In a systematic review of 950 patients across 13 eligible studies, not limited to tumor size or number, downstaging had a pooled success rate of 0.48 (95% CI, 0.39–0.58) and a pooled post-LT HCC recurrence rate of 0.16 (95% CI, 0.11–0.23) [7]. Explant histologic characteristics among successfully downstaged patients (UNOS-DS) are generally favorable, with low rates of poorly differentiated grade (0–8.1%) and microvascular invasion (1.6–15.4%) [5, 8, 9]. Conversely, inferior outcomes are observed among those who experience tumor progression despite LRT [4].

In 2017, the United Network for Organ Sharing implemented a national downstaging protocol (UNOS-DS), by which patients beyond the MC can be eligible for priority LT listing via HCC exception points (Table 22.1). The initial selection criteria are based on the downstaging criteria proposed by UCSF: a single lesion less than or equal to 8 cm, or 2 or 3 lesions smaller than 5 cm, or 4 or 5 nodules all smaller than 3 cm, with a total tumor diameter of less than 8 cm [5]. There is a minimum observation period of 3 months of disease stability after successful downstaging to within the MC, and macrovascular tumor invasion and extrahepatic disease are contraindications to LT. In this experience, post-LT survival outcomes have been comparable to those within conventional MC: Three-year post-LT survival was 79.1% for UNOS-DS, compared to 83.2% for the MC ($P = 0.17$) [9]. Three-year HCC recurrence rates, however, were higher in the UNOS-DS groups (12.8%), compared to LT recipients within the MC (6.9%).

Notably, only 42–69% of patients in early downstaging protocols were successfully downstaged to within the MC—meaning that a substantial proportion do not progress to LT candidacy, whether because of tumor progression, inadequate response to LRT, or decompensation after LRT [10]. Previous studies have identified AFP of greater than 1000 ng/mL and Child-Pugh score class B or C (vs A) as statistically significant predictors of unsuccessful downstaging [5, 8]. As LRT in this population is not without risk, early identification of those who are unlikely to be successfully downstaged may avoid unnecessary LRT and unrealistic expectations, as well as facilitate timely oncologic care and consideration of systemic therapy.

Beyond United Network for Organ Sharing Downstaging Protocol

LT may be a viable treatment option even for those with initial tumors beyond UNOS-DS. Several single-center studies have explored downstaging for LT in patients exceeding both conventional and expanded UNOS-DS, with no upper limit on tumor number, size, or total HCC tumor diameters. In this population, downstaging to the MC has been successful in

Table 22.1 United network for organ sharing downstaging protocols in patients with hepatocellular carcinoma presenting beyond the Milan criteria

UNOS-DS protocol	All comers protocol
Inclusion criteria	
1. HCC exceeding UNOS T2 criteria but meeting one of the following: (a) Single lesion ≤ 8 cm (b) 2 or 3 lesions each ≤ 5 cm with sum of maximal tumor diameters ≤ 8 cm (c) 4 or 5 lesions each ≤ 3 cm with sum of maximal tumor diameters ≤ 8 cm 2. Absence of vascular invasion or extrahepatic disease based on cross-sectional imaging	1. HCC exceeding UNOS-DS by any of the following: (a) HCC tumor number (b) HCC tumor size (c) Total HCC tumor diameter 2. Absence of vascular invasion or extrahepatic disease based on cross-sectional imaging
Criteria for successful downstaging	
Residual tumor(s) within the MC	
Criteria for downstaging failure and exclusion from LT	
1. Progression of tumor(s) beyond inclusion criteria for downstaging based on tumor size and number 2. Any evidence of extrahepatic, lymphatic, or vascular tumor spread	1. Progression of tumor burden beyond MC after initial successful downstaging 2. Development of any new HCC lesion(s) (not including residual/recurrent disease at previous LRT site) 3. Any evidence of extrahepatic, lymphatic, or vascular tumor spread
Additional guidelines	
Minimum observation period of 3 mo between downstaging and LT	Minimum observation period of 6 mo between downstaging and LT

Abbreviations: *DS* downstaging, *HCC* hepatocellular carcinoma, *LRT* locoregional therapy, *LT* liver transplant, *MC* Milan criteria, *UNOS* United network for organ sharing

24–65% of patients, albeit with mixed results regarding post-LT outcomes when compared to patients with HCC within the MC [11–14]. Barakat et al. [13] identified noninfiltrating tumor as a predictor of successful downstaging in this population, whereas size, number of tumors, and AFP were not statistically significant. In the UCSF cohort, the sum of the number of lesions and the largest lesion diameter were the only significant predictors of successful downstaging [14]. Notably, although downstaging was successful in the majority of patients in this cohort, only 13.5% eventually underwent LT, reflecting the shortage of available donor organs and long waitlist times in the region.

In the US experience, there were 121 LTs performed between 2012 and 2015 for patients initially exceeding the MC and UNOS-DS criteria. Guidelines for downstaging and listing these patients via the "All comers" (AC) protocol are shown in Table 22.1. Three-year post-LT survival rates of 71.4% were observed, compared to 83.2% for those always within the MC ($P = 0.04$) [9]. These survival rates are inferior to those within the MC yet still within what is considered an acceptable outcome for deceased-donor LT. HCC recurrence rates at 3 years were higher in the AC group (16.7%) compared to the UNOS-DS (12.8%) and Milan (6.9%) groups.

Beyond Tumor Size and Number

Surrogate markers of tumor biology have been identified to predict outcomes after LT for HCC, independent of tumor size and number. Higher AFP levels have been consistently identified as a negative predictor of post-LT outcome, at all stages of HCC [9, 15–17]. In the downstaging population in particular, AFP greater than 100 ng/mL was identified as a particular subgroup with a higher risk of HCC recurrence and death, with 3-year survival of 59.5%, compared with 80.6% among patients with AFP less than 20 ng/mL, and 72.1% among patients with AFP of 20–99 ng/mL [9]. Other serum biomarkers such as the neutrophil-to-lymphocyte ratio, AFP slope, AFP-L3, and des-gamma carboxyprothrombin (DCP), as well as radiographic features including ^{18}F-fludeoxyglucose (FDG) avidity on positron emission tomography, have also been shown to predict HCC recurrence and post-LT mortality [15, 18–21]. In addition, tumor biology can be inferred from dynamic tumor behavior, such as response to LRT or tumor progression while awaiting LT [22]. The observation that patients with HCC outside the MC who were successfully downstaged achieved similar outcomes after a period of waiting time inspired the "ablate and wait" strategy, which requires a period of observation before LT during which the tumor biology can be clarified [23].

It is increasingly clear that tumor size and number alone are insufficient markers of tumor aggressiveness and recurrence. If criteria for initial tumor size and number are liberalized, we will need to rely on these surrogate markers to assess tumor biology and define the selection criteria for LT. Combining biomarkers with tumor number and size may help to more accurately risk-stratify this population. In the Metroticket 2.0 analysis, the addition of AFP to the number and size of tumors outperformed the MC, UCSF, and up-to-7 criteria in the prediction of 5-year post-LT HCC-specific survival [24]. Similarly, the Kyoto criteria for living-donor LT (tumor number ≤ 10, total tumor diameter ≤ 5 cm, serum DCP ≤ 400 mAU/mL) incorporates DCP to represent tumor biology and has been reported to result in excellent post-LT outcomes: a 5-year survival rate of 82% and HCC recurrence rate of 7%.

Liver allocation policy in the United States now considers some of the aforementioned variables, implementing a 6-month observation time before LT, an AFP cutoff of 500 ng/mL after LRT in patients with AFP greater than 1000 ng/mL, and a formalized downstaging protocol to ensure response to LRT. Further study is needed to identify appropriate biomarker thresholds, the utility of specific radiographic features such as FDG avidity on positron emission tomography, and the optimal observation time before LT to identify those who would benefit from LT.

These variables may be particularly useful to help risk stratify LT candidates with an initial tumor burden beyond the MC. This higher-risk group may need a stricter AFP cutoff to select those who will benefit most from LT. We propose that patients undergoing downstaging should demonstrate response to LRT—by both imaging and AFP of less than 100 ng/mL—to be eligible for LT.

Survival Benefit of Liver Transplantation

In the United States, all patients with HCC are assigned the same amount of exception points and thus similar allocation priority and probability of undergoing LT, regardless of tumor characteristics. The diverse risks of waitlist dropout and post-LT HCC recurrence are currently not well represented by the US organ allocation system.

The concept of survival benefit in LT considers the expected survival without LT against expected survival after LT, balancing the two-fold risks of waitlist mortality and LT. The HCC population is heterogeneous, and so the survival benefit of LT can be quite variable. In an intention-to-treat analysis, Lai et al. [25] demonstrate the largest LT survival benefit for patients with HCC outside the MC, with low AFP and incomplete response or inability to undergo LRT. In contrast, patients with low LT survival benefit are of 2 categories: poor radiological response and high AFP, with too advanced or aggressive malignancy to benefit from LT, and complete radiologic response and low Model for End-Stage Liver Disease score with well-treated tumor and thus no clear benefit from LT.

Patients presenting with HCC beyond the MC and UNOS-DS may have higher-risk disease or may be disadvantaged by a delayed diagnosis. A proportion of these patients may still benefit greatly from LT, whereby a higher urgency for LT may mitigate, or justify, the slightly lower yet still excellent post-LT outcomes. For example, a patient with HCC exceeding UNOS-DS criteria with a 9 cm tumor, who is able achieve reduction in viable tumor to 3 cm and normal AFP after radioembolization, has much to gain from LT and should undergo LT before the tumor progresses beyond transplantable criteria. By contrast, a patient with a single, well-treated 3-cm tumor without residual or recurrent tumor may be at lower risk for waitlist dropout with no imminent need for LT and will derive a lower survival benefit from undergoing LT. Mazzaferro [26] has proposed an adaptive approach accounting for response to LRT, in which higher waitlist priority is granted to patients with larger tumors that are downstaged, as well as those with more decompensated liver disease who may be unable to tolerate bridging LRT. Radiologic response to the MC remains an essential criterion for the downstaging pathway, selecting for patients with more favorable tumor biology and response to LRT and selecting out those with an unacceptably high risk of HCC recurrence and mortality after LT.

Organ Utility

Organ allocation systems seek to make the best use of donated organs, currently a limited resource in many countries. In the context of organ shortage, expanding access to LT for HCC may have collateral effects on the LT waitlist. Already, in the United States, patients with HCC make up 20% of the waitlist and benefit from a higher transplant rate compared to non-HCC patients. This gap has narrowed as incremental policy changes over the years have adjusted the prioritization for HCC through the amount of exception points granted [27].

Under the guiding principles of equity and justice, implementation of more liberal HCC criteria should aim for a survival threshold at which the benefit of transplanting patients with HCC outweighs the harm for not transplanting patients without HCC. A previous Markov model addressing this question suggests that a 5-year survival rate of at least 61% among patients beyond the MC must be achieved to outweigh the harm to others awaiting LT and loss of utility, in terms of quality-adjusted life years [28]. Current LT survival outcomes—for HCC within the MC, UNOS-DC, and AC—exceed proposed utility thresholds and standard metrics for post-LT survival. These outcomes should only improve as we refine our selection criteria, develop more accurate biomarkers, and improve the effectiveness of bridging LRT.

Considering the escalating risk of dropout with longer waitlist times, and thus greater urgency for LT, patients with advanced HCC may be appropriate for living-donor LT. Selection criteria and acceptable survival thresholds may be different for living donation with regard to acceptable post-LT survival and outcome, although this would have to be balanced against the added risk to a potential donor.

Conclusion

LT may be a viable treatment option for those with larger tumors beyond UNOS-DS criteria who are successfully downstaged to the MC. By nature, patients with initially more advanced HCC will be a higher risk of waitlist dropout as a result of tumor progression and inability to downstage to within transplantable criteria. Considering that a minority of patients eventually undergo LT and the low intent-to-treat survival, management of expectations at the time of initial evaluation before attempted downstaging is essential. However, with careful selection practices, it is clear that certain patients with larger tumors can do well with LT and should not be excluded a priori based on tumor size or number alone.

The authors recommend selected patients with initial tumors beyond UNOS downstaging criteria may benefit from LT, and current criteria for LT priority should be broadened to include larger tumors. We propose that patients initially outside the MC should demonstrate not only response to LRT by imaging but also AFP of less than 100 ng/mL to be eligible for LT.

References

1. Mazzaferro V, Regalia E, Doci R, et al. Liver transplantation for the treatment of small hepatocellular carcinomas in patients with cirrhosis. N Engl J Med. 1996;334(11):693–9. https://doi.org/10.1056/NEJM199603143341104.
2. Clavien PA, Lesurtel M, Bossuyt PMM, Gores GJ, Langer B, Perrier A. Recommendations for liver transplantation for hepatocellular carcinoma: an international consensus conference report. Lancet Oncol. 2012;13(1):e11–22. https://doi.org/10.1016/S1470-2045(11)70175-9.
3. Duffy JP, Vardanian A, Benjamin E, et al. Liver transplantation criteria for hepatocellular carcinoma should be expanded: a 22-year experience with 467 patients at UCLA. Ann Surg. 2007;246(3):502–9. https://doi.org/10.1097/SLA.0b013e318148c704.
4. Otto G, Herber S, Heise M, et al. Response to transarterial chemoembolization as a biological selection criterion for liver transplantation in hepatocellular carcinoma. Liver Transpl. 2006;12(8):1260–7. https://doi.org/10.1002/lt.20837.
5. Yao FY, Kerlan RK Jr, Hirose R, et al. Excellent outcome following down-staging of hepatocellular carcinoma prior to liver transplantation: an intention-to-treat analysis. Hepatology. 2008;48(3):819–27. https://doi.org/10.1002/hep.22412.
6. Ravaioli M, Grazi GL, Piscaglia F, et al. Liver transplantation for hepatocellular carcinoma: results of down-staging in patients initially outside the Milan selection criteria. Am J Transplant. 2008;8(12):2547–57. https://doi.org/10.1111/j.1600-6143.2008.02409.x.
7. Parikh ND, Waljee AK, Singal AG. Downstaging hepatocellular carcinoma: a systematic review and pooled analysis. Liver Transpl. 2015;21(9):1142–52. https://doi.org/10.1002/lt.24169.
8. Mehta N, Guy J, Frenette CT, et al. Excellent outcomes of liver transplantation following down-staging of hepatocellular carcinoma to within Milan criteria: a multicenter study. Clin Gastroenterol Hepatol. 2018;16(6):955–64. https://doi.org/10.1016/j.cgh.2017.11.037.
9. Mehta N, Dodge JL, Grab JD, Yao FY. National experience on down-staging of hepatocellular carcinoma before liver transplant: influence of tumor burden, alpha-fetoprotein, and wait time. Hepatology. 2020;71(3):943–54. https://doi.org/10.1002/hep.30879.
10. Yao FY, Fidelman N. Reassessing the boundaries of liver transplantation for hepatocellular carcinoma: where do we stand with tumor down-staging? Hepatology. 2016;63(3):1014–25. https://doi.org/10.1002/hep.28139.
11. Chapman WC, Garcia-Aroz S, Vachharajani N, et al. Liver transplantation for advanced hepatocellular carcinoma after downstaging without up-front stage restrictions. J Am Coll Surg. 2017;224(4):610–21. https://doi.org/10.1016/j.jamcollsurg.2016.12.020.
12. Lewandowski RJ, Kulik LM, Riaz A, et al. A comparative analysis of transarterial downstaging for hepatocellular carcinoma: chemoembolization versus radioembolization. Am J Transplant. 2009;9(8):1920–8. https://doi.org/10.1111/j.1600-6143.2009.02695.x.
13. Barakat O, Wood RP, Ozaki CF, et al. Morphological features of advanced hepatocellular carcinoma as a predictor of downstaging and liver transplantation: an intention-to-treat analysis. Liver Transpl. 2010;16(3):289–99. https://doi.org/10.1002/lt.21994.
14. Sinha J, Mehta N, Dodge JL, Poltavskiy E, Roberts J, Yao F. Are there upper limits in tumor burden for down-staging of hepatocellular carcinoma to liver transplant? Analysis of the all-comers protocol. Hepatology. 2019;70(4):1185–96. https://doi.org/10.1002/hep.30570.
15. Mehta N, Heimbach J, Harnois DM, et al. Validation of a Risk Estimation of Tumor Recurrence After Transplant (RETREAT) score for hepatocellular carcinoma recurrence after liver transplant. JAMA Oncol. 2017;3(4):493–500. https://doi.org/10.1001/jamaoncol.2016.5116.
16. Hameed B, Mehta N, Sapisochin G, Roberts JP, Yao FY. Alpha-fetoprotein level > 1000 ng/mL as an exclusion criterion for liver transplantation in patients with hepatocellular carcinoma meeting the Milan criteria. Liver Transpl. 2014;20(8):945–51. https://doi.org/10.1002/lt.23904.
17. Berry K, Ioannou GN. Comparison of liver transplant–related survival benefit in patients with versus without hepatocellular carcinoma in the United States. Gastroenterology. 2015;149(3):669–80. https://doi.org/10.1053/j.gastro.2015.05.025.
18. Halazun KJ, Najjar M, Abdelmessih RM, et al. Recurrence after liver transplantation for hepatocellular carcinoma: a new MORAL to the story. Ann Surg. 2017;265(3):557–464. https://doi.org/10.1097/SLA.0000000000001966.
19. Chaiteerakij R, Zhang X, Addissie BD, et al. Combinations of biomarkers and Milan criteria for predicting hepatocellular carcinoma recurrence after liver transplantation. Liver Transpl. 2015;21(5):599–606. https://doi.org/10.1002/lt.24117.
20. Yang SH, Suh KS, Lee HW, et al. The role of (18)F-FDG-PET imaging for the selection of liver transplantation candidates among hepatocellular carcinoma patients. Liver Transpl. 2006;12(11):1655–60. https://doi.org/10.1002/lt.20861.
21. Lee JW, Paeng JC, Kang KW, et al. Prediction of tumor recurrence by 18F-FDG PET in liver transplantation for hepatocellular carcinoma. J Nucl Med. 2009;50(5):682–27. https://doi.org/10.2967/jnumed.108.060574.
22. Lai Q, Avolio AW, Graziadei I, et al. Alpha-fetoprotein and modified response evaluation criteria in solid tumors progression after locoregional therapy as predictors of hepatocellular cancer recurrence and death after transplantation. Liver Transpl. 2013;19(10):1108–18. https://doi.org/10.1002/lt.23706.

23. Roberts JP, Venook A, Kerlan R, Yao F. Hepatocellular carcinoma: ablate and wait versus rapid transplantation. Liver Transpl. 2010;16(8):925–9. https://doi.org/10.1002/lt.22103.
24. Mazzaferro V, Sposito C, Zhou J, et al. Metroticket 2.0 model for analysis of competing risks of death after liver transplantation for hepatocellular carcinoma. Gastroenterology. 2018;154(1):128–39. https://doi.org/10.1053/j.gastro.2017.09.025.
25. Lai Q, Vitale A, Iesari S, et al. Intention-to-treat survival benefit of liver transplantation in patients with hepatocellular cancer. Hepatology. 2017;66(6):1910–9. https://doi.org/10.1002/hep.29342.
26. Mazzaferro V. Squaring the circle of selection and allocation in liver transplantation for HCC: an adaptive approach. Hepatology. 2016;63(5):1707–17. https://doi.org/10.1002/hep.28420.
27. Kim WR, Lake JR, Smith JM, et al. OPTN/SRTR 2017 annual data report: liver. Am J Transplant. 2019;19(Suppl 2):184–283. https://doi.org/10.1111/ajt.15276.
28. Volk ML, Vijan S, Marrero JA. A novel model measuring the harm of transplanting hepatocellular carcinoma exceeding Milan criteria. Am J Transplant. 2008;8(4):839–46. https://doi.org/10.1111/j.1600-6143.2007.02138.x.

Should All Tyrosine Kinase Inhibitors Be Considered Equal for First-Line Hepatocellular Carcinoma Treatment?

23

Naim Alkhouri and Sukeshi Patel Arora

Both tyrosine kinase inhibitors (TKIs) approved for first-line treatment of advanced hepatocellular carcinoma (HCC) have their position in our treatment armamentarium and no clear winner can be declared at this point.

Main Points

1. Patients with advanced HCC, such as those with macrovascular tumor invasion or extrahepatic metastases and those with progression after locoregional therapy, are classical candidates for systemic treatment according to current guidelines.
2. There are 2 TKI therapies approved for first-line use in HCC: lenvatinib and sorafenib.
3. Lenvatinib may have a more potent vascular endothelial growth factor receptor (VEGFR)-inhibiting activity than sorafenib, and it targets other receptors that are not targeted by sorafenib such as FGFR1 to 4, KIT, and RET.
4. Lenvatinib was noninferior to sorafenib in terms of overall survival (OS) in the REFLECT study.
5. Until data on their real-world efficacy and tolerability are available, it is not known which compound should be selected in a first-line setting.
6. Equally important, data demonstrating efficacy for second-line treatments such as regorafenib or cabozantinib have been acquired only in sorafenib-experienced patients.
7. Data gaps for lenvatinib remain, especially in patients with 50% or more liver involvement, clear invasion of the bile duct, or main portal vein invasion.

Patients with advanced HCC, such as those with macrovascular tumor invasion or extrahepatic metastases and those with progression after locoregional therapy, are classical candidates for systemic treatment according to current guidelines. With newer, emerging immunotherapies showing efficacy over sorafenib in the first-line setting, more prospective data is needed for the use of the traditional "first-line TKIs" as "second-line" treatments after front-line immunotherapy. Nonetheless, for patients with contraindications to immunotherapy, such as transplantation and uncontrolled autoimmune conditions, lenvatinib and sorafenib remain the mainstay in the first-line setting for the treatment of advanced HCC. The first treatment to show survival benefit in patients with advanced HCC was sorafenib, a TKI that inhibits cancer proliferation and angiogenesis. Since 2007, sorafenib has been considered standard of care for HCC in the first-line setting [1]. Although other therapeutics have been investigated since then, these studies have been negative when compared to sorafenib. This was until the REFLECT study showed that lenvatinib was noninferior to sorafenib in terms of OS [2]. However, this study demonstrated that these 2 TKIs have a distinct toxicity profile as well as other outcomes that were dissimilar, thus, asking the question: Are all TKIs created equal in HCC? This chapter will provide a concise review of available clinical data on both sorafenib and lenvatinib as the only Food and Drug Administration–approved first-line therapies for advanced HCC and then discuss similarities and differences in their efficacy and safety profiles.

N. Alkhouri
Department of Hepatology, Arizona Liver Health, Chandler, AZ, USA

S. P. Arora (✉)
Division of Hematology/Oncology, Mays Cancer Center, University of Texas (UT) Health San Antonio, San Antonio, TX, USA
e-mail: aroras@uthscsa.edu

N. S. Reau, A. G. Singal (eds.), *Controversies in Hepatocellular Carcinoma*, https://doi.org/10.1007/978-3-031-90225-3_23

Sorafenib

Sorafenib (Nexavar, Bayer-Onyx Pharmaceuticals) is an oral multikinase inhibitor that inhibits tumor angiogenesis and tumor-cell proliferation through activity VEGFRs 1–3, platelet-derived growth factor receptor β (PDGFR-β), and the serine–threonine kinases Raf-1 and B-Raf [3].

Phase 3 Trials with Sorafenib

The SHARP Trial

Trial Design, End Points, and Patient Characteristics

The Hepatocellular Carcinoma Assessment Randomized Protocol (SHARP) trial [1] established sorafenib as the standard of care in advanced HCC in a multicenter, phase 3, double-blind, placebo-controlled trial [1]. In this landmark trial, 602 patients with advanced HCC were randomly assigned to receive sorafenib 400 mg twice daily vs placebo [1]. Baseline characteristics were well balanced in both arms, with the majority of patients having an Eastern Cooperative Oncology Group (ECOG) score of 0 to 1, Child-Pugh score (CPS) class A cirrhosis, and HCC at Barcelona Clinic Liver Cancer (BCLC) stage C [1]. This study mainly recruited from the United States and Europe, and the most common causes of cirrhosis were hepatitis C virus (HCV; 29% vs 27%), alcohol-associated liver disease only (26% vs 26%), and hepatitis B virus (HBV; 19% vs 18%) for the sorafenib and placebo groups, respectively [1].

Efficacy

The median OS was 10.7 months with sorafenib vs 7.9 months with placebo (hazard ratio [HR] = 0.69; 95% CI, 0.55–0.87; $P < 0.001$) [1] (Table 23.1). The survival rates at 1 year were 44% in the sorafenib arm and 33% in the placebo arm, representing a 31% relative reduction in the risk of death. The disease control rate was 43% with sorafenib vs 32% with placebo ($P = 0.002$) [1]. Of note, the median time to progression (TTP) was statistically significantly longer in those treated with sorafenib compared to placebo (5.5 vs 2.8 months; HR = 0.58; 95% CI, 0.45–0.74; $P < 0.001$) and the estimated rate of progression-free survival (PFS) at 4 months was 62% with sorafenib and 42% with placebo.

Safety and Tolerability

In the sorafenib group, 80% had adverse events (AEs), with the most common being diarrhea (39%), fatigue (22%), palmar-plantar erythrodysesthesia (21%), and anorexia (14%) [1] (Table 23.2). There was a 37% to 38% rate of discontinuation of drug in both arms [1]. Since the approval of this drug, we have learned that optimal doses for tolerability vary among patients and titrating to tolerable doses is critical to decrease side effects, increase adherence, and maintain efficacy [4]. With the goal to find the highest tolerable dose, strategies such as starting at 200 mg twice daily with dose escalation allow dose optimization.

The Asia-Pacific Study

Trial Design, End Points, and Patient Characteristics

The Asia-Pacific Study [5] is a multinational, phase 3, double-blind, placebo-controlled trial that focused on an Asia-Pacific population, which predominantly has cirrhosis caused by HBV [5]. In this study, 271 patients with advanced HCC were randomly assigned 2:1 to receive sorafenib 400 mg twice daily vs placebo [5]. Baseline characteristics were well balanced in both arms, with the majority of patients having ECOG 0 to 1, CPS A cirrhosis, and HCC at BCLC stage C [5]. The most common causes of cirrhosis were HCV (1% vs 4%) and HBV (71% vs 77%) for the sorafenib and placebo groups, respectively [5].

Efficacy

Patients with advanced HCC who received sorafenib had a median OS of 6.5 months vs 4.2 months with placebo (HR = 0.68 [95% CI, 0.50–0.93]; $P = 0.014$) [5]. The median TTP also improved from 1.4 months with placebo to 2.8 months with sorafenib (HR = 0.57 [95% CI, 0.42–0.79]; $P = 0.0005$) [5].

Safety and Tolerability

Overall, in the sorafenib group, any grade toxicity in patients was 81.9% vs 38.7% in the placebo group [5]. With sorafenib, the most common toxicities with sorafenib were palmar-plantar erythrodysesthesia (45%), diarrhea 26%]), alopecia (25%),

Table 23.1 Clinical outcomes of first-line tyrosine kinase inhibitors in advanced hepatocellular carcinoma

	Overall survival, mo	Time to progression, mo	Time to radiological progression, mo	Objective response rate, %	Median duration on treatment, mo
SHARP: Sorafenib vs placebo [1]	10.7 vs 7.9 (HR 0.69; 95% CI, 0.55–0.87; $P < 0.001$	4.1 vs 4.9	5.5 vs 2.8 ($P < 0.001$)	–	5.3
REFLECT: Lenvatinib vs sorafenib (noninferiority) [2]	13.6 vs 12.3 (HR 0.92, 95% CI, 0.79–1.06)	8.9 vs 3.7 (HR 0.63; 95% CI, 0.53–0.73)	–	24.1 vs 9.2	5.7

Abbreviations: *HR* hazard ratio, *SHARP hepatocellular carcinoma assessment randomized protocol*, *mo* months

Table 23.2 Adverse events for first-line treatments based on Phase 3 clinical trials

	Sorafenib vs placebo		Lenvatinib vs sorafenib	
	Any grade (%)	Grade 3–4 (%)	Grade 1–2 (%)	Grade 3 (%)
Gastroenterological				
(a) Anorexia	14	< 1	34 vs 27	8 vs 3
(b) Diarrhea	39	8	39 vs 46	4 vs 4
(c) Nausea	11	< 1	20 vs 14	1 vs 1
(d) Vomiting	5	1	16% vs	1 vs 1
Constitutional				
(a)Fatigue	22	4	30 vs 25	4 vs 4
(b) Weight loss	9	2	31 vs 22	8 vs 3
Dermatologic				
(a) Alopecia	14	0	3 vs 25	0 vs 0
(b) Dry skin	8	0	–	–
(c) Hand-foot skin reaction	21	8	27 vs 52	3 vs 11
(d) Pruritis	8		–	–
(e) Rash or desquamation	16	1%	–	–
(f) Others	5	1	–	7 vs 5
(g) Increased blood bilirubin			15 vs 13	5 vs 8
(h) Hypothyroidism			16 vs 2	0 vs 0
Voice changes	6	0	24 vs 12	< 1 vs 0
Hypertension	5	2	42 vs 30	23 vs 14
Liver dysfunction	< 1	< 1	14 vs 17	5 vs 8
Abdominal pain not otherwise specified	8	2	17 vs 18	2 vs 3
Bleeding	7	1	–	–

fatigue (20%), rash or desquamation (20%), hypertension (19%), and anorexia (13%) [5]. Most of these toxicities were predominantly grade 1 or 2 AEs [5] (see Table 23.2).

Prospective, Noninterventional Study: The Global Investigation of Therapeutic Decisions in Hepatocellular Carcinoma and of its Treatment with Sorafenib (GIDEON) Study

In clinical practice, dosing of sorafenib varies from the recommended daily dose based on CPS, performance status, comorbidities, and/or physician preference. A global, prospective, noninterventional study evaluated the safety of sorafenib in patients with HCC in real-life practice, showing a variation in patient and disease characteristics and treatment patterns for US patients compared to the rest of the world [6, 7]. In the United States, a higher percentage of CPS class B patients receive sorafenib than in European and Asian regions [6]. The median OS was higher for CPS class A than CPS class B, with a median OS of 13.6 months vs 5.2 months, respectively [8]. Although a similar proportion of patients received an 800 mg initial dose, the median daily dose was 721 mg in CPS class B vs 680 mg in CPS class A [7]. Despite the higher average dose received in CPS class B patients, their OS was worse than CPS class A patients, possibly because of the shorter median duration of sorafenib in patients with CPS class B (8.6 weeks) vs CPS class A (13.7 weeks) [7].

Lenvatinib

Lenvatinib (Lenvima, Eisai Inc) is an oral TKI that has multiple targets including VEGFRs 1 to 3, FGF receptors 1 to 4, PDGF receptor α, RET, and KIT leading to suppression of tumor angiogenesis and proliferation [9]. A phase 2 trial that was conducted in Asia and included patients with advanced HCC demonstrated that lenvatinib at 12 mg daily had potent clinical activity with a manageable AE profile [10]. Further analysis of the pharmacokinetic data showed that the ideal dose is 8 mg daily for patients that weigh less than 60 kg and 12 mg daily for those that weigh 60 kg or more, informing the design of the phase 3 REFLECT trial.

Phase 3 Trial with Lenvatinib

The REFLECT Trial

Trial Design, End Points, and Patient Characteristics

The REFLECT trial [2] was a multicenter, global, randomized, open-label, phase 3 noninferiority trial that enrolled systemic chemotherapy-naive patients with advanced HCC and randomly assigned them to receive weight-based lenvatinib (12 mg for body weight ≥ 60 kg and 8 mg for body weight < 60 kg) once daily or sorafenib 400 mg twice daily

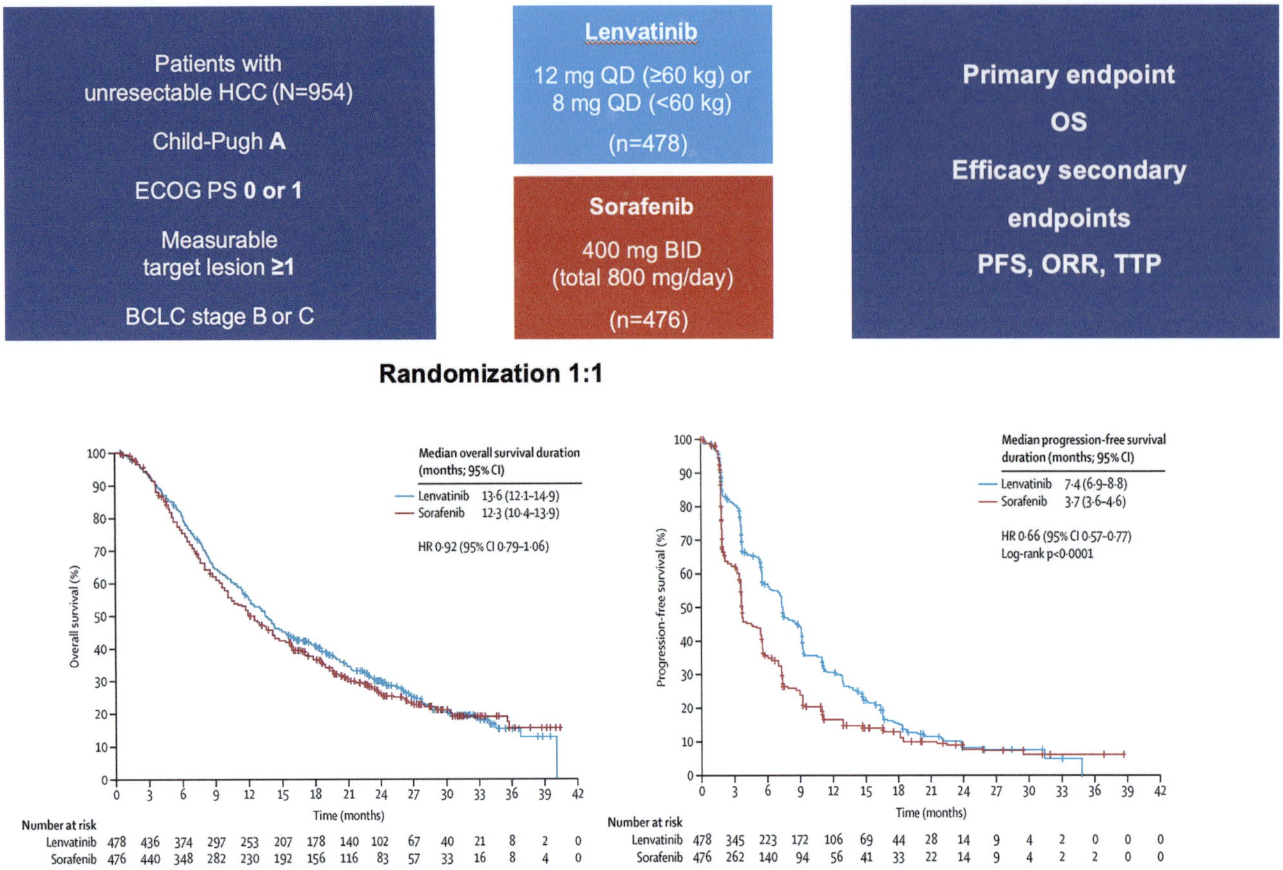

Fig. 23.1 REFLECT trial design and main end points. Patients with advanced hepatocellular carcinoma (HCC) were randomly assigned to receive lenvatinib or sorafenib in a 1:1 ratio. Lenvatinib was noninferior to sorafenib in overall survival (OS) and statistically significantly superior in improving progression-free survival (PFS). BCLC, Barcelona Clinic Liver Cancer; BID, twice a day; ECOG, Eastern Cooperative Oncology Group; HR, hazard ratio; ORR, objective response rate; QD, every day; TTP, time to progression

in a 1:1 ratio (Fig. 23.1). Dose reduction/interruption for lenvatinib-related AEs (to 8 and 4 mg daily, or 4 mg every other day) was allowed. It is important to note the fact that patients with 50% or more liver involvement, clear invasion of the bile duct, or main portal vein invasion were excluded from the trial. The primary end point was OS, with the main secondary end points being PFS, TTP, and objective response rate. The trial enrolled 954 patients (478 received lenvatinib and 476 received sorafenib) with balanced baseline characteristics between the groups in terms of age, sex, race, weight, performance status, and BCLC stage. Compared to the sorafenib arm, patients in the lenvatinib arm were more likely to have HBV (53% vs 48%) and alpha fetoprotein greater than 200 ng/mL (46% vs 39%) and less likely to have HCV (19% vs 26%).

Efficacy

Lenvatinib demonstrated noninferiority to sorafenib in achieving the primary end point, with a median OS duration of 13.6 months in the lenvatinib group compared to 12.3 months in the sorafenib group (HR = 0.92; 95% CI, 0.79–1.06), making the REFLECT trial the first positive trial against an active comparator in first-line unresectable HCC treatment. Importantly, lenvatinib was statistically significantly superior to sorafenib in achieving secondary efficacy end points including PFS (7.3 months vs 3.6 months; HR = 0.64; 95% CI, 0.55–0.75) and objective response rate (40.6% vs 12.4%; HR = 5.01; 95% CI, 3.59–7.01) based on masked independent imaging review that used the mRECIST criteria. The use of mRECIST may be a limitation of this study, as it depends on both the quality of the imaging modalities and the ability of the reviewer. However, mRECIST is an established tool in HCC, [11, 12] and exploratory post hoc analysis confirmed that PFS based on investigator mRECIST assessment was similar to those observed on independent imaging review using both RECIST 1.1 [13].

Safety and Tolerability

Treatment-emergent AEs occurred in the majority of patients who received either lenvatinib or sorafenib, consistent with the previous experience with TKIs. AEs that were more frequent in the lenvatinib group included systemic hypertension, hypothyroidism, proteinuria, and dysphonia (see Table 23.2). On the other hand, palmar-plantar erythrodysesthesia, alopecia, and diarrhea were more common with sorafenib. The median treatment duration with lenvatinib was 5.7 months compared to 3.7 months with sorafenib and the median time to first dose reduction, for patients requiring a dose reduction, was 10 weeks in the lenvatinib arm and 3.7 weeks in the sorafenib arm, indicating potentially better tolerability with lenvatinib.

Future Perspectives

With the advent of immune checkpoint inhibitor trials in HCC, we have two therapies with Food and Drug Administration approval, atezolizumab plus bevacizumab and tremelimumab plus durvalumab, which are preferred first-line options based on improved survival when compared to sorafenib [14, 15]. As a result, sorafenib or lenvatinib have shifted to second-line treatments, but still remain first-line options for patients who cannot receive immunotherapy (i.e., patients with uncontrolled autoimmune disorders, patients with solid organ liver transplantation, patients with contraindications to immunotherapy). Future studies should investigate the efficacy of these traditional "first-line" TKIs in the "second-line." setting for the treatment of advanced HCC.

Conclusion

Despite the positive results of the REFLECT trial, it is not clear which compound should be selected between sorafenib and lenvatinib in a first-line setting. It is expected that in the near future, data on their real-world efficacy and tolerability will help inform clinicians' therapeutic decisions. Importantly, most experience with second-line treatments such as regorafenib or cabozantinib has been only in sorafenib-experienced patients, [15, 16] and data on their efficacy cannot be extrapolated if given after lenvatinib. Furthermore, data are needed with lenvatinib on patients with 50% or more liver involvement, clear invasion of the bile duct, or main portal vein invasion. On the other hand, lenvatinib may have a more potent VEGFR-inhibiting activity than sorafenib and it targets other receptors that are not targeted by sorafenib such as FGFR1 to 4, KIT, and RET, although it remains unknown to what extent these differences may affect overall efficacy.

In conclusion, and to answer the question that we started with, both TKIs approved for first-line treatment of advanced HCC have their position in our treatment armamentarium and no clear winner can be declared at his point pending further data, especially now that TKIs are used after first-line immunotherapy for the majority of patients with HCC.

References

1. Llovet JM, Ricci S, Mazzaferro V, et al. Sorafenib in advanced hepatocellular carcinoma. N Engl J Med. 2008;359(4):378–90. https://doi.org/10.1056/NEJMoa0708857.
2. Kudo M, Finn RS, Qin S, et al. Lenvatinib versus sorafenib in first-line treatment of patients with unresectable hepatocellular carcinoma: a randomised phase 3 non-inferiority trial. Lancet. 2018;391(10126):1163–73. https://doi.org/10.1016/S0140-6736(18)30207-1.
3. Vogel A, Cervantes A, Chau I, et al. Hepatocellular carcinoma: ESMO clinical practice guidelines for diagnosis, treatment and follow-up. Ann Oncol. 2018;29(Suppl 4):iv238–55. https://doi.org/10.1093/annonc/mdy308.
4. Reiss KA, Yu S, Mamtani R, et al. Starting dose of sorafenib for the treatment of hepatocellular carcinoma: a retrospective, multi-institutional study. J Clin Oncol. 2017;35(31):3575–81. https://doi.org/10.1200/JCO.2017.73.8245.
5. Cheng AL, Kang YK, Chen Z, et al. Efficacy and safety of sorafenib in patients in the Asia-Pacific region with advanced hepatocellular carcinoma: a phase III randomised, double-blind, placebo-controlled trial. Lancet Oncol. 2009;10(1):25–34. https://doi.org/10.1016/S1470-2045(08)70285-7.
6. Lencioni R, Kudo M, Ye SL, et al. First interim analysis of the GIDEON (global investigation of therapeutic decisions in hepatocellular carcinoma and of its treatment with sorafeNib) non-interventional study. Int J Clin Pract. 2012;66(7):675–83. https://doi.org/10.1111/j.1742-1241.2012.02940.x.
7. Lencioni R, Kudo M, Ye SL, et al. GIDEON (global investigation of therapeutic DEcisions in hepatocellular carcinoma and of its treatment with sorafeNib): second interim analysis. Int J Clin Pract. 2014;68(5):609–17. https://doi.org/10.1111/ijcp.12352.
8. Marrero JA, Lencioni R, Ye SL, et al. Final analysis of GIDEON (global investigation of therapeutic decisions in hepatocellular carcinoma [HCC] and of its treatment with Sorafenib [sor]) in > 3000 sor-treated patients (pts): clinical findings in pts with liver dysfunction. J Clin Oncol. 2013;31(Suppl):abstr 4126.
9. Yamamoto Y, Matsui J, Matsushima T, et al. Lenvatinib, an angiogenesis inhibitor targeting VEGFR/FGFR, shows broad antitumor activity in human tumor xenograft models associated with microvessel density and pericyte coverage. Vasc Cell. 2014;6:18. https://doi.org/10.1186/2045-824X-6-18.
10. Ikeda K, Kudo M, Kawazoe S, et al. Phase 2 study of lenvatinib in patients with advanced hepatocellular carcinoma. J Gastroenterol. 2017;52(4):512–9. https://doi.org/10.1007/s00535-016-1263-4.
11. Lencioni R, Montal R, Torres F, et al. Objective response by mRECIST as a predictor and potential surrogate end-point of overall survival in advanced HCC. J Hepatol. 2017;66(6):1166–72. https://doi.org/10.1016/j.jhep.2017.01.012.
12. Meyer T, Palmer DH, Cheng AL, Hocke J, Loembé AB, Yen CJ. mRECIST to predict survival in advanced hepatocellular carcinoma: analysis of two randomised phase II trials comparing nintedanib vs sorafenib. Liver Int. 2017;37(7):1047–55. https://doi.org/10.1111/liv.13359.
13. Eisenhauer EA, Therasse P, Bogaerts J, et al. New response evaluation criteria in solid tumours: revised RECIST guideline (version 1.1). Eur J Cancer. 2009;45(2):228–47. https://doi.org/10.1016/j.ejca.2008.10.026.
14. Finn RS, Qin S, Ikeda M, Galle PR, Ducreux M, Kim TY, Kudo M, Breder V, Merle P, Kaseb AO, Li D, Verret W, Xu DZ, Hernandez S, Liu J, Huang C, Mulla S, Wang Y, Lim HY, Zhu AX, Cheng AL. Atezolizumab plus bevacizumab in unresectable hepatocellular carcinoma. N Engl J Med. 2020;382(20):1894–905. https://doi.org/10.1056/NEJMoa1915745.

15. Abou-Alfa GK, Lau G, Kudo M, Chan SL, Kelley RK, Furuse J, Sukeepaisarnjaroen W, Kang YK, Van Dao T, De Toni EN, Rimassa L, Breder V, Vasilyev A, Heurgué A, Tam VC, Mody K, Thungappa SC, Ostapenko Y, Yau T, Azevedo S, Varela M, Cheng AL, Qin S, Galle PR, Ali S, Marcovitz M, Makowsky M, He P, Kurland JF, Negro A, Sangro B. Tremelimumab plus durvalumab in unresectable hepatocellular carcinoma. NEJM Evid. 2022;1(8):EVIDoa2100070. https://doi.org/10.1056/EVIDoa2100070.
16. Kudo M. Lenvatinib may drastically change the treatment landscape of hepatocellular carcinoma. Liver Cancer. 2018;7(1):1–19. https://doi.org/10.1159/000487148.

Should All Patients Be Considered for Locoregional Therapy Before Systemic Therapy?

24

Jordan Tasse and Bulent Arslan

All patients with very-early, early-, and intermediate-stage hepatocellular carcinoma (HCC) should be considered for locoregional therapy (LRT) before systemic therapy.

Main Points

1. LRT is the minimally invasive targeted tumor treatment using either a percutaneous ablative or transarterial embolic approach.
2. Patients diagnosed with very-early to early-stage HCC should be considered for curative therapy. This includes surgical resection, LRT with percutaneous ablation or transarterial radiation segmentectomy, and liver transplantation (LT).
3. If the patient is being considered for LT, LRT is preferred.
4. In patients with intermediate-stage HCC, LRT with transarterial radioembolization (TARE) or transarterial chemoembolization (TACE) should be the first consideration. The goal in these patients is to downstage to potential curative therapy with LT, resection, or ablation.
5. Those with advanced-stage HCC should be evaluated both for systemic therapy and LRT. However, patients who are candidates for LRT should undergo this before systemic therapy because of better tumor response rates and toxicity.

LRT including percutaneous thermal ablation, TACE, and TARE have become the mainstay of treatment for patients with early- to intermediate-stage (Barcelona Clinic Liver Cancer [BCLC]) HCC. Systemic therapies have traditionally shown limited tumor response in HCC. With the introduction of sorafenib showing modest survival benefits, this therapy has been recommended in patients with advanced-stage disease. Newer systemic therapies have recently been introduced that may potentially change the paradigm in the treatment of patients presenting with all stages of HCC. This chapter will address current recommendations and future considerations regarding LRT and systemic therapy in the treatment of HCC.

HCC is a primary malignancy of the liver, usually occurring in patients with chronic underlying liver disease and cirrhosis. It is the ninth leading cause of cancer death in the United States and, despite advances in prevention, screening, and treatment, incidence and mortality remain high. Multiple treatment modalities are available; however, given the progressive nature of liver cirrhosis and the propensity for the cirrhotic liver to develop HCC, orthotopic liver transplantation (LT) remains the primary treatment option considered for long-term cure. In addition, surgical resection and percutaneous thermal ablation can be considered for a curative goal in patients with limited disease burden and Child-Pugh score (CPS) class A or B cirrhosis. Currently, the BCLC is the most widely used system to stage HCC that includes factors such as patient Eastern Conference Oncology Group performance status, CPS classification of underlying liver disease, and extent of tumor involvement of the liver with and without extrahepatic spread of disease.

Hepatocellular Carcinoma Staging

BCLC is the most commonly used staging system for HCC and is routinely used to guide treatment decisions. In patients with early- and very-early-stage disease, curative intent therapies for long-term survival including resection, LRT, and/or orthotopic LT should be employed. In patients with intermediate-stage disease, LRT with chemoembolization or radioembolization should be used to prolong survival.

J. Tasse (✉) · B. Arslan
Rush University Medical Center, Chicago, USA
e-mail: Jordan_C_Tasse@rush.edu;
BULENT_ARSLAN@rush.edu

N. S. Reau, A. G. Singal (eds.), *Controversies in Hepatocellular Carcinoma*, https://doi.org/10.1007/978-3-031-90225-3_24

Systemic therapy is not currently recommended as first-line therapy in these populations.

Currently, the discussion between whether a patient should undergo LRT or systemic therapy begins with advanced-stage HCC. In some of these patients, systemic therapy may be the best option to prolong survival. However, LRT should also be considered to potentially downstage disease and allow for curative intent therapies such as LT, ablation, or resection. Terminal-stage HCC patients should be given the best supportive care without LRT or systemic therapy. Available literature and current recommendations regarding LRT and systemic therapies for all stages of HCC will be discussed.

Locoregional Therapy

LRT is a minimally invasive targeted tumor treatment using either a percutaneous thermal ablative or transarterial embolic approach. Percutaneous ablation includes radiofrequency ablation (RFA), microwave ablation (MWA), or cryoablation. Transarterial therapies include chemoembolization, radioembolization, and bland embolization.

Thermal ablation is a widely accepted treatment option for patients with early-stage HCC and has been shown to be both safe and effective with multiple studies. Ablation has advantages that include its minimally invasive nature; it can be curative and is easily repeatable in cases of tumor recurrence. Ablation is typically considered for very-early and early-stage disease in tumors less than 3 cm in diameter. Tumor response rates approach greater than 90% with newer ablative technologies and adverse events (AEs) are low (< 5%). Most patients go home the same day or the next.

There have been 4 randomized controlled trials (RCTs) comparing thermal ablation with RFA to surgical resection for HCC. Three of these showed no statistically significant difference in overall survival (OS) between the 2 cohorts, [1–3] while AEs were higher in the resection group. One showed improved OS and progression-free survival (PFS) with surgical resection [4]. Additionally, nonrandomized comparative studies have shown varying results. Lee et al. [5] showed improved OS for patients undergoing RFA vs surgical resection after propensity score analysis. Another evaluation demonstrated lower AEs and improved OS with RFA compared to surgery for tumors less than 2 cm in diameter [6]. Some studies have reported similar outcomes between RFA and surgical resection [7–14]. Several other studies have found better OS with surgical resection vs percutaneous ablation [15–21]. Even in those studies, when evaluating tumors 2 cm or smaller in diameter [19] or 3 cm or smaller, [15, 19, 20] there was no statistically significant difference in OS between the groups. For very-early and early-stage HCC, with tumors less than or equal to 3 cm in diameter, ablation shows a greater than 90% tumor response and survival outcomes similar to surgical resection.

Transarterial therapies including TACE and TARE aim to induce tumor necrosis by injection of embolic material (particles, chemotherapy, radiation) directly into the tumor blood supply without the need for systemic circulation of the material. This derives from the fact that HCC tumor neovascularity occurs almost exclusively from the arterial supply, while the surrounding liver gains the majority of its perfusion from the portal vein. In 2002 and 2003, 2 RCTs showed statistically significant OS benefits with TACE compared to best supportive care for unresectable HCC [22, 23]. More recently TARE has shown statistically significantly improved time to progression (TTP) with comparable OS to chemoembolization in the treatment of patients with BCLC stage A and B disease [24].

TACE and TARE are currently recommended for patients with intermediate-stage HCC. A newer technique involving radioembolization, termed *radiation segmentectomy*, involves selective injection of yttrium 90 (Y90) microspheres into a targeted tumor liver segmental branch with high doses of radiation. Outcomes for early-stage HCC have shown similar tumor response rates, tumor control, and survival benefit when compared to curative intent therapies [25].

Patients with HCC being listed for LT often have to wait an extended period of time, sometimes more than 1 year, for a suitable liver to become available. TACE and TARE have both been evaluated in this setting and shown efficacy at preventing tumor burden from progressing to beyond transplantation criteria while the patient awaits a suitable liver, termed *bridging* [26–31]. One study showed superiority of TARE compared to TACE [28] in terms of fewer side effects and longer TTP (13.3 months vs 9.4 months).

TACE and TARE have also shown efficacy in downstaging HCC tumor burden, that is, treating patients with more advanced disease and shrinking the tumor to within criteria to be considered for curative therapy such as LT, resection, or ablation [31–40].

For patients with BCLC stage C disease, including portal invasion and/or extrahepatic disease with preserved liver function, sorafenib was previously recommended as the therapy of choice. More recently newer immunotherapy-based combinations have replaced Sorafenib as the first-line therapy of choice in patients with advanced-stage disease. These include Atezolizumab-Bevacizumab and Durvalumab-Tremilimumab. TACE has been shown to be noninferior to sorafenib [41, 42] in this setting. However, TARE is used in this setting because of its nonembolic nature, especially in cases of portal vein tumor thrombus. A meta-analysis showed improved survival with TARE for patients with HCC and portal vein invasion. Two phase 3 RCTs were published in 2017 comparing Y90 to sorafenib: the SARAH trial and SIRveNIB trial [43, 44]. It should be noted that both trials were designed to show the superiority of TARE over sorafenib and both used resin-based microsphere Y90 as opposed to glass microspheres. Glass microspheres are used at many centers and are approved by the Food and Drug Administration for the treatment of HCC, while resin microspheres are approved for the treatment of metastatic colorectal carcinoma in the liver. The SARAH trial randomly assigned 467 patients to receive either Y90 or sorafenib, and SIRveNIB randomly assigned 360 patients. The primary end point in each trial was median OS; reported survival for Y90 was 8.8 months and 8.0 months, respectively, and survival for sorafenib was 10 months and 9.9 months ($P = 0.36$ and $P = 0.18$, respectively). In each study, Y90 did not demonstrate superiority in terms of OS; however, Y90 scored statistically significantly higher on quality of life metrics and was noted to have significantly fewer serious AEs. In addition, there was increased tumor response with TARE compared to sorafenib. The reason why better tumor response did not translate into a benefit in OS nor in PFS should be further investigated in terms of patient selection.

Systemic Therapy

First Line

Systemic therapies have historically proven ineffective in the treatment of HCC. In 2008 sorafenib, a multityrosine kinase inhibitor, was the first medication shown to improve OS in the treatment of patients with advanced-stage HCC [45]. In this multicenter, phase-3 RCT, 602 patients with advanced HCC were randomly assigned to receive sorafenib or placebo. Medial OS was 10.7 months in the sorafenib group vs 7.9 months in the placebo group ($P \leq 0.001$). AEs were higher in the sorafenib group, reported at 80%. This survival benefit of sorafenib was also confirmed in the Asia-Pacific trial the same year [46].

Subsequently, lenvatinib ($n = 478$), a kinase inhibitor, was compared with sorafenib ($n = 476$) in a multicenter, randomized noninferiority trial for the first-line treatment of patients with unresectable HCC [47]. Lenvatinib did show promising results in terms of TTP (lenvatinib 8.9 months; sorafenib 3.7 months); however, OS (lenvatinib, 13.6 months; sorafenib, 12.3 months) and treatment-related AEs were not statistically significantly different between the 2 groups. Based on the noninferior results of this study, lenvatinib was approved for first-line therapy in the United States in 2018. Current National Comprehensive Cancer Network recommendations are for its use in patients with no worse than CPS class A cirrhosis. This represents a reasonable alternative systemic therapy to sorafenib.

More recently, investigations of immunotherapy-based combinations have occurred. Atezolizumab, a programmed death-ligand 1 inhibitor plus bevacizumab, a vascular endothelial growth factor inhibitor, in the first-line treatment of advanced HCC were published in 2020. In the phase 3 IMBrave 150 trial, 501 previously untreated patients with advanced HCC were randomly assigned to receive atezolizumab plus bevacizumab vs sorafenib. This combination significantly improved overall survival versus Sorafenib (median 19.2 months versus 13.4 months, $p = <0.001$). Additionally, PFS was also improved with Atezolizumab-Bevacizumab compared to Sorafenib at 6.8 versus 4.3 months ($p = <0.001$) [48] Grade 3 and 4 AEs were comparable between the 2 groups at 57% in the combined therapy group vs 55% with sorafenib. Subsequently in 2022, the results of the open-label, phase 3 HIMALAYA trial were published comparing the combination of Tremilimumab and Durvalumab (STRIDE) versus Sorafenib [49]. Median overall survival with STRIDE was 16.43 months versus 13.77 months with Sorafenib OS at 36 months was 30.7% with STRIDE compared to 20.2 months with Sorafenib. With the publication of these results, both the Atezolizumab-Bevacizumab and the STRIDE combinations became the recommended first-line systemic therapy options in unresectable HCC. If these options are not feasible, then Sorafenib or Lenvatinib or Durvalumab are the current alternative first-line recommendations in the most recent BCLC guidelines.

Second Line

Second-line therapy is considered for patients who progress on first-line therapy and whose performance status and liver function are sufficient to tolerate it. There are ongoing clinical trials evaluating the effectiveness of various second-line therapies in the treatment of HCC. Regorafenib [50] and nivolumab [51] have shown a survival benefit in this setting. Other considerations include pembrolizumab, lenvatinib, carbozantinib, and ramucirumab.

Combination Locoregional Therapy with Systemic Therapy

It has been hypothesized that administering sorafenib in conjunction with TACE may accentuate angiogenic factors and improve outcomes. However, 4 out of 5 trials comparing this approach have failed to demonstrate OS or PFS benefits compared to TACE alone [52]. Larger phase 3 trials are necessary to further evaluate this approach.

The addition of radioembolization using resin microspheres to sorafenib did not show a survival benefit in the phase 3 SORAMIC trial [53]. This trial evaluated 424 patients not eligible for TACE and median OS (12.1 vs 11.4 months), and the combined therapy group showed higher frequency of grade 3 or 4 AEs. The STOP HCC trial is a phase 3 trial comparing the addition of TARE with glass microspheres to sorafenib and is ongoing [54]. The first 2 interim analyses supported continuation of the trial. Additional trials combining TACE and TARE with immunotherapy-based systemic regimens are ongoing in hopes of synergistic effects without overlapping toxicities.

All Patients Should Be Considered for Locoregional Treatment Before Systemic Therapy

There has been a resurgence of interest in systemic therapy for HCC with the demonstration that the immunotherapy-based combinations improve survival compared to Sorafenib in advanced-stage disease. Both the Atezolizumab-Bevacizumab and the Durvalumab-Tremilimumab combinations have shown improved overall survival.

Patients diagnosed with very-early to early-stage HCC should be considered for curative therapy. This includes surgical resection, LRT with percutaneous ablation, and LT. If the patient is being considered for LT, LRT is preferred. This allows bridging to LT with a minimally invasive approach. Systemic therapy is not recommended in this setting.

In patients with intermediate-stage HCC, LRT with TARE or TACE should be the first consideration. The goal in these patients is to downstage to potential curative therapy with LT, resection, or ablation. If this is not feasible, TACE or TARE should be used to prolong survival. Systemic therapy can be considered in patients who are not candidates for LRT.

For patients with advanced-stage HCC, both LRTs and systemic options should be considered. Despite the lack of survival benefit for TARE over sorafenib demonstrated in the SARAH and SirVeNib trials, they did reveal some important findings. Patients had higher tumor response rates with TARE. Because of this, starting treatment with LRT first may give the patient the best chance to achieve downstaging to potentially curative therapies. If the patient with advanced-stage disease starts treatment with systemic therapy and the tumor progresses or liver function declines, they will likely no longer be candidates for LRT. Although the converse is also true with progression or decompensation after LRT potentially making systemic therapy contraindicated, dramatic tumor response is more likely to occur after LRT. In addition, patients experienced lower toxicity with TARE and quality of life metrics were higher. Given the more favorable AE profile with LRT, this may be preferable to systemic therapy for the initial treatment of patients with locally advanced disease and without extrahepatic metastases. LRTs usually employ single treatment sessions with TARE and at most several with TACE. This compared to the frequent dosing of systemic therapy may be a better option for many patients.

Conclusion

All patients with very-early, early-, and intermediate-stage HCC should be considered for LRT before systemic therapy based on the current literature. Those with advanced-stage HCC should be evaluated for both. However, patients who are candidates for LRT should undergo this before systemic therapy because of better tumor response rates and toxicity. Patients with metastatic disease should typically receive systemic therapy alone.

References

1. Chen MS, Li JQ, Zheng Y, et al. A prospective randomized trial comparing percutaneous local ablative therapy and partial hepatectomy for small hepatocellular carcinoma. Ann Surg. 2006;243(3):321–8. https://doi.org/10.1097/01.sla.0000201480.65519.b8.
2. Feng K, Yan J, Li X, et al. A randomized controlled trial of radiofrequency ablation and surgical resection in the treatment of small hepatocellular carcinoma. J Hepatol. 2012;57(4):794–802. https://doi.org/10.1016/j.jhep.2012.05.007.
3. Fang Y, Chen W, Liang X, et al. Comparison of long-term effectiveness and complications of radiofrequency ablation with hepatectomy for small hepatocellular carcinoma. J Gastroenterol Hepatol. 2014;29(1):193–200. https://doi.org/10.1111/jgh.12441.
4. Huang J, Yan L, Cheng Z, et al. A randomized trial comparing radiofrequency ablation and surgical resection for HCC conforming to the Milan criteria. Ann Surg. 2010;252(6):903–12. https://doi.org/10.1097/SLA.0b013e3181efc656.
5. Lee YH, Hsu CY, Chu CW, et al. Radiofrequency ablation is better than surgical resection in patients with hepatocellular carcinoma within the Milan criteria and preserved liver function: a retrospective study using propensity score analyses. J Clin Gastroenterol. 2015;49(3):242–9. https://doi.org/10.1097/MCG.0000000000000133.
6. Peng ZW, Lin XJ, Zhang YJ, et al. Radiofrequency ablation versus hepatic resection for the treatment of hepatocellular carcinomas 2 cm or smaller: a retrospective comparative study. Radiology. 2012;262(3):1022–33. https://doi.org/10.1148/radiol.11110817.
7. Hong SN, Lee SY, Choi MS, et al. Comparing the outcomes of radiofrequency ablation and surgery in patients with a single small hepatocellular carcinoma and well-preserved hepatic function. J Clin Gastroenterol. 2005;39(3):247–52. https://doi.org/10.1097/01.mcg.0000152746.72149.31.
8. Yamagiwa K, Shiraki K, Yamakado K, et al. Survival rates according to the cancer of the liver Italian program scores of 345 hepatocellular carcinoma patients after multimodality treatments during a 10-year period in a retrospective study. J Gastroenterol Hepatol. 2008;23(3):482–90. https://doi.org/10.1111/j.1440-1746.2007.05262.x.
9. Yamakado K, Nakatsuka A, Takaki H, et al. Early-stage hepatocellular carcinoma: radiofrequency ablation combined with chemoembolization versus hepatectomy. Radiology. 2008;247(1):260–6. https://doi.org/10.1148/radiol.2471070818.
10. Kim GA, Shim JH, Kim MJ, et al. Radiofrequency ablation as an alternative to hepatic resection for single small hepatocellular

carcinomas. Br J Surg. 2016;103(1):126–35. https://doi.org/10.1002/bjs.9960.
11. Jiang L, Yan L, Wen T, et al. Comparison of outcomes of hepatic resection and radiofrequency ablation for hepatocellular carcinoma patients with multifocal tumors meeting the Barcelona-clinic liver cancer stage a classification. J Am Coll Surg. 2015;221(5):951–61. https://doi.org/10.1016/j.jamcollsurg.2015.08.00.
12. Kang TW, Kim JM, Rhim H, et al. Small hepatocellular carcinoma: radiofrequency ablation versus nonanatomic resection – propensity score analyses of long-term outcomes. Radiology. 2015;275(3):908–19. https://doi.org/10.1148/radiol.15141483.
13. Yang H, Lee J, Lee DH, et al. Small single-nodule hepatocellular carcinoma: comparison of transarterial chemoembolization, radiofrequency ablation, and hepatic resection by using inverse probability weighting. Radiology. 2014;271(3):909–18. https://doi.org/10.1148/radiol.13131760.
14. Cucchetti A, Piscaglia F, Cescon M, et al. An explorative data-analysis to support the choice between hepatic resection and radiofrequency ablation in the treatment of hepatocellular carcinoma. Dig Liver Dis. 2014;46(3):257–63. https://doi.org/10.1016/j.dld.2013.10.015.
15. Hung HH, Chiou YY, Hsia CY, et al. Survival rates are comparable after radiofrequency ablation or surgery in patients with small hepatocellular carcinomas. Clin Gastroenterol Hepatol. 2011;9(1):79–86. https://doi.org/10.1016/j.cgh.2010.08.018.
16. Vivarelli M, Guglielmi A, Ruzzenente A, et al. Surgical resection versus percutaneous radiofrequency ablation in the treatment of hepatocellular carcinoma on cirrhotic liver. Ann Surg. 2004;240(1):102–7. https://doi.org/10.1097/01.sla.0000129672.51886.44.
17. Guglielmi A, Ruzzenente A, Valdegamberi A, et al. Radiofrequency ablation versus surgical resection for the treatment of hepatocellular carcinoma in cirrhosis. J Gastrointest Surg. 2008;12(1):192–8. https://doi.org/10.1007/s11605-007-0392-8.
18. Gory I, Fink M, Bell S, et al. Radiofrequency ablation versus resection for the treatment of early stage hepatocellular carcinoma: a multicenter Australian study. Scand J Gastroenterol. 2015;50(5):567–76. https://doi.org/10.3109/00365521.2014.953572.
19. Hasegawa K, Kokudo N, Makuuchi M, et al. Comparison of resection and ablation for hepatocellular carcinoma: a cohort study based on a Japanese nationwide survey. J Hepatol. 2013;58:724–9. https://doi.org/10.1016/j.jhep.2013.05.027.
20. Yun WK, Choi MS, Choi D, et al. Superior long-term outcomes after surgery in child-Pugh class a patients with single small hepatocellular carcinoma compared to radiofrequency ablation. Hepatol Int. 2011;5(2):722–9. https://doi.org/10.1007/s12072-010-9237-8.
21. Desiderio J, Trastulli S, Pasquale R, et al. Could radiofrequency ablation replace liver resection for small hepatocellular carcinoma in patients with compensated cirrhosis? A 5-year follow-up. Langenbeck's Arch Surg. 2013;398(1):55–62. https://doi.org/10.1007/s00423-012-1029-2.
22. Llovet JM, Real MI, Montaña X, et al. Arterial embolisation or chemoembolisation versus symptomatic treatment in patients with unresectable hepatocellular carcinoma: a randomised controlled trial. Lancet. 2002;359:1734–9. https://doi.org/10.1016/S0140-6736(02)08649-X.
23. Lo CM, Ngan H, Tso WK, et al. Randomized controlled trial of transarterial lipiodol chemoembolization for unresectable hepatocellular carcinoma. Hepatology. 2002;35(5):1164–71. https://doi.org/10.1053/jhep.2002.33156.
24. Salem R, Gordon AC, Mouli S, et al. Y90 radioembolization significantly prolongs time to progression compared with chemoembolization in patients with hepatocellular carcinoma. Gastroenterology. 2016;151(6):1155–63. https://doi.org/10.1053/j.gastro.2016.08.029.
25. Lewandowski R, Gabr A, Abouchaleh N, et al. Radiation segmentectomy: potential curative therapy for early hepatocellular carcinoma. Radiology. 2018;287(3):1050–8. https://doi.org/10.1148/radiol.2018171768.
26. Bouchard-Fortier A, Lapointe R, Perreault P, Bouchard L, Pomier-Layrargues G. Transcatheter arterial chemoembolization of hepatocellular carcinoma as a bridge to liver transplantation: a retrospective study. Int J Hepatol. 2011;2011:974514. https://doi.org/10.4061/2011/974514.
27. Graziadei IW, Sandmueller H, Waldenberger P, et al. Chemoembolization followed by liver transplantation for hepatocellular carcinoma impedes tumor progression while on the waiting list and leads to excellent outcome. Liver Transpl. 2003;9(6):557–63. https://doi.org/10.1053/jlts.2003.50106.
28. Salem R, Lewandowski R, Kulik L, et al. Radioembolization results in longer time-to-progression and reduced toxicity compared with chemoembolization in patients with hepatocellular carcinoma. Gastroenterology. 2011;140(2):497–507. https://doi.org/10.1053/j.gastro.2010.10.049.
29. Biederman DM, Titano JJ, Korff RA, et al. Radiation segmentectomy versus selective chemoembolization in the treatment of early-stage hepatocellular carcinoma. J Vasc Interv Radiol. 2018;29(1):30–37.e2. https://doi.org/10.1016/j.jvir.2017.08.026.
30. Ettorre GM, Levi Sandri GB, Laurenzi A, et al. Yttrium-90 radioembolization for hepatocellular carcinoma prior to liver transplantation. World J Surg. 2017;41(1):241–9. https://doi.org/10.1007/s00268-016-3682-z.
31. Tohme S, Sukato D, Chen HW, et al. Yttrium-90 radioembolization as a bridge to liver transplantation: a single-institution experience. J Vasc Interv Radiol. 2013;24(11):1632–8. https://doi.org/10.1016/j.jvir.2013.07.026.
32. Yao FY, Mehta N, Flemming J, et al. Downstaging of hepatocellular cancer before liver transplant: long-term outcome compared to tumors within Milan criteria. Hepatology. 2015;61(6):1968–77. https://doi.org/10.1002/hep.27752.
33. Yao FY, Hirose R, LaBerge JM, et al. A prospective study on downstaging of hepatocellular carcinoma prior to liver transplantation. Liver Transpl. 2005;11(12):1505–14. https://doi.org/10.1002/lt.20526.
34. Lei J, Wang W, Yan L. Downstaging advanced hepatocellular carcinoma to the Milan criteria may provide a comparable outcome to conventional Milan criteria. J Gastrointest Surg. 2013;17(8):1440–6. https://doi.org/10.1007/s11605-013-2229-y.
35. Yu CY, Out HY, Huang TL, et al. Hepatocellular carcinoma downstaging in liver transplantation. Transplant Proc. 2012;44(2):412–4. https://doi.org/10.1016/j.transproceed.2012.01.043.
36. Chapman WC, Doyle MBM, Stuart JE, et al. Outcomes of neoadjuvant transarterial chemoembolization to downstage hepatocellular carcinoma before liver transplantation. Ann Surg. 2008;248(4):617–25. https://doi.org/10.1097/SLA.0b013e31818a07d4.
37. Green TJ, Rochon PJ, Chang S, et al. Downstaging disease in patients with hepatocellular carcinoma outside of Milan criteria: strategies using drug-eluting bead chemoembolization. J Vasc Interv Radiol. 2013;24:1613–22. https://doi.org/10.1016/j.jvir.2013.07.024.
38. Riaz A, Kulik L, Lewandowski RJ, et al. Radiologic-pathologic correlation of hepatocellular carcinoma treated with internal radiation using yttrium-90 microspheres. Hepatology. 2009;49(4):1185–93. https://doi.org/10.1002/hep.22747.
39. Kulik L, Vouche M, Koppe S, et al. Prospective randomized pilot study of Y90+/-sorafenib as bridge to transplantation in hepatocellular carcinoma. J Hepatol. 2014;61(2):309–17. https://doi.org/10.1016/j.jhep.2014.03.023.
40. Abdelfattah MR, Al-Sebayel M, Broering D, Alsuhaibani H. Radioembolization using yttrium-90 microspheres as bridg-

ing and downstaging treatment for unresectable hepatocellular carcinoma before liver transplantation: initial single-center experience. Transplant Proc. 2015;47(2):408–11. https://doi.org/10.1016/j.transproceed.2014.11.004.
41. Kirstein MM, Voigtländer T, Schweitzer N, et al. Transarterial chemoembolization versus sorafenib in patients with hepatocellular carcinoma and extrahepatic disease. United European Gastroenterol J. 2018;6(2):238–46. https://doi.org/10.1177/2050640617716597.
42. Pinter M, Hucke F, Graziadei I, et al. Advanced-stage hepatocellular carcinoma: transarterial chemoembolization versus sorafenib. Radiology. 2012;263(2):590–9. https://doi.org/10.1148/radiol.12111550.
43. Vilgrain V, Pereira H, Assenat E, et al. Efficacy and safety of selective internal radiotherapy with yttrium-90 resin microspheres compared with sorafenib in locally advanced and inoperable hepatocellular carcinoma (SARAH): an open-label randomized controlled phase 3 trial. Lancet Oncol. 2017;18(12):1624–36. https://doi.org/10.1016/S1470-2045(17)30683-6.
44. Chow P, Gandhi M, Tan SB, et al. SIRveNIB: selective internal radiation therapy versus sorafenib in Asia-Pacific patients with hepatocellular carcinoma. J Clin Oncol. 2018;36(19):1913–21. https://doi.org/10.1200/JCO.2017.76.0892.
45. Llovet JM, Ricci S, Mazzaferro V, et al. Sorafenib in advanced hepatocellular carcinoma. N Engl J Med. 2008;359(4):378–90. https://doi.org/10.1056/NEJMoa0708857.
46. Cheng AL, Kang YK, Chen Z, et al. Efficacy and safety of sorafenib in patients in the Asia-Pacific region with advanced hepatocellular carcinoma: a phase III randomised, double-blind, placebo-controlled trial. Lancet Oncol. 2009;10(1):25–34. https://doi.org/10.1016/S1470-2045(08)70285-7.
47. Kudo M, Finn R, Qin S, et al. Lenvatinib versus sorafenib in first line treatment of patients with unresectable hepatocellular carcinoma: a randomized phase-3 non-inferiority trial. Lancet. 2018;391(10126):1163–73. https://doi.org/10.1016/S0140-6736(18)30207-1.
48. Finn R, Shukui Q, Masafumi I, Galle P, Ducreux M, et al. Atezolizumab plus bevacizumab in unresectable hepatocellular carcinoma. N Engl J Med. 2020;382:1894–905.
49. Abou-Alfa G, Lau G, Kudo M, Chan S, et al. Tremelimumab plus durvalumab in unresectable hepatocellular carcinoma. NEJM Evidence. 2022;1(8)
50. Bruix J, Qin S, Merle P, et al. Regorafenib for patients with hepatocellular carcinoma who progressed on sorafenib treatment (RESORCE): a randomised, double-blind, placebo-controlled, phase 3 trial. Lancet. 2017;389(10064):56–66. https://doi.org/10.1016/S0140-6736(16)32453-9.
51. Crocenzi TS, El-Khoueiry AB, Cheung T, et al. Nivolumab (nivo) in sorafenib (sor)-naive and -experienced pts with advanced hepatocellular carcinoma (HCC): CheckMate 040 study [abstract]. J Clin Oncol. 2017;35(Suppl):4013. https://doi.org/10.1200/JCO.2017.35.15_suppl.4013.
52. Liu L, Chen H, Wang M, et al. Combination therapy of sorafenib and TACE for unresectable HCC: a systematic review and meta-analysis. PLoS One. 2014;9(3):e91124. https://doi.org/10.1371/journal.pone.0091124.
53. Ricke AU, Klümpen HJ, Amthauer H, et al. Impact of combined selective internal radiation therapy and sorafenib on survival in advanced hepatocellular carcinoma. J Hepatol. 2019;71(6):1164–74. https://doi.org/10.1016/j.jhep.2019.08.006.
54. Chauhan N, Bukovcan J, Boucher E, et al. Intra-arterial TheraSphere yttrium-90 glass microspheres in the treatment of patients with unresectable hepatocellular carcinoma: protocol for the STOP-HCC phase 3 randomized controlled trial. JMIR Res Protoc. 2018;7(8):e11234. https://doi.org/10.2196/11234.

Should Direct-Acting Antivirals Be Given to Patients with Hepatocellular Carcinoma?

25

Nicole E. Rich and Amit G. Singal

Direct-acting antivirals (DAAs) appear to be beneficial in patients with hepatitis C virus (HCV) and concomitant hepatocellular carcinoma (HCC) and can be considered on a case-by-case basis.

Main Points

1. DAAs have dramatically altered the landscape for care of patients with HCV, including those with concomitant HCC.
2. DAAs appear to be safe and not associated with increased risk of recurrence in patients with a history of complete response (i.e., no viable tumor) following HCC-directed therapy.
3. DAAs appear to be safe in patients with active (i.e., viable) HCC and can be considered in select patients.
4. In patients with early HCC eligible for curative therapy, we recommend initiating DAA therapy after HCC complete response given the higher response rates than patients with active HCC.
5. In patients who are liver transplant (LT) candidates, decisions on the timing of DAA treatment should be made on a case-by-case basis, considering patient preferences and donor organ availability.
6. There are limited data evaluating DAA therapy in patients with intermediate- and advanced-stage HCC, but treatment can be considered in select cases based on overall life expectancy, comorbid conditions, and response to HCC therapy.

N. E. Rich (✉) · A. G. Singal
Department of Medicine, University of Texas Southwestern Medical Center, Dallas, TX, USA
e-mail: Nicole.Rich@UTSouthwestern.edu; Amit.Singal@UTSouthwestern.edu

HCV is the most common cause of HCC in the United States, and patients with HCV-related cirrhosis have an annual risk of developing HCC between 2% and 4% [1]. DAAs have revolutionized HCV treatment because they are well tolerated, safe, and efficacious with rates of sustained virologic response (SVR) exceeding 95%, even in populations once considered difficult to treat. Moreover, HCV eradication is cost-effective, [2] with several established benefits including improvement in all-cause mortality, [3] liver function, [4] portal hypertension, [5] and extrahepatic manifestations of HCV, [6] as well as reduced risk of de novo HCC [7, 8]. However, there has been controversy regarding the safety and benefit of DAAs in patients with HCC. In this chapter, we review the available literature on the use of DAAs in patients with HCC.

Patients with a History of Hepatocellular Carcinoma with Complete Response

While a large body of literature has consistently demonstrated that DAAs reduce the risk of incident HCC in patients with HCV, [7–11] potential concerns about increased risk of HCC recurrence led to controversy and extensive debate about the safety of DAAs in patients with a history of HCC with complete response (i.e., treated HCC with no viable tumor). In 2016, Reig and colleagues [12] observed a higher than expected proportion of patients with HCC recurrence (27.6%) after a median of 5.7 months following DAA treatment in an observational study of 58 patients with history of HCC, prompting concern for increased risk (and aggressiveness) of HCC recurrence following DAA treatment. The proposed mechanism was thought to be related to a decrease of immune surveillance of microscopic tumor clones following rapid clearance of HCV and decreased hepatic inflammation [13]. This finding resulted in substantial debate, varying practice patterns among providers, [14] and led to several

N. S. Reau, A. G. Singal (eds.), *Controversies in Hepatocellular Carcinoma*, https://doi.org/10.1007/978-3-031-90225-3_25

subsequent studies evaluating the effect of DAAs on HCC recurrence risk.

Several published studies have since refuted the findings by Reig et al. [12] First, a study including 3 prospective French cohorts showed lower recurrence rates (< 13%) and no increased risk of recurrence between DAA-treated and untreated patients [15]. Later, a meta-analysis of 24 studies (n = 1840 patients) found the proportion of patients who developed HCC recurrence varied widely following DAA therapy, ranging from 0% to 59% within a 2-year period [16]. Overall, the pooled estimate for recurrence rate was 24.4% (95% CI, 16.2–28.3%), and more than three-fourths (77.8%) of patients with HCC recurrence were detected at an early tumor stage. Among included studies, HCC recurrence was associated with a history of prior HCC recurrence and a shorter time interval between complete response to HCC treatment and initiation of DAA therapy. Finally, data from a large, multicenter cohort in North America confirmed these findings and found no difference in recurrence risk (hazard ratio [HR] = 0.90; 95% CI, 0.70–1.16) between DAA-treated and DAA-untreated patients [17]. In addition, among patients who developed recurrent HCC, the patterns of recurrence including timing, tumor burden, and response to HCC treatment were similar between those who were treated with DAA and those who were not. While these studies share some methodological limitations, including nonstandard HCC surveillance and treatment protocols between sites and lack of a comparator group, they collectively suggest that DAA treatment is likely safe and not associated with increased HCC recurrence.

The effect of DAAs on overall survival (OS) in patients with HCC, independent of the risk of HCC recurrence, has also been investigated in several studies. The ITA.LI.CA group [18] demonstrated that hepatic decompensation was a more important determinant of prognosis than HCC recurrence in 328 patients with a history of complete response to curative HCC therapy (e.g., resection or ablation). Similarly, in a study of nearly 800 HCC patients at 31 North American health systems, Singal et al. [19] found DAAs were associated with lower mortality compared to untreated patients (HR = 0.54; 95% CI, 0.33–0.90) in patients with complete response to HCC-directed therapy. As expected, the proportion of HCC-related deaths was similar in DAA-treated compared to untreated patients (30.2% vs 29.1%, $P = 0.89$), but there were fewer liver-related deaths in the DAA-treated group (16.3% vs 34.0%, $P = 0.03$). In subgroup analysis, the mortality reduction was observed in those with SVR (HR = 0.29; 95% CI, 0.18–0.47) but not those without SVR, reinforcing that this association was likely causal instead of related to selection bias of which patients underwent DAA treatment. In a follow-up study, this North American group found post-SVR patients had significantly lower odds of hepatic decompensation compared to viremic patients (OR = 0.18, 95% CI 0.06–0.59). This finding was consistent across subgroups of patients with Child-Pugh class A cirrhosis (OR = 0.22, 95% CI 0.04–0.77), intermediate/advanced HCC (OR = 0.20, 95% CI 0.04–0.65), and those treated with nonablative HCC therapies (OR = 0.21, 95% CI 0.07–0.67) [20].

These studies all have a few limitations, including their retrospective nature as well as the possibility of immortal time bias and selection bias, as DAAs may be used more selectively in patients perceived to have better prognosis. However, authors attempted to adjust for these factors using techniques such as landmark analyses, inverse probability weighting, and propensity score matching. Collectively, these data suggest that DAAs likely improve OS of patients with a history of complete response to HCC therapy by preserving underlying liver function, allowing for reduced cirrhosis-related mortality.

Patients with Active Hepatocellular Carcinoma

Several studies have demonstrated a decrease in SVR rates among patients with active HCC (i.e., untreated or previously treated HCC with viable tumor). An early single-center study by Prenner et al. [21] among 421 patients with HCV cirrhosis found higher rates of DAA failure among patients with HCC compared to those without HCC (21% vs 12%, $P = 0.009$); nearly all (93%) DAA failures in the HCC group occurred among patients with active HCC, whereas patients who received curative treatment with resection or transplant achieved excellent SVR rates, similar to those observed in patients without HCC. Subsequently, a large Veterans Affairs study [22] found more than three-fourths of patients with HCC achieved SVR; however, the researchers observed lower SVR rates among HCC patients compared to those without HCC and those with HCC treated post-transplant (74.4% vs 91.1% and 94.0%, respectively). In a meta-analysis of 49 studies from 15 countries, comprising 3341 and 35,701 patients with and without HCC, respectively, pooled SVR rates were lower in patients with HCC compared to those without HCC (89.6% vs 93.3%, $P = 0.001$), though there was statistically significant heterogeneity across studies ($I^2 > 75\%$) [23]. Overall, SVR rates were lower among patients with active HCC compared to those with complete response to HCC treatment (73.1% vs 92.6%, $P < 0.001$). The authors found SVR rates differed by DAA regimen, with similar SVR rates between HCC and non-HCC patients who were treated with non–sofosbuvir-based regimens (e.g., daclatasvir/asunaprevir [91.7% vs 89.8%, $P = 0.66$] and ombitasvir/paritaprevir/ritonavir [97.2% vs 94.8%, $P = 0.79$]), while lower SVR rates were noted among HCC patients treated with sofosbuvir-based regimens (e.g., ledipasvir/sofosbuvir) compared to those without HCC (92.6% vs 97.8%, $P = 0.03$). SVR rates also differed by HCC treatment modality, with patients receiving potentially curative HCC treatments (e.g., LT, resection, local ablation) having higher rates of SVR compared to those who received transarterial chemoembolization (91.3% vs 70.0%, $P < 0.001$). Though lower SVR rates have been observed consistently among patients with active HCC, the mechanism remains unclear. Several hypotheses have been proposed, including immune dysregulation among HCC patients, [24] HCC tumor cells serving as a reservoir for HCV, and alterations in the vascular supply to HCV-infected tumor cells affecting DAA uptake [25].

There have been few data evaluating the potential benefit of DAA therapy in patients with active HCC. It is possible that DAA therapy may improve liver dysfunction to reduce liver-related mortality as well as potentially allow more aggressive HCC-directed therapy; however, these patients would continue to have a high competing risk of HCC-related mortality—particularly those with intermediate- or advanced-stage HCC, who typically do not have curative therapy options. In a multicenter, retrospective study by Bruno and colleagues, [26] patients treated with interferon who achieved SVR had better 5-year OS after development of HCC compared to those with active viremia (65.9% vs 31.9%, $P = 0.004$). Moreover, end-stage liver disease (rather than HCC) was deemed to be the cause of death in 37% of patients in the viremic group, whereas none of the patients in the SVR group died of cirrhosis-related complications. A subsequent multinational cohort study [27] including 2 US and 6 Asian centers demonstrated that patients with active HCC who were treated with DAAs and achieved SVR had a statistically significantly lower risk of 5-year all-cause (HR = 0.37; 95% CI, 0.16–0.83) and liver-related (HR = 0.34; 95% CI, 0.13–0.88) mortality compared to those who did not receive DAA. Notably, all patients in this study were eligible for (and received) HCC treatment, with approximately 70% of patients in both groups having early-stage HCC and receiving curative HCC treatment.

Data remain limited on the benefits of DAA therapy in patients with intermediate- or advanced-stage HCC. The most recent American Association for the Study of Liver Diseases and the Infectious Diseases Society of America guidance [28] recommends HCV treatment for all patients, except those with a short life expectancy, that cannot be improved by HCV treatment, LT, or other therapy, and are thus otherwise unlikely to benefit from HCV eradication. Though patients with advanced-stage HCC were previously considered to have a median life expectancy of 8–13 months, [29] the emergence of several new, promising first- and second-line therapies [30, 31] for advanced HCC and improvements in OS raises questions as to if and when DAA therapy should now be considered in patients with advanced HCC. Though data are limited in HCC, the concomitant use of DAAs and sorafenib appears to be safe, [32] while the use of newer agents including immune checkpoint inhibitors for HCC during DAA treatment has not been well studied. Further data are needed to examine the effect of DAA treatment on OS among patients with non–early-stage HCC.

Patients with Hepatocellular Carcinoma Who Are Candidates for Liver Transplantation

There are other important considerations regarding the timing of HCV treatment in LT candidates, including those with HCC. Unfortunately, rates of LT waitlist dropout among patients with HCC have continued to increase over time, with considerable geographic disparities [33]. Deferring HCV treatment until the posttransplant period is cost-effective [34] and can expand the donor pool, [35] allowing access to organs from HCV-positive donors, thereby decreasing time on the waiting list and the potential for waitlist dropout and mortality. Further, DAAs are safe and highly effective in the post-LT setting, including in patients transplanted for HCC [22, 36]. Finally, some authors have suggested deferring HCV treatment until post LT is prudent given lower SVR rates (though still > 70%) in the pre-LT period compared to after LT [22].

An Italian study by Zanetto et al. [37] showed no increased rate of waiting list dropout among HCC patients treated with DAAs before LT. Recently, a large international, multicenter intention-to-treat analysis of 1012 patients with HCV-related HCC listed for LT found that neither pre- nor post-LT DAA treatment was associated with increased risk of HCC recurrence (HR = 0.44; 95% CI, 0.19–1.00 and HR = 0.62; 95% CI, 0.33–1.16, respectively) [37]. Further, there was a mortality benefit with improved 5-year recurrence-free survival in the DAA-treated group compared to the DAA-naive group (93.4% vs 73.9%, $P < 0.001$). Last, there was no difference in HCC recurrence patterns between DAA-treated and -untreated patients in this study. In sum, these data suggest there is no contraindication to treating patients with DAAs either pre or post transplant.

Overall, DAA treatment decisions in the setting of transplant are complex and should be individualized; a discussion should be had with the patient using shared decision-making based on access to LT (including geographic region as well as availability of a living donor), as well as comorbid conditions, quality of life considerations, and patient preferences.

Timing of Direct-Acting Antiviral Treatment in Patients with Hepatocellular Carcinoma

Although DAAs have demonstrated benefits in patients with HCC, data are limited regarding the optimal timing of DAA treatment in this population—should patients be treated at the time of HCC diagnosis before HCC treatment, or after documented complete response? As there are no conclusive data that DAA therapy is associated with increased risk of recurrence, differential time to recurrence, or aggressive tumor behavior in patients with HCC, the benefits of DAA treatment in patients with complete response to HCC treatment likely outweigh any potential risks. In light of the meta-analysis published by Ji et al. [23] showing a pooled SVR rate for 73% for patients with active HCC compared to 93% for patients with inactive HCC or no HCC, it is reasonable to defer DAA treatment until after HCC treatment. In a 2019 American Gastroenterological Association clinical guidance document, Singal et al. [38] recommended waiting to initiate DAA therapy for 4–6 months after HCC complete response to allow for confirmation of complete and durable response, which appears consistent with the current practice of most hepatology providers [14]. There are fewer data informing the optimal timing of DAA treatment in patients with intermediate- or advanced-stage HCC. Given the heterogeneity in tumor biology and treatment response, [39] it is reasonable to first initiate HCC-directed therapy with locoregional or systemic therapy. In patients with an objective response (partial or complete response) or stable disease, DAA therapy could be considered. In patients with tumor progression, prognosis will likely be driven by HCC-related mortality and the benefit of DAA therapy is likely limited.

Conclusion

The use of highly effective DAAs has dramatically altered the landscape for care of patients with HCV, including those with concomitant HCC. Though the currently published data have some limitations, DAAs appear to be safe and not associated with increased risk of recurrence in patients with a history of complete response (i.e., no viable tumor) following HCC-directed therapy. Similarly, DAAs appear to be safe in patients with active (i.e., viable) HCC and can be considered in select patients. In patients with early HCC eligible for curative therapy, we recommend initiating DAA therapy after HCC complete response given higher response rates than patients with active HCC. In patients who are LT candidates, decisions on the timing of DAA treatment should be made on a case-by-case basis, considering patient preferences and donor organ availability. There are limited data evaluating DAA therapy in patients with intermediate- and advanced-stage HCC, but treatment can be considered in select cases based on overall life expectancy, comorbid conditions, and response to HCC therapy. Overall, DAAs appear to be beneficial in patients with HCV and concomitant HCC and can be considered on a case-by-case basis.

References

1. Kulik L, El-Serag HB. Epidemiology and management of hepatocellular carcinoma. Gastroenterology. 2019;156(2):477–91. https://doi.org/10.1053/j.gastro.2018.08.065.
2. Chhatwal J, Kanwal F, Roberts MS, Dunn MA. Cost-effectiveness and budget impact of hepatitis C virus treatment with sofosbuvir and ledipasvir in the United States. Ann Intern Med. 2015;162(6):397–406. https://doi.org/10.7326/M14-1336.
3. Backus LI, Belperio PS, Shahoumian TA, et al. Direct-acting antiviral sustained virologic response: impact on mortality in patients without advanced liver disease. Hepatology. 2018;68(3):827–38. https://doi.org/10.1002/hep.29811.
4. Foster GR, Irving WL, Cheung MCM, et al. Impact of direct acting antiviral therapy in patients with chronic hepatitis C and decompensated cirrhosis. J Hepatol. 2016;64(6):1224–31. https://doi.org/10.1016/j.jhep.2016.01.029.
5. Lens S, Alvarado-Tapias E, Mariño Z, et al. Effects of all-oral antiviral therapy on HVPG and systemic hemodynamics in patients with hepatitis C virus-associated cirrhosis. Gastroenterology. 2017;153(5):1273–1283.e1. https://doi.org/10.1053/j.gastro.2017.07.016.
6. El-Serag HB, Christie IC, Puenpatom A, Castillo D, Kanwal F, Kramer JR. The effects of sustained virological response to direct-acting anti-viral therapy on the risk of extrahepatic manifestations of hepatitis C infection. Aliment Pharmacol Ther. 2019;49(11):1442–7. https://doi.org/10.1111/apt.15240.
7. Kanwal F, Kramer J, Asch SM, Chayanupatkul M, Cao Y, El-Serag HB. Risk of hepatocellular cancer in HCV patients treated with direct-acting antiviral agents. Gastroenterology. 2017;153(4):996–1005.e1. https://doi.org/10.1053/j.gastro.2017.06.012.
8. Ioannou GN, Green PK, Berry K. HCV eradication induced by direct-acting antiviral agents reduces the risk of hepatocellular carcinoma. J Hepatol. 2018;68:25–32. https://doi.org/10.1016/j.jhep.2017.08.030.
9. Li DK, Ren Y, Fierer DS, et al. The short-term incidence of hepatocellular carcinoma is not increased after hepatitis C treatment with direct-acting antivirals: an ERCHIVES study. Hepatology. 2018;67(6):2244–53. https://doi.org/10.1002/hep.29707.
10. Romano A, Angeli P, Piovesan S, et al. Newly diagnosed hepatocellular carcinoma in patients with advanced hepatitis C treated with DAAs: a prospective population study. J Hepatol. 2018;69(2):345–52. https://doi.org/10.1016/j.jhep.2018.03.009.
11. Calvaruso V, Cabibbo G, Cacciola I, et al. Incidence of hepatocellular carcinoma in patients with HCV-associated cirrhosis treated with direct-acting antiviral agents. Gastroenterology. 2018;155(2):411–421.e4. https://doi.org/10.1053/j.gastro.2018.04.008.
12. Reig M, Mariño Z, Perelló C, et al. Unexpected high rate of early tumor recurrence in patients with HCV-related HCC undergoing interferon-free therapy. J Hepatol. 2016;65(4):719–26. https://doi.org/10.1016/j.jhep.2016.04.008.
13. Villani R, Vendemiale G, Serviddio G. Molecular mechanisms involved in HCC recurrence after direct-acting antiviral therapy. Int J Mol Sci. 2018;20(1):49. https://doi.org/10.3390/ijms20010049.
14. Rich NE, Yang JD, Perumalswami PV, et al. Provider attitudes and practice patterns for direct-acting antiviral therapy for patients with hepatocellular carcinoma. Clin Gastroenterol Hepatol. 2020;18(4):974–83. https://doi.org/10.1016/j.cgh.2019.07.042.

15. ANRS Collaborative Study Group on Hepatocellular Carcinoma (ANRS CO22 HEPATHER, CO12 CirVir and CO23 CUPILT cohorts). Lack of evidence of an effect of direct-acting antivirals on the recurrence of hepatocellular carcinoma: data from three ANRS cohorts. J Hepatol. 2016;65(4):734–40. https://doi.org/10.1016/j.jhep.2016.05.045.
16. Saraiya N, Yopp AC, Rich NE, Odewole M, Parikh ND, Singal AG. Systematic review with meta-analysis: recurrence of hepatocellular carcinoma following direct-acting antiviral therapy. Aliment Pharmacol Ther. 2018;48(2):127–37. https://doi.org/10.1111/apt.14823.
17. Singal AG, Rich NE, Mehta N, et al. Direct-acting antiviral therapy not associated with recurrence of hepatocellular carcinoma in a multicenter north American cohort study. Gastroenterology. 2019;156(6):1683–1692.e1. https://doi.org/10.1053/j.gastro.2019.01.027.
18. Cabibbo G, Petta S, Barbara M, et al. Hepatic decompensation is the major driver of death in HCV-infected cirrhotic patients with successfully treated early hepatocellular carcinoma. J Hepatol. 2017;67(1):65–71. https://doi.org/10.1016/j.jhep.2017.01.033.
19. Singal AG, Rich NE, Mehta N, et al. Direct-acting antiviral therapy for hepatitis C virus infection is associated with increased survival in patients with a history of hepatocellular carcinoma. Gastroenterology. 2019;157(5):1253–1263.e2. https://doi.org/10.1053/j.gastro.2019.07.040.
20. Parikh ND, Mehta N, Hoteit MA, Yang JD, John BV, Moon AM, Salgia RJ, Pillai A, Kassab I, Saeed N, Thyssen E, Nathani P, McKinney J, Chan W, Durkin C, Connor M, Alsudaney M, Konjeti R, Durand B, Nissen NN, Kim HP, Paknikar R, Rich NE, Schipper MJ, Singal AG. Association between sustained virological response and clinical outcomes in patients with hepatitis C infection and hepatocellular carcinoma. Cancer. 2022;128(19):3470–8. https://doi.org/10.1002/cncr.34378.
21. Prenner SB, VanWagner LB, Flamm SL, Salem R, Lewandowski RJ, Kulik L. Hepatocellular carcinoma decreases the chance of successful hepatitis C virus therapy with direct-acting antivirals. J Hepatol. 2017;66(6):1173–81. https://doi.org/10.1016/j.jhep.2017.01.020.
22. Beste LA, Green PK, Berry K, Kogut MJ, Allison SK, Ioannou GN. Effectiveness of hepatitis C antiviral treatment in a USA cohort of veteran patients with hepatocellular carcinoma. J Hepatol. 2017;67(1):32–9. https://doi.org/10.1016/j.jhep.2017.02.027.
23. Ji F, Yeo YH, Wei MT, et al. Sustained virologic response to direct-acting antiviral therapy in patients with chronic hepatitis C and hepatocellular carcinoma: a systematic review and meta-analysis. J Hepatol. 2019;71(3):473–85. https://doi.org/10.1016/j.jhep.2019.04.017.
24. Sachdeva M, Chawla YK, Arora SK. Immunology of hepatocellular carcinoma. World J Hepatol. 2015;7(17):2080–90. https://doi.org/10.4254/wjh.v7.i17.2080.
25. Harrod E, Moctezuma-Velazquez C, Gurakar A, Ala A, Dieterich D, Saberi B. Management of concomitant hepatocellular carcinoma and chronic hepatitis C: a review. Hepatoma Res. 2019;5:28. https://doi.org/10.20517/2394-5079.2019.15.
26. Bruno S, Di Marco V, Iavarone M, et al. Improved survival of patients with hepatocellular carcinoma and compensated hepatitis C virus-related cirrhosis who attained sustained virological response. Liver Int. 2017;37(10):1526–34. https://doi.org/10.1111/liv.13452.
27. Dang H, Yeo YH, Yasuda S, et al. Cure with interferon-free direct-acting antiviral is associated with increased survival in patients with hepatitis C virus–related hepatocellular carcinoma from both east and west. Hepatology. 2020;71(6):1910–22. https://doi.org/10.1002/hep.30988.
28. Ghany MG, Morgan TR. Hepatitis C guidance 2019 update: American Association for the study of liver diseases–infectious diseases society of America recommendations for testing, managing, and treating hepatitis C virus infection. Hepatology. 2020;71(2):686–721. https://doi.org/10.1002/hep.31060.
29. Villanueva A. Hepatocellular carcinoma. N Engl J Med. 2019;380(15):1450–62. https://doi.org/10.1056/NEJMra1713263.
30. Bangaru S, Marrero JA, Singal AG. Review article: new therapeutic interventions for advanced hepatocellular carcinoma. Aliment Pharmacol Ther. 2020;51(1):78–89. https://doi.org/10.1111/apt.15573.
31. Finn RS, Ducreux M, Qin S, et al. IMbrave150: a randomized phase III study of 1L atezolizumab plus bevacizumab vs sorafenib in locally advanced or metastatic hepatocellular carcinoma. J Clin Oncol. 2018;36(15 Suppl):TPS4141. https://doi.org/10.1200/JCO.2018.36.15_suppl.TPS4141.
32. Revuelta-Herrero JL, Giménez-Manzorro A, Matilla-Peña A, Herranz-Alonso A, Sanjurjo-Sáez M. Concomitant use of sorafenib with ombitasvir/paritaprevir/ritonavir and dasabuvir: effectiveness and safety in clinical practice. J Clin Pharm Ther. 2018;43(6):906–9. https://doi.org/10.1111/jcpt.12727.
33. Mehta N, Dodge JL, Hirose R, Roberts JP, Yao FY. Increasing liver transplantation wait-list dropout for hepatocellular carcinoma with widening geographical disparities: implications for organ allocation. Liver Transpl. 2018;24(10):1346–56. https://doi.org/10.1002/lt.25317.
34. Salazar J, Saxena V, Kahn JG, et al. Cost-effectiveness of direct-acting antiviral treatment in hepatitis C–infected liver transplant candidates with compensated cirrhosis and hepatocellular carcinoma. Transplantation. 2017;101(5):1001–8. https://doi.org/10.1097/TP.0000000000001605.
35. Crismale JF, Ahmad J. Expanding the donor pool: hepatitis C, hepatitis B and human immunodeficiency virus–positive donors in liver transplantation. World J Gastroenterol. 2019;25(47):6799–812. https://doi.org/10.3748/wjg.v25.i47.6799.
36. Liu J, Ma B, Cao W, et al. Direct-acting antiviral agents for liver transplant recipients with recurrent genotype 1 hepatitis C virus infection: systematic review and meta-analysis. Transpl Infect Dis. 2019;21(2):e13047. https://doi.org/10.1111/tid.13047.
37. Zanetto A, Shalaby S, Vitale A, et al. Dropout rate from the liver transplant waiting list because of hepatocellular carcinoma progression in hepatitis C virus–infected patients treated with direct-acting antivirals. Liver Transpl. 2017;23(9):1103–12. https://doi.org/10.1002/lt.24790.
38. Singal AG, Lim JK, Kanwal F. AGA clinical practice update on interaction between oral direct-acting antivirals for chronic hepatitis C infection and hepatocellular carcinoma: expert review. Gastroenterology. 2019;156(8):2149–57. https://doi.org/10.1053/j.gastro.2019.02.046.
39. Rich NE, John BV, Parikh ND, et al. Hepatocellular carcinoma demonstrates heterogeneous growth patterns in a multi-center cohort of patients with cirrhosis. Hepatology. 2020;72(5):1654–65. https://doi.org/10.1002/hep.31159.

Should Locoregional Therapy Be Given in Combination with Systemic Therapy?

26

Jennifer Wang and Anjana Pillai

Locoregional therapy (LRT) given in combination with systemic therapy may improve time to progression (TTP) and overall survival (OS), however, at a price of increased side effects. It is still not clear how and which treatments should be combined for optimal outcome.

Main Points

1. Retrospective studies demonstrate lower incidence of recurrence, decreased TTP, and better OS at 1 year in patients treated with sorafenib plus radiofrequency ablation (RFA) vs. RFA alone.
2. A meta-analysis of transarterial chemoembolization (TACE) in combination with sorafenib found that combination therapy statistically significantly increased objective response rate (ORR), disease control rate, and OS compared with TACE alone, although the rate of adverse events (AEs) was also higher in TACE plus sorafenib group. Other meta-analyses have shown that the combination of sorafenib with TACE has improved TTP, but not OS.
3. Well-designed, randomized controlled, prospective trials are still needed before TACE-tyrosine kinase inhibitor combination therapy can be recommended as routine practice.
4. The SORAfenib in combination with local MICro-therapy guided by gadolinium-EOB-DTPA-enhanced MRI (SORAMIC) trial compared Transarterial Radioembolization (TARE) (yttrium 90 [Y90] resin) plus sorafenib to sorafenib alone in patients with advanced hepatocellular carcinoma (HCC) and showed that the combination therapy did not result in a statistically significant improvement in OS, though the results could have been affected by patient selection.
5. EMERALD-1 and LEAP-012 trials demonstrated that combining Transaterial Chemoembolization (TACE) with systemic therapy improved progression-free survival (PFS), but OS data are not currently available.

HCC is the most common form of primary liver cancer and is the fourth leading cause of cancer-related death worldwide [1]. In the United States, HCC mortality increased by 35% between 2002 and 2012, while the incidence continues to rise [2, 3], LRT can be considered as curative for those patients with early-stage HCC who are not surgical resection candidates [4, 5] for bridging or downstaging in the setting of liver transplantation, or as palliative therapy in those with more advanced disease [6–13]. However, because the majority of patients with HCC present in the intermediate/advanced stages [4, 5, 14], treatment options include LRT, systemic therapy, and in select patients, a combination of both. In this chapter, we will focus on current data and examine the benefits of combination of LRT with systemic therapy in a subgroup of patients with unresectable HCC.

Radiofrequency Ablation Plus Systemic Therapy

The synergism between LRT and systemic therapy has led to several studies on the safety and benefits of combination therapy, although the results have been conflicting (Table 26.1).

There are a few retrospective studies that showed lower incidence of recurrence (56.7% vs. 87.5%), decreased time to progression (TTP) (17 vs. 6.1 months) and overall survival (OS) (85.6% vs. 80.1% at 1 year) in patients treated with sorafenib plus RFA vs. RFA alone [15–17]. In a retrospective multicenter study, comparison between 64 patients who received sorafenib plus RFA and 64 patients treated with RFA alone showed that the 1-, 2-, 3-, and 4-year overall survival rates were 85.6%, 64.0%, 58.7%, and 50.3%,

J. Wang · A. Pillai (✉)
Division of Gastroenterology and Hepatology, University of Illinois Chicago, Chicago, IL, USA
e-mail: jwang022@uic.edu; Anjana.Pillai@bsd.uchicago.edu

N. S. Reau, A. G. Singal (eds.), *Controversies in Hepatocellular Carcinoma*, https://doi.org/10.1007/978-3-031-90225-3_26

Table 26.1 Positive trials in favor of LRT + systemic therapy

Trial (author, year published)	Treatment regimen	Stage of HCC	Patients (n)	Duration and daily dose of systemic therapy	Efficacy (combined vs. monotherapy)	References
Phase II (Sansonno, 2012)	c-TACE + sorafenib (sequential, 30 days after) vs. c-TACE + placebo	HCV-infected, BCLC stage B, CTP class A	31 vs. 31	N/A	TTP: 9.2 vs. 4.9 months; $P < 0.01$	[23]
Retrospective (Bai, 2013)	TACE + sorafenib (concurrent) vs. TACE alone	Intermediate or advanced HCC	82 vs. 164	N/A	TTP: 6.3 vs. 4.3 months; $P = 0.04$ OS: 7.5 vs. 5.1 months, $P = 0.009$	[40]
Phase II single-arm (Cosgrove, 2015)	DEB-TACE + sorafenib (1 week before and concurrent)	BCLC stage B/C, CTP class A-B7, ECOG 0-1	50	N/A	TTP 13.9 months OS 20.4 months	[41]
Phase II single-arm (Chao, 2016)	On-demand c-TACE + sorafenib (interrupted, discontinued for 4 days before and after TACE)	BCLC stage A/B, CTP class A/B, ECOG 0-1	192	N/A; 697 mg daily	PFS: 384 days TTP: 415 days	[42]
Observational registry study (Geschwind, 2016)	c-TACE or DEB-TACE with prior or concomitant sorafenib;	Unresectable HCC with a life expectancy >8 weeks	3202	15 weeks/688 mg	OS: Prior TACE 12.7 months; No prior TACE 9.2 months; TACE + concomitant sorafenib 21.6 months; TACE + sequential sorafenib 9.7 months	[43]
Phase III (Kudo, 2019)	c-TACE + sorafenib (started 2–3 weeks prior to TACE, discontinued 2 days before and after each TACE session vs. c-TACE alone	Unresectable HCC (BCLC stage A/B/C), CTP≤ 7, ECOG 0-1	80 vs. 76	38.7 weeks/355.2 mg	PFS: 25.2 vs. 13.5 months, $P = 0.006$ OS: not met	[26]
Phase II (Feng, 2014)	RFA followed by sorafenib vs. RFA alone	BCLC stage 0-B1 HCC	64 vs. 64	N/A	The 1-, 2-, 3-, and 4-year OS rates were 85.6%, 64.0%, 58.7%, and 50.3%, respectively, in the sorafenib-RFA group, and 80.7%, 47.2%, 30.9%, and 30.9%, respectively, in the RFA group	[15]
Retrospective (Fukuda, 2014)	RFA followed by sorafenib vs. RFA alone	Single HCC <3 cm	16 vs. 30	N/A	RFA-induced ablated area: 46.3 mm ± 10.3 and 33.0 mm ± 6.9 vs. 32.9 mm ± 7.6 and 25.6 mm ± 5.7.	[16]
Phase II (Kan, 2015)	RFA followed by sorafenib vs. RFA alone	HBV-related medium-sized HCC (3.1–5 cm)	30 vs. 32	N/A	Recurrence rate: 56.7 vs. 87.5%; $P < 0.01$ TTP 17 vs. 6.1 months; $P < 0.05$	[17]
Phase III (Sangro, 2025)	Durvalumab + bevacizumab + TACE vs. durvalumab + TACE vs. TACE alone	Unresectable HCC amenable to embolization, BCLC A/B/C, CTP A-B7, excludes Vp3 and Vp4	204 vs. 207 vs. 205	13.8 vs. 7.2 vs. 8.7 months	Median PFS: 15 vs. 10 vs. 8.2 months	[31]

(continued)

Table 26.1 (continued)

Trial (author, year published)	Treatment regimen	Stage of HCC	Patients (n)	Duration and daily dose of systemic therapy	Efficacy (combined vs. monotherapy)	References
Phase III (Kudo, 2025)	Pembrolizumab + lenvatinib + TACE vs. TACE alone	Unresectable HCC amenable to TACE in 1–2 sessions, CTP A	237 vs. 243	12.4 vs. 8.4 months	Median PFS: 14.6 vs. 10 months. 24-months survival: 75% vs. 69% (HR 0.80, *P* 0.0867	[32]
Negative trials of LRT + systemic therapy						
Phase III RCT (Kudo, 2011)	TACE followed by sorafenib (sequential) vs. TACE + placebo	Unresectable HCC, CTP class A, >25% tumor necrosis after 1 or 2 TACE	229 vs. 227	17.1 weeks/386 mg	TTP: 5.4 vs. 3.7 months; $P = 0.252$ OS: 29.7 days vs. not reached	[22]
Phase II (Lencioni, 2016)	DEB-TACE + sorafenib (concurrent) vs. DEB-TACE + placebo	Intermediate stage HCC (BCLC stage B), CTP class A, ECOG 0	154 vs. 153	21 weeks/566 mg	TTP: 169 vs. 166 days; $P = 0.07$ OS: Not reached	[24]
Phase III (Meyer, 2017)	DEB-TACE + sorafenib (1 week before, concurrent) vs. DEB-TACE + placebo	Intermediate stage HCC, CTP class A, ECOG ≤ 1	157 vs. 156	17.1 weeks/660 mg	PFS: 238 vs. 235 days; $P = 0.94$	[25]
Phase III (Park, 2019)	On-demand c-TACE + concurrent sorafenib vs. sorafenib alone	BCLC stage B/C, CTP≤ 7, ECOG 0-1	170 vs. 169	23.7 weeks vs. 18.3 weeks/ 557.9 mg vs. 584.7 mg	OS 12.8 vs. 10.8 months; $P = 0.29$ TTP 5.3 vs. 3.5 months; $P = 0.003$ PFS 5.2 vs. 3.6 months; $P = 0.01$	[44]
Retrospective (Gramenzi, 2015)	TARE vs. sorafenib	BCLC stage B/C, CTP class A/B, ECOG 0-1	63 vs. 74	16 weeks/608 mg	OS: 13.2 vs. 14.4 months; $P = 0.392$	[35]
Phase III (SARAH trial, Vilgrain, 2017)	TARE (Y90 resin) vs. sorafenib	Unresectable, BCLC stage B/C, CTP class A/B, ECOG 0-1	237 vs. 222	11.2 weeks/800 mg	OS: 8.0 vs. 9.9 months; $P = 0.18$	[37]
Phase III (*SIRveNIB trial*, Chow, 2018)	TARE (Y90 resin) vs. sorafenib	BCLC stage B/C without extrahepatic disease	182 vs. 178	13.8 weeks/644.5 mg	OS: 8.8 vs. 10.0 months; $P = 0.36$ Grade 3-4 AEs:27.7% vs.50.6%; $P < 0.001$	[36]
Phase II (SORAMIC trial, Ricke, 2019)	TARE (Y90 resin) followed by sorafenib vs. sorafenib	BCLC stage B/C, CTP ≤ B7, ECOG <2, poor candidate for TACE	216 vs. 208	29.4 weeks/474.4 mg vs. 19.3 weeks/523.1 mg	OS: 12.1 vs. 11.4 months; $P = 0.95$	[39]

respectively, in the sorafenib-RFA group, and 80.7%, 47.2%, 30.9%, and 30.9%, respectively, in the RFA group [15]. The very low rate of drug interruption (6%) in the study likely resulted in positive outcome. Prospective and randomized trials are needed to validate the efficacy and safety of the combined therapy.

Transarterial Chemoembolization Plus Systemic Therapy

TACE causes acute hypoxia within the tumor, leading to upregulation of local and systemic vascular endothelial growth factor, which could induce tumor progression and metastasis. The rationale to add sorafenib, a VEGF inhibitor, to prevent revascularization and tumor proliferation seems clear. A number of phase II and retrospective studies have investigated the efficacy and safety of TACE in combination with sorafenib with inconsistent results (see Table 26.1). The studies varied in timing of TACE with sorafenib (sequential vs. interrupted vs. concurrent), dosage and duration of sorafenib, as well as stages of HCC included.

In the first prospective study to examine the safety and efficacy of combining sorafenib and drug-eluting beads (DEB)-TACE in patients with unresectable HCC, the toxicity profile of the combination therapy was noted to be similar to that of sorafenib alone, and manageable with dose adjustment of sorafenib [18]. The study also demonstrated that the combined modality may also be efficacious with preliminary analysis, showing an objective response rate of 58%. Subsequent cohort studies affirmed the safety and tolerability of combined therapy in addition to its potential efficacy

[19, 20]. However, following the promising results of the above studies, a small prospective study of 15 patients who received doxorubicin-based TACE in combination with sorafenib was terminated early due to safety concerns [21]. There were 32 serious adverse events (grade $\geq$ 3) and 4 deaths considered to be TACE related. In addition to concerns about the safety of combination therapy, its efficacy also became controversial. Kudo and colleagues completed a phase III, randomized controlled trial of Japanese and Korean patients who received sorafenib 1-3 months after TACE with drug-eluting beads (DEB-TACE) and compared them to patients who received placebo after TACE [22]. The study did not show a difference in TTP, but the daily dose of sorafenib given was very low (median 386 mg) with 73% of patients undergoing dose reductions and 91% experiencing dose interruptions in the combination group. The authors noted there may be a race and geographic variation in trial outcomes.

In a well-designed RCT, enrolling only patients with hepatitis C (HCV) who received conventional TACE (cTACE) , the authors found that sorafenib administered 30 days after TACE as an adjuvant therapy showed promise in improving TTP (9.2 months in the cTACE + sorafenib group vs. 4.9 months in the cTACE alone group) [23]. The biggest benefits of the study were in reducing metachronous, multicentric progression of HCC in addition to local progression of TACE-treated areas. The authors postulated that the nontumoral areas in HCV-related HCC has high carcinogenic risk that is effectively influenced by adjuvant sorafenib. The promising outcomes suggest that there may also be a difference in response to therapy depending on the etiology and duration of the underlying cirrhosis, timing and dosage of sorafenib administered and the type of TACE treatment utilized.

The SPACE trial was the first large global RCT testing the efficacy of combining DEB-TACE with sorafenib in patients with intermediate-stage HCC [24]. The trial attempted to standardize TACE by using exclusively DEB-TACE. In total, 307 patients were randomized to DEB-TACE plus sorafenib or DEB-TACE plus placebo. The authors did not find a difference in TTP but noted manageable toxicity profile with the combination therapy. The negative result was felt to be due to the short duration of sorafenib therapy (mean 21 weeks). The decision to terminate sorafenib treatment based on the criteria for TACE-untreatable progression may not have been appropriate. The authors noted a regional difference in outcomes as patients from Asian countries had a greater improvement in TTP than patients from non-Asian countries—a key distinction was the duration of sorafenib administered (30 vs. 17.4 weeks, respectively).

The SPACE trial was followed by another negative study—TACE-2 study [25]. Twenty institutes in England participated in this study to clarify whether combining DEB-TACE and sorafenib improved outcome but did not find difference in progression-free survival or overall survival. The negative result raised the question of whether progression based on RESIST 1.1 (Response Evaluation Criteria in Solid Tumors) or mRECIST (Modified Response Evaluation Criteria in Solid Tumors) was adequate in trials of TACE combination therapies. The authors noted that sorafenib may have been prematurely discontinued based on appearance of new intrahepatic lesions, which does not always indicate disease progression given intrahepatic metastasis is common in HCC. This led to the proposal for new definition of disease progression.

After nearly a decade of negative trials, the TACTICS trial by Kudo and colleagues was the first phase 3 trial to meet its primary end point. It redefined disease progression as unTACEable progression or TACE failure as opposed to Response Evaluation Criteria in Cancer of the Liver (RECICL) or mRECIST criteria [26]. The study showed that sorafenib plus concurrent DEB-TACE has significantly better median progression-free survival (PFS) than the DEB-TACE only group (25.2 vs. 13.5 months) with comparable safety outcomes. The combination group also had longer interval between TACE sessions. It was hypothesized that the administration of sorafenib 2–3 weeks prior to TACE helped with normalization of tumor vessels, leading to improved efficacy of TACE. Sorafenib also helped stabilize disease progression between TACE sessions. The duration of sorafenib treatment was longer than prior trials, which could explain the positive results.

Meta-analyses of TACE in combination with sorafenib have also been conducted. A meta-analysis including 14 studies and a total of 1670 patients with advanced HCC showed that combination therapy significantly increased objective response rate, disease control rate and OS compared with TACE alone, although the rate of adverse events was also higher in TACE plus sorafenib group [27]. Other meta-analyses have shown that the combination of sorafenib with TACE has improved TTP, but not OS [28, 29].

The landmark IMbrave150 trial in 2019 established atezolizumab plus bevacizumab as the standard first-line treatment for unresectable HCC, ushering in a new era of combination therapy with the addition of immunotherapy or anti-VEGF therapy to TACE [30]. The EMERALD-1 trial is a multiregional, randomized, double-blind, placebo-controlled, phase 3 study that met its primary endpoint with a significant PFS improvement in participants with unresectable HCC who received the combination of durvalumab, bevacizumab and TACE. The median PFS in this group reached 15 months (95% CI 11.1–18.9) compared to 10 months with durvalumab plus TACE (95% CI 9–12.7) and 8.2 months with TACE alone (95% CI 6.9–11.1) [31]. The PFS benefit was consistent across subgroups of a heterogeneous population of patients. The number and timing of TACE was at the investigator's discretion but up to four TACE procedures were allowed within 16 weeks. Durvalumab was given 1 week after the first TACE procedure, whereas bevacizumab was not given until 2 weeks after the last TACE procedure due to concern for side effects with concurrent treatment. Serious adverse events and adverse events leading to discontinuation were more common in the group that received durvalumab and bevacizumab, which was consistent with a longer duration of exposure in this group. The most common grade maximum 3–4 adverse events were hypertension, anemia, acute kidney injury and proteinuria in the group that received combination therapy. EMERLAD-1 is ongoing for the final overall survival analysis, which could potentially change the standard of care in this population.

Similarly, LEAP-012 is positive phase 3 trial that demonstrated improved PFS in patients who received combination therapy [32]. In total, 480 patients with intermediate-stage HCC were randomized to receive TACE plus lenvatinib plus pembrolizumab or TACE plus dual placebo for up to 2 years. Max of two TACE procedures were allowed before the first 9-week evaluation imaging. The study demonstrated a median PFS of 14.6 months (95% CI 12.6–16.7) in the combination group compared with 10 months in the TACE alone group (95% CI 8.1–12.2). There was a favorable trend for improved two-year survival in the lenvatinib plus pembrolizumab group compared to the placebo (75% vs. 69%, HR 0.80, $P = 0.087$), but longer follow-up is needed. In terms of safety, grade 3 or higher treatment-related adverse events were more common in the combination group (71%) compared to placebo (32%), and deaths due to adverse events occurred in four (2%) of participants who received lenvatinib plus pembrolizumab.

The positive results of EMERALD-1 and LEAP-012 trials are encouraging and suggest that combining TACE with systemic therapy may improve outcomes of patients with intermediate-stage HCC. However, it remains uncertain whether a PFS benefit alone without a confirmed overall survival benefit, especially in the context of increased toxicity, is sufficient to warrant a change in the current standard of care. Notably, both trials were TACE based with fixed sessions, which diverge from real-world practice, where TACE is often performed on an on-demand basis according to radiological response. This discrepancy raises further questions about optimal timing and sequencing of systemic therapy. Although meaningful advances in the management of intermediate-stage HCC have been limited in the past decade, these trials represent a hopeful step forward.

Transarterial Radioembolization Plus Systemic Therapy

TARE is another LRT frequently used in patients with intermediate-stage HCC. It selectively delivers radioactive microspheres loaded with Y90, iodine-131, or rhenium-188 to a tumor-feeding artery to cause radiation-induced cell necrosis at the tumor level [32–34].

In a proof-of-concept, retrospective study, Gramenzi et al. compared 74 patients on sorafenib with 63 patients in the TARE arm for the treatment of intermediate-locally advanced HCC. The authors noted comparable survival outcomes between the two groups (median OS: 13.2 months in TARE patients and 14.4 months in sorafenib group) [35].

Since that study, there have been two large phase 3 RCTs comparing TARE to sorafenib in patients with locally advanced HCC: SARAH and SIRveNIB trials [36, 37]. Both trials failed to show survival difference but did show improved toxicity profile of TARE over sorafenib. The European SARAH trial (SorAfenib versus Radioembolisation in Advanced Hepatocellular carcinoma) enrolled 467 patients from multiple centers across France and randomized them to continuous sorafenib or TARE 2–5 weeks after randomization. The study showed that TARE did not lead to a significant improvement in survival but did reduce the severity of side effects and was significantly better tolerated with better quality of life [37]. The SIRveNIB (selective internal radiation therapy v sorafenib) was conducted at multiple centers in the Asia-Pacific region, with 360 patients, comparing the efficacy of a single delivery of radioembolization versus continuous sorafenib dosing. The study yielded similar findings as the SARAH trial in terms of similar overall survival between the two groups and better tolerability of TARE group. Furthermore, it showed significantly improved PFS and TTP in the TARE-treated population [36].

The reasons for negative results of SARAH and SIRveNIB were felt to be due to poor study design, lack of center experience with TARE, and no use of boosted radiation [38]. A subgroup analysis of the SIRveNIB trial did find survival benefit with TARE over sorafenib in patients with BCLC stage C HCC (9.2 vs. 5.8 months; $P = 0.0475$), highlighting the importance of patient selection [36].

The SORAMIC trial was the first large RCT comparing TARE (Y-90 resin) plus sorafenib ($n = 216$) to sorafenib alone ($n = 208$) in patients with advanced HCC and showed that the combination therapy did not result in a significant improvement in overall survival compared with sorafenib alone [39]. Subgroup analysis suggested that patient selection may have contributed to the negative results. There was a survival benefit for combination therapy in patients who were ≤ 65 years (HR 0.65; 95% CI 0.43–1.00; $P = 0.046$); non-cirrhotic patients (HR 0.46; 95% CI 0.25–0.86); $P = 0.013$), and patients with HCC of nonalcoholic etiology (HR 0.63; 95% CI 0.45–0.89); $P = 0.009$).

Following the encouraging results from the EMERALD-1 study, EMERALD-Y90 is a phase 2 study to evaluate the efficacy and safety of TARE with durvalumab monotherapy, followed by durvalumab plus bevacizumab [NCT06040099]. Additional ongoing clinical trials investigating the response rate of TARE with atezolizumab and bevacizumab (STRATUM, NCT05377034), TARE followed by durvalumab plus tremelimumab (ROWAN, NCT05063565) and TARE with concurrent durvalumab plus tremelimumab (IMMUWIN, NCT04522544) are highly anticipated. The results of these trials could contribute to the establishment of the role of combination therapy in the treatment of HCC.

Future Directions

HCC is a complex disease that requires a multidisciplinary approach to develop the best possible individualized treatment plan. At present, the role of combination therapy with LRT plus systemic therapy is not well established in the treatment algorithm of HCC, although its safety and tolerability are well established in multiple trials (see Table 26.1). The trials comparing survival outcomes of combination therapy versus either modality alone have shown inconsistent results, largely due to heterogenous nature of study design and patient population. The most recent trials have shown improved progression-free survival with combination therapy over TACE alone in patients with advanced HCC. Better understanding of molecular hepatocarcinogenesis of HCC and development of predictive biomarkers would be invaluable to guide optimal choice of therapy and help predict the response to currently available treatment modalities.

References

1. Kanwal F, Singal AG. Surveillance for hepatocellular carcinoma: current best practice and future direction. Gastroenterology. 2019;157(1):54–64. https://doi.org/10.1053/j.gastro.2019.02.049.
2. Bertuccio P, Turati F, Carioli G, et al. Global trends and predictions in hepatocellular carcinoma mortality. J Hepatol. 2017;67(2):302–9. https://doi.org/10.1016/j.jhep.2017.03.011.
3. Cancer Today. International Agency for Research on Cancer, World Health Organization. Published 2022. Accessed 28 April 2025. https://gco.iarc.fr/today/en/data-sources-methods#incidence
4. Llovet J, Brú C, Bruix J. Prognosis of hepatocellular carcinoma: the BCLC staging classification. Semin Liver Dis. 1999;19(03):329–38. https://doi.org/10.1055/s-2007-1007122.
5. Maida M, Orlando E, Cammà C, Cabibbo G. Staging systems of hepatocellular carcinoma: a review of literature. World J Gastroenterol. 2014;20(15):4141. https://doi.org/10.3748/wjg.v20.i15.4141.
6. Ibrahim SM, Kulik L, Baker T, et al. Treating and downstaging hepatocellular carcinoma in the caudate lobe with Yttrium-90 radioembolization. Cardiovasc Intervent Radiol. 2012;35(5):1094–101. https://doi.org/10.1007/s00270-011-0292-x.
7. Kulik LM, Atassi B, van Holsbeeck L, et al. Yttrium-90 microspheres (TheraSphere®) treatment of unresectable hepatocellular carcinoma: downstaging to resection, RFA and bridge to transplantation. J Surg Oncol. 2006;94(7):572–86. https://doi.org/10.1002/jso.20609.
8. Yao FY, Hirose R, LaBerge JM, et al. A prospective study on downstaging of hepatocellular carcinoma prior to liver transplantation. Liver Transpl. 2005;11(12):1505–14. https://doi.org/10.1002/lt.20526.
9. Lai Q, Vitale A, Iesari S, et al. The intention-to-treat effect of bridging treatments in the setting of Milan criteria–in patients waiting for liver transplantation. Liver Transpl. 2019;25(7):1023–33. https://doi.org/10.1002/lt.25492.
10. Sapisochin G, Barry A, Doherty M, et al. Stereotactic body radiotherapy vs. TACE or RFA as a bridge to transplant in patients with hepatocellular carcinoma. An intention-to-treat analysis. J Hepatol. 2017;67(1):92–9. https://doi.org/10.1016/j.jhep.2017.02.022.
11. Green TJ, Rochon PJ, Chang S, et al. Downstaging disease in patients with hepatocellular carcinoma outside of Milan criteria: strategies using drug-eluting bead chemoembolization. J Vasc Interv Radiol. 2013;24(11):1613–22. https://doi.org/10.1016/j.jvir.2013.07.024.
12. Galle PR, Forner A, Llovet JM, et al. EASL clinical practice guidelines: management of hepatocellular carcinoma. J Hepatol. 2018;69(1):182–236. https://doi.org/10.1016/j.jhep.2018.03.019.
13. Marrero JA, Kulik LM, Sirlin CB, et al. Diagnosis, staging, and management of hepatocellular carcinoma: 2018 practice guidance by the American Association for the Study of Liver Diseases. Hepatology. 2018;68(2):723–50. https://doi.org/10.1002/hep.29913.
14. Saraswat VA, Pandey G, Shetty S. Treatment algorithms for managing hepatocellular carcinoma. J Clin Exp Hepatol. 2014;4:S80–9. https://doi.org/10.1016/j.jceh.2014.05.004.
15. Feng X, Xu R, Du X, et al. Combination therapy with sorafenib and radiofrequency ablation for BCLC stage 0–B1 hepatocellular carcinoma: a multicenter retrospective cohort study. Am J Gastroenterol. 2014;109(12):1891–9. https://doi.org/10.1038/ajg.2014.343.
16. Fukuda H, Numata K, Moriya S, et al. Hepatocellular carcinoma: concomitant sorafenib promotes necrosis after radiofrequency ablation – propensity score matching analysis. Radiology. 2014;272(2):598–604. https://doi.org/10.1148/radiol.14131640.
17. Kan X, Jing Y, Wan QY, et al. Sorafenib combined with percutaneous radiofrequency ablation for the treatment of medium-sized hepatocellular carcinoma. Eur Rev Med Pharmacol Sci. 2015;19(2):247–55. http://www.ncbi.nlm.nih.gov/pubmed/25683938
18. Pawlik TM, Reyes DK, Cosgrove D, Kamel IR, Bhagat N, Geschwind JFH. Phase II trial of sorafenib combined with concurrent transarterial chemoembolization with drug-eluting beads for hepatocellular carcinoma. J Clin Oncol. 2011;29(30):3960–7. https://doi.org/10.1200/JCO.2011.37.1021.
19. Cabrera R, Pannu DS, Caridi J, et al. The combination of sorafenib with transarterial chemoembolisation for hepatocellular carcinoma. Aliment Pharmacol Ther. 2011;34(2):205–13. https://doi.org/10.1111/j.1365-2036.2011.04697.x.
20. Park JW, Koh YH, Kim HB, et al. Phase II study of concurrent transarterial chemoembolization and sorafenib in patients with unresectable hepatocellular carcinoma. J Hepatol. 2012;56(6):1336–42. https://doi.org/10.1016/j.jhep.2012.01.006.
21. Sieghart W, Pinter M, Reisegger M, et al. Conventional transarterial chemoembolisation in combination with sorafenib for patients with hepatocellular carcinoma: a pilot study. Eur Radiol. 2012;22(6):1214–23. https://doi.org/10.1007/s00330-011-2368-z.
22. Kudo M, Imanaka K, Chida N, et al. Phase III study of sorafenib after transarterial chemoembolisation in Japanese and Korean patients with unresectable hepatocellular carcinoma. Eur J Cancer. 2011;47(14):2117–27. https://doi.org/10.1016/j.ejca.2011.05.007.
23. Sansonno D, Lauletta G, Russi S, Conteduca V, Sansonno L, Dammacco F. Transarterial chemoembolization plus sorafenib: a sequential therapeutic scheme for HCV-related intermediate-stage hepatocellular carcinoma: a randomized clinical trial. Oncologist. 2012;17(3):359–66. https://doi.org/10.1634/theoncologist.2011-0313.

24. Lencioni R, Llovet JM, Han G, et al. Sorafenib or placebo plus TACE with doxorubicin-eluting beads for intermediate stage HCC: the SPACE trial. J Hepatol. 2016;64(5):1090–8. https://doi.org/10.1016/j.jhep.2016.01.012.
25. Meyer T, Fox R, Ma YT, et al. Sorafenib in combination with transarterial chemoembolisation in patients with unresectable hepatocellular carcinoma (TACE 2): a randomised placebo-controlled, double-blind, phase 3 trial. Lancet Gastroenterol Hepatol. 2017;2(8):565–75. https://doi.org/10.1016/S2468-1253(17)30156-5.
26. Kudo M, Ueshima K, Ikeda M, et al. Randomised, multicentre prospective trial of transarterial chemoembolisation (TACE) plus sorafenib as compared with TACE alone in patients with hepatocellular carcinoma: TACTICS trial. Gut. 2020;69(8):1492–501. https://doi.org/10.1136/gutjnl-2019-318934.
27. Cai R, Song R, Pang P, et al. Transcatheter arterial chemoembolization plus sorafenib versus transcatheter arterial chemoembolization alone to treat advanced hepatocellular carcinoma: a meta-analysis. BMC Cancer. 2017;17(1):714. https://doi.org/10.1186/s12885-017-3707-5.
28. Li L, Zhao W, Wang M, et al. Transarterial chemoembolization plus sorafenib for the management of unresectable hepatocellular carcinoma: a systematic review and meta-analysis. BMC Gastroenterol. 2018;18(1):138. https://doi.org/10.1186/s12876-018-0849-0.
29. Wang G, Liu Y, Feng ZS, et al. Sorafenib combined with transarterial chemoembolization in patients with hepatocellular carcinoma: a meta-analysis and systematic review. Hepatol Int. 2016;10(3):501–10. https://doi.org/10.1007/s12072-015-9700-7.
30. Finn RS, Qin S, Ikeda M, et al. Atezolizumab plus bevacizumab in unresectable hepatocellular carcinoma. N Engl J Med. 2020;382(20):1894–905. https://doi.org/10.1056/NEJMoa1915745.
31. Sangro B, Kudo M, Erinjeri JP, et al. Durvalumab with or without bevacizumab with transarterial chemoembolisation in hepatocellular carcinoma (EMERALD-1): a multiregional, randomised, double-blind, placebo-controlled, phase 3 study. Lancet. 2025;405(10474):216–32. https://doi.org/10.1016/S0140-6736(24)02551-0.
32. Kudo M, Ren Z, Guo Y, et al. Transarterial chemoembolisation combined with lenvatinib plus pembrolizumab versus dual placebo for unresectable, non-metastatic hepatocellular carcinoma (LEAP-012): a multicentre, randomised, double-blind, phase 3 study. Lancet. 2025;405(10474):203–15. https://doi.org/10.1016/S0140-6736(24)02575-3.
33. Sangro B, Salem R. Transarterial chemoembolization and radioembolization. Semin Liver Dis. 2014;34(04):435–43. https://doi.org/10.1055/s-0034-1394142.
34. Salem R, Lewandowski RJ, Mulcahy MF, et al. Radioembolization for hepatocellular carcinoma using Yttrium-90 microspheres: a comprehensive report of long-term outcomes. Gastroenterology. 2010;138(1):52–64. https://doi.org/10.1053/j.gastro.2009.09.006.
35. Gramenzi A, Golfieri R, Mosconi C, et al. Yttrium-90 radioembolization vs sorafenib for intermediate-locally advanced hepatocellular carcinoma: a cohort study with propensity score analysis. Liver Int. 2015;35(3):1036–47. https://doi.org/10.1111/liv.12574.
36. Chow PKH, Gandhi M, Tan SB, et al. SIRveNIB: selective internal radiation therapy versus sorafenib in Asia-Pacific patients with hepatocellular carcinoma. J Clin Oncol. 2018;36(19):1913–21. https://doi.org/10.1200/JCO.2017.76.0892.
37. Vilgrain V, Pereira H, Assenat E, et al. Efficacy and safety of selective internal radiotherapy with yttrium-90 resin microspheres compared with sorafenib in locally advanced and inoperable hepatocellular carcinoma (SARAH): an open-label randomised controlled phase 3 trial. Lancet Oncol. 2017;18(12):1624–36. https://doi.org/10.1016/S1470-2045(17)30683-6.
38. Sposito C, Mazzaferro V. The SIRveNIB and SARAH trials, radioembolization vs. sorafenib in advanced HCC patients: reasons for a failure, and perspectives for the future. HepatoBiliary Surg Nutr. 2018;7(6):487–9. https://doi.org/10.21037/hbsn.2018.10.06.
39. Ricke J, Klümpen HJ, Amthauer H, et al. Impact of combined selective internal radiation therapy and sorafenib on survival in advanced hepatocellular carcinoma. J Hepatol. 2019;71(6):1164–74. https://doi.org/10.1016/j.jhep.2019.08.006.
40. Bai W, Wang YJ, Zhao Y, et al. Sorafenib in combination with transarterial chemoembolization improves the survival of patients with unresectable hepatocellular carcinoma: a propensity score matching study. J Dig Dis. 2013;14(4):181–90. https://doi.org/10.1111/1751-2980.12038.
41. Cosgrove DP, Reyes DK, Pawlik TM, Feng AL, Kamel IR, Geschwind JFH. Open-label single-arm phase II trial of sorafenib therapy with drug-eluting bead transarterial chemoembolization in patients with unresectable hepatocellular carcinoma: clinical results. Radiology. 2015;277(2):594–603. https://doi.org/10.1148/radiol.2015142481.
42. Chao Y, Chung Y, Han G, et al. The combination of transcatheter arterial chemoembolization and sorafenib is well tolerated and effective in a Sian patients with hepatocellular carcinoma: final results of the START trial. Int J Cancer. 2015;136(6):1458–67. https://doi.org/10.1002/ijc.29126.
43. Geschwind JF, Kudo M, Marrero JA, et al. TACE treatment in patients with sorafenib treated unresectable hepatocellular carcinoma in clinical practice: final analysis of GIDEON. Radiology. 2016;279(2):630–40. https://doi.org/10.1148/radiol.2015150667.
44. Park JW, Kim YJ, Kim DY, et al. Sorafenib with or without concurrent transarterial chemoembolization in patients with advanced hepatocellular carcinoma: the phase III STAH trial. J Hepatol. 2019;70(4):684–91. https://doi.org/10.1016/j.jhep.2018.11.029.

Should Patients with Decompensated Cirrhosis Receive Systemic Therapy?

27

Anthony Bejjani and Richard S. Finn

Main Points

1. For patients with hepatocellular carcinoma (HCC), there are two major prognostic factors, tumor burden and degree of liver dysfunction.
2. Select studies provide evidence that overall survival (OS) is predominantly limited by CPS class B status and not antitumor efficacy [1, 2].
3. To minimize the effect of cirrhosis on survival outcomes, the majority of phase 3 studies evaluating systemic agents for the treatment of HCC have excluded patients with CPS scores greater than A5 and A6, except for a minority that allowed up to B7.
4. It can be hypothesized that in some cases treating the advanced HCC may improve liver function and improve survival. Now, with newer combinations that have higher response rates, it might be reasonable to consider these options for the treatment of select patients.

HCC usually arises in the background of chronic liver disease, which is a competing risk for death for patients with all stages of HCC. Given that the primary purpose of treatment for patients with HCC is to extend OS, consideration of a patient's underlying liver function must be considered to achieve this goal and to limit treatment-related toxicity that may be increased in the setting of end-stage liver disease. To minimize the effect of cirrhosis on survival outcomes, the majority of phase 3 studies evaluating systemic agents for the treatment of HCC have excluded patients with CPS scores greater than A5 and A6, except for a minority that allowed up to B7. As a result, the true efficacy of these agents in patients with more advanced, decompensated cirrhosis is unknown. There have been small, single-arm, and observational studies that have given us a sense of the safety and tolerability of some of these agents in the less-compensated patients. Survival data from these studies are difficult to interpret without a control arm, but consistently the prognostic significance of CPS score is confirmed. The most experience to date comes from the use of sorafenib based on its approval in 2007, and current data suggest sorafenib likely is a safe option for patients with CPS class B cirrhosis. Other more recently approved systemic agents, including lenvatinib, regorafenib, cabozantinib, and ramucirumab, have much fewer data regarding safety and efficacy outside the CPS class A group. Limited data with immune checkpoint inhibitors suggests they have a favorable safety profile in patients with CPS class B cirrhosis. In this chapter, we will review the available evidence for the use of extending treatment to patients with decompensated cirrhosis.

A. Bejjani · R. S. Finn (✉)
Department of Medicine, Division of Hematology/Oncology, Geffen School of Medicine at University of California Los Angeles, Jonsson Comprehensive Cancer Center, Los Angeles, CA, USA
e-mail: abejjani@mednet.ucla.edu; RFinn@mednet.ucla.edu

Staging Hepatocellular Carcinoma: The Importance of Liver Function in Outcomes of HCC

Unlike other malignancies in which survival is dominated by tumor characteristics (i.e., tumor, nodes, metastasis as used in the TNM system), for patients with HCC there are two major prognostic factors, tumor burden and degree of liver dysfunction. The Barcelona Clinic Liver Cancer (BCLC) staging system includes an assessment of both characteristics (as well as performance status), has been incorporated into most prospective studies in HCC, and is advocated by global HCC guidelines [3, 4]. The BCLC uses the CPS, which includes both objective and subjective measurements (degree of ascites and encephalopathy) of liver dysfunction, but there have been more recent efforts to include assessment-based objective data such as the Model for End-Stage Liver Disease or albumin-bilirubin scores [5–7]. Regardless, it is

N. S. Reau, A. G. Singal (eds.), *Controversies in Hepatocellular Carcinoma*, https://doi.org/10.1007/978-3-031-90225-3_27

very clear that in clinical trial design and daily practice, ignoring the degree of liver function will increase toxicity and alter expected survival. The most common etiologies of cirrhosis include chronic viral hepatitis (hepatitis B and C), alcohol use, and nonalcohol steatohepatitis. These all represent chronic inflammatory processes that lead to cirrhosis and clinical sequelae of portal hypertension, impaired hepatocyte function, and liver nodularity (cirrhosis). These patients are susceptible to acute decompensations of liver function that may manifest as encephalopathy, coagulopathy, ascites, and/or gastrointestinal bleeding, accompanied by high risk of further clinical deterioration [8]. All-cause mortality considerably worsens in this state, especially in patients who concurrently have HCC [9].

The CPS for cirrhosis has been widely used to easily classify the severity of cirrhosis at the bedside and time of evaluation. The score serves as a good prognostic factor as mortality rates are quite distinct for CPS class A patients compared to CPS class C patients [9]. Additional staging systems derived through a metanalysis also delineate the presence of ascites and/or bleeding esophageal varices as markers of decompensation, in which mortality markedly rises compared to a compensated state [9]. Given the direct relationship between increased morbidity and mortality and degree of liver dysfunction, it is often beneficial to engage hepatology patients to optimize their condition before initiation of treatment for patients with HCC. It is important to remember that the presence of HCC also worsens the prognosis of decompensated patients, complicating the ability to demonstrate clinical benefits of anticancer therapies [9].

Expert consensus guidelines for the treatment of earlier stages of HCC also emphasize the selection of patients with compensated cirrhosis to reduce the likelihood of decompensation. CPS class B8 or higher and refractory ascites are largely regarded as absolute contraindications to conventional transarterial chemoembolization [10]. Multiple single-institution studies have shown radiofrequency ablation to be safe in CPS class B patients, with one study identifying pretreatment bilirubin as a risk factor for decompensation [1, 2, 11]. These studies also provide further evidence that OS is predominantly limited by CP class B status and not antitumor efficacy [1, 2, 11].

Sorafenib in Advanced Hepatocellular Carcinoma

The phase 3 development of sorafenib, a small-molecule inhibitor of vascular endothelial growth factor receptors (VEGFRs) 1, 2, and 3, platelet-derived growth factor receptor β (PDGFR-β), and Raf-1/B-Raf, came after a phase 2 single-arm study in patients with advanced HCC and both CPS class A and CPS class B liver disease. While there were no statistically significant pharmacokinetic differences observed between the CPS class A and CPS class B groups, there was a decrease in time to progression (TTP) and survival between the 2 groups. TTP for CPS class A patients was 21 weeks (95% CI, 16–25 weeks) and for CPS class B patients was 13 weeks (95% CI, 9–18 weeks) and OS time for CPS class A patients was 41 weeks (95% CI, 36.6–63.6 weeks) and 14 weeks for CPS class B patients (95% CI, 11.6–25.7 weeks). In addition, CPS class B patients had more frequent worsening of their cirrhosis [12]. The pivotal SHARP trial compared sorafenib to placebo. The study met its primary end point of OS in the study population (10.7 months vs 7.9 months; hazard ratio [HR] = 0.69; 95% CI, 0.55–0.87) [13]. To avoid masking the benefit of sorafenib by patients dying of advanced liver disease, the trial required CPS class A liver disease, and accrued only 14 patients (5%) with CPS class B [13]. A concurrent phase 3 trial conducted predominantly in the Asia-Pacific region demonstrated a very similar magnitude of benefit with an HR of 0.68 for the benefit of sorafenib compared to placebo in advanced HCC, but with considerably lower OS (6.5 months vs 4.2 months) compared to the SHARP population [14]. This difference most likely stems from the Asia-Pacific population having statistically significantly more extrahepatic spread, BCLC stage C, and poorer performance status patients compared to the SHARP population, as this trial also had quite few CPS class B patients (2.7% of patient population) [14]. Of note, both trials excluded patients with clinically significant gastrointestinal bleeding within 30 days before study entry. Both of these trials, in excluding CPS class B patients, were unable to provide prospective evidence of efficacy and safety of sorafenib in advanced HCC.

In the decade following the pivotal SHARP trial, many agents were compared to sorafenib in the first-line setting and did not demonstrate superiority or noninferiority. As a result, with sorafenib remaining the only agent for first-line systemic treatment of advanced HCC, numerous cohort studies, retrospective analyses, and prospective trials were conducted to expand treatment beyond CPS class A patients.

In one retrospective cohort of sorafenib-treated patients through the Austrian registry with CPS class A/B/C cirrhosis, TTP remained similar across groups (median 2.2 months CPS class A vs median 2.9 months CPS class B vs 2 CPS class C patients with 2.8/5.2-month TTP) [15]. However, OS was quite poor with advanced cirrhosis (median OS at least 8.3 months in CPS class A vs 4.3 months in CPS class B vs 1.5 months in CPS class C; P = 0.0001) [15]. One small study attempted to evaluate the safety and efficacy of sorafenib in the CPS class B7 and CPS class B8 populations exclusively. The study had to be terminated before completing accrual because there was a large dropout from patients who were decompensated during the screening period and could not go on to study treatment, highlighting the fragility of this population of CPS class B patients [16]. Another prospective study by Guedes Leal et al. [17] further delineated safety and efficacy for CPS class B patients by cirrhosis severity (i.e., B7, B8, and B9), with the overall CPS class B group having statistically significantly worse OS compared to CPS class A patients (12 months vs 6 months). There were no statistically significant OS differences between CPS class B7 (8 months), B8 (5 months), and B9 patients (6 months) in this study, but a similar numerical trend of worse OS with worsening CPS status seen in other single-institution studies was observed [17–19]. Of note, the frequency of adverse events (AEs) for the patients in the Guedes Leal study [17] leading to interruption was similar across CPS class A patients (30%) and CPS class B patients (28%). CPS class B patients had more AEs related to liver decompensation (ascites, encephalopathy) as compared to CP A patients (hand-foot syndrome) [17]. A meta-analysis of 30 trials (4 randomized controlled, 26 cohort studies) of patients who were treated with sorafenib for advanced HCC showed CPS class B status as a negative predictor for OS, with a median OS of 8.8 months for CPS class A patients vs 4.6 months for CPS class B patients ($P = 0.001$), but with similar tolerability of sorafenib [20]. A large retrospective cohort of Veterans Affairs patients established noninferiority (prespecific HR = 1.10) of sorafenib doses of less than 800 mg/day in CPS class A and CPS class B patients, thus providing more dosing options [21].

Beyond the data from these individual institutions and studies, 2 major registries, GIDEON and INSIGHT, have provided key data regarding real-world experience with sorafenib with patients with advanced cirrhosis.

The GIDEON (Global Investigation of therapeutic DEcisions in HCC and Of its treatment with sorafeNib) trial, conducted in more than 40 countries and involving more than 3000 patients, was a prospective, observational registry study undertaken to evaluate the use of sorafenib in HCC patients, especially to gather more comprehensive data on the use of sorafenib in patients with CPS class B cirrhosis [22]. Marrero et al. [23] conducted an analysis of safety and efficacy of sorafenib stratified by CPS class status (A, B7/8/9, all B, and C). Median OS for CPS class A cirrhosis (1968 patients, 61% of study population) was 13.8 months, statistically significantly higher than CPS class B patients (666 patients, 21% of study population). Initial starting sorafenib dosages were 800 mg in most patients in both groups (72% CPS class A and 70% CPS class B) with fewer dose reductions in the CPS class B patients (29%) as compared to CPS class A (40%). Notably, there were more drug discontinuations among the CPS class B patients (40%) compared to CPS class A (29%). However, drug-related AEs were similar (17% CPS class A vs 21% CPS class B) with more events classified as liver dysfunction in CPS class B (42 patients, 6%), most often within the first 24 weeks of treatment (37/42 patients).

The INSIGHT study was a noninterventional, prospective, multicenter, observational study performed in 124 sites across Austria and Germany [24]. Similar to the GIDEON trial, patients with BCLC stage C HCC and CPS class A cirrhosis had a median OS of 16.8 months compared to 7.5 months with CPS class B cirrhosis. Notably, there were statistically significant findings between CPS class B7/8 and B9 patients, with median OS of 8.1 months/9.5 months and only 2.4 months, respectively ($P = 0.0032$). The median OS in CPS class B7/8 and B9 patients mirrored the median TTP, which was 3.7 months/3.3 months and 1.2 months, respectively. The safety profiles and reasons for discontinuation of sorafenib were similar to the GIDEON trial. These findings allude to observations from several studies that advanced HCC is one of the most statistically significant negative prognostic factors in patients with decompensated cirrhosis [9]. Given the low overall response rate of sorafenib, therapies that induce responses are needed for patients with decompensated cirrhosis to avert a likely rapid deterioration in clinical status.

Despite this prior experience showing poor outcomes with sorafenib in CPS class B patients, more recent trials have shown improved OS in all CPS classes, in part thanks to improved supportive care and experience with managing patients on sorafenib. As previously noted, in the GIDEON and INSIGHT studies, recruiting patients from January 2009 to April 2012 and 2008 to 2014, respectively, median OS in CPS class A patients was 13.8 months and 16.8 months, respectively, both exceeding the SHARP population. A smaller registry in Italy, SOFIA, comprised 259 patients (88% of population) with CPS class A and 37 patients (12%) with CPS class B cirrhosis [25]. Median survival of the CPS class A patients was 12.7 months vs 7.7 months for CPS class B patients [25]. In a small Japanese study in 2018, median OS for CPS class B patients treated with sorafenib for advanced HCC was 7.4 months, comparing favorably to prior studies [26]. The median OS for the CPS class A population was 13.4 months, also exceeding the SHARP study [26].

Overall, CPS class B status predicts for worse OS compared to CPS class A patients with similar tolerability and AE profile attributable to sorafenib. Survival has improved over time with more access to treatments and improvements in supportive care. A concern for treating CPS class B patients, beyond limited survival due to underlying cirrhosis, is that more hepatic dysfunction events have been noted that may be drug related. Indeed, in the GIDEON trial, 6% of patients discontinued because of hepatic dysfunction [23]. In the SOFIA registry, 16–17% of patients required dose reduction or interruption because of hepatic decompensation [25]. A post hoc analysis showed a statistically significant improvement in OS for the 77 patients who received 400 mg/day of sorafenib for more than 70% of the treatment period compared with the 219 patients who were either fully dosed at 800 mg/day or received 400 mg/day of sorafenib for less than 70% of the treatment period (21.6 vs 9.6 months). This likely reflects different durations of treatment (6.8 vs 3 months), underscoring the need for careful follow-up and dosing to ensure patients continue to receive therapy.

The Evolving Landscape of Systemic Therapy in Advanced Hepatocellular Carcinoma

Since its approval, there have been substantial clinical efforts to develop agents that are more effective in the front-line setting and active after progression of sorafenib in the second-line setting. Only in the past few years have there been several successes in both these settings. However, like sorafenib, the phase 3 development of these agents has concentrated on the well-compensated, CPS class A population. All the currently approved agents and available data from their pivotal studies are listed in Table 27.1. Regorafenib, a multikinase inhibitor targeting VEGFR, PDGFR, RET, KIT, FGFR1, and TIE-2, was the first drug approved in the second line after progression on sorafenib. In the phase 3 RESORCE trial, regorafenib demonstrated improved median OS compared to placebo (10.6 months vs 7.8 months) [27]. All patients on trial had tolerated doses of at least 400 mg of sorafenib/day for at least 20 days of the last 28-day cycle and had documented radiographic progression, with almost all patients (97–98%) having CPS class A cirrhosis.

With expanding treatment options, it remains crucial to identify patients who may be candidates for continued therapy. One retrospective cohort created a scoring system, NEXT, which identifies poor prognosis of patients after progression on sorafenib if the patients had an intermediate score (1–2) or high score (3–4) [34]. While CPS class B or C predicted poor OS, the wide CIs in intermediate risk suggest subsets of patients may benefit from continued therapy [34].

Terashima and colleagues [35] studied patients treated with sorafenib at a university hospital in Japan, assessing liver functional reserve after discontinuing sorafenib to assess candidacy for regorafenib. While 23 of 90 (25.6%) patients transitioned to CPS class B when they were CPS class A at the start of sorafenib treatment, 10 of 35 (28.6%) patients improved from CPS class B before starting sorafenib to CPS class A after stopping sorafenib, and 15 of 35 (42.9%) remained stable at CPS class B7/8 [35]. This provides some support for the use of sorafenib in CPS class B patients and indicates that a subset will be candidates for second-line treatment with reogorafenib at progression. However, there is limited published experience using regorafenib in patients with CPS class B liver function. One small case series included only 3 patients with statistically significant grade 3/4 aspartate transaminase/alanine transaminase elevations in that group [36].

For patients that are candidates, the sequential use of systemic therapy correlates with OS. Secondary analyses of the RESORCE trial showed a median OS of 26 months in patients from the time of sorafenib initiation to death on the trial in the regorafenib arm vs 19.2 months in the placebo arm [37]. Lenvatinib, a small-molecule tyrosine kinase inhibitor that inhibits VEGFRs 1 to 3, fibroblast growth factor receptor (FGFR) 1–4, PDGFRα, RET, and KIT, demonstrated noninferiority in first-line treatment compared to sorafenib in the phase 3 REFLECT trial (median OS 13.6 months vs 12.3 months with sorafenib) [28]. The dosing of lenvatinib in HCC differs from other malignancies, relying on a weight-based adjustment, as it was discovered from earlier phase 2 studies that lenvatinib's pharmacokinetic does differ in the setting of cirrhosis [38]. In a small retrospective cohort in Japan, sequential therapy with sorafenib/regorafenib/lenvatinib or sorafenib/lenvatinib (if intolerant to sorafenib) showed a strong correlation ($r = 0.946$; 95% CI, 0.918–0.965, $P < 0.001$) between duration of therapy and OS [39].

To better understand the optimal dose intensity of lenvatinib, especially given lower starting doses of lenvatinib in

Table 27.1 Pivotal studies of currently approved agents in hepatocellular carcinoma with available data for Child-Pugh score class B patients

Systemic agent	Trial name for approval	No. of CPS class B patients in treatment arm	Median OS of CPS class B patients
Sorafenib	SHARP [13]	14	Not reported
Lenvatinib	REFLECT [28]	3	Not reported
Regorafenib	RESORCE [27]	5	Not reported
Cabozantinib	CELESTIAL [29]	7	Not reported
Ramucirumab	REACH-2 [30]	0	No patients
	REACH (efficacy/safety for CPS class B, AFP >400 ng/mL) [31][a]	23 ramucirumab 16 placebo	3.65 mo vs 2.76 mo placebo (HR = 0.67; 95% CI, 0.33–1.37)
Nivolumab	CheckMate 040 [32]	49	7.6 mo
Pembrolizumab	KEYNOTE 224 [33]	6	Not reported

[a]Ramucirumab approval based on REACH-2 results
Abbreviations: *AFP* alpha fetoprotein, *CPS* Child-Pugh score, *HR* hazard ratio, *OS* overall survival

HCC as compared to thyroid cancer, Sasaki et al. [40, 41] conducted a retrospective analysis of patients with CPS class A and CPS class B cirrhosis with HCC treated with lenvatinib. Fourteen of eighty one patients were CPS class B and were given lenvatinib 8 mg [41]. CPS class B patients had a 0% response rate and a statistically significantly worse OS compared to CPS class A patients ($P < 0.001$). Besides CPS as a prognosticator for outcome, the authors felt these results were in part explained by statistically significantly lower relative dose intensity of lenvatinib [41]. Their work suggests that achieving a sufficient relative dose intensity of lenvatinib is necessary to achieve responses that were highly correlated with survival [41]. A total of 84% of the whole patient population required dose reduction or withdrawal because of AEs, a substantial challenge for CPS class B patients who have less reserve and already start at lower doses of lenvatinib [41]. Indeed, the recommended phase 2 maximally tolerated dose of lenvatinib in CPS class B patients was 8 mg, potentially limiting efficacy in light of a narrow therapeutic index [42]. Initial real-world safety data from Japan for patients treated with full, weight-based dose lenvatinib suggests a high rate of liver dysfunction within the first 4 weeks of treatment (number of patients with CPS class score 5, 6, 7, 8, 9, ≥ 10 of 65, 50, 8, 0, 0, 0 before treatment vs 50, 39, 22, 8, 0, 4 patients after treatment, respectively, $P < 0.001$). These data suggest that lenvatinib use in CPS class B patients at baseline needs to be administered with caution [43].

Cabozantinib, a tyrosine kinase inhibitor of VEGFR 1-3, MET, and AXL, has demonstrated improved median OS in the phase 3 CELESTIAL trial compared to placebo after progression on sorafenib [29]. No statistically significant published data exist for safety for cabozantinib in CPS class B patients. Ramucirumab, a monoclonal antibody against the VEGF receptor-2, demonstrated improved median OS in patients with alpha fetoprotein (AFP) greater than 400 ng/mL vs placebo in the phase 3 REACH-2 trial in an exclusively CPS class A patient population [30]. An exploratory analysis of the original REACH trial, which did not use elevated AFP as an inclusion criterion, did include CPS class B7/8 patients, 39 of whom had an AFP greater than 400 ng/mL. The CPS class A5 and A6 subgroups with an AFP greater than 400 ng/mL showed a benefit for ramucirumab vs placebo (CPS class A5 median OS of 8.61 months vs 4.76 months with placebo, HR = 0.61; 95% CI, 0.43–0.87; $P = 0.01$; CPS class A6 median OS of 5.72 months vs 3.61 months with placebo, HR = 0.64; 95% CI, 0.42–0.98; $P = 0.04$) [44]. Patients in the CPS class B7/8 subgroup with AFP greater than 400 ng/mL had a median OS of 3.65 months vs 2.76 months (HR 0.67; 95% CI, 0.33–1.37) [44, 45]. Ninety-two percent of patients with CPS class B7/8 treated with ramucirumab had any grade 3 or greater treatment-emergent AEs compared to 63.2% with placebo, and compared to 59–64% with CPS class A5/6 patients treated with ramucirumab [41]. A total of 15.8% of CPS class B7/8 patients treated with ramucirumab and 18.4% of CPS class B7/8 patients treated with placebo had grade 3 or greater ascites compared to 6.6% of CPS class A6 patients treated with ramucirumab and 4.1% of CPS class A6 patients treated with placebo [44].

The immune checkpoint inhibitors nivolumab and pembrolizumab have both received accelerated approval by the Food and Drug Administration for treatment of advanced HCC after progression on sorafenib. These approvals were based on single-arm studies that demonstrated an acceptable safety profile and efficacy in this population, the so-called CheckMate 040 and KEYNOTE-224 studies, with nivolumab and pembrolizumab, respectively [32, 33]. With both agents, these phase 1/2 studies demonstrated overall response rates of 15–20% and prolonged median duration of response of more than 1 year in mostly CPS class A patients [32, 33]. The combination of the checkpoint inhibitors nivolumab and ipilimumab, anti–PD-1 and anti–cytotoxic T-lymphocyte antigen 4, respectively, recently received accelerated approval by the Food and Drug Administration based on an objective response rate of 31% with a median duration of 17 months [44]. At this time there are insufficient data on this combination in CPS class B patients. A separate cohort of the CheckMate 040 study evaluated single-agent nivolumab in CPS class B7/8 patients (N = 49) and demonstrated an objective response rate of 10% and a median OS of 7.6 months [46]. Importantly, treatment-related AEs were reported in 25 (51%) patients with 4 (8.2%) patients having select hepatic treatment-related AEs. Overall, these data compared favorably to the CPS class A cohort and did not lead to an increase in discontinuations compared to CPS class A as well. Similar to the small molecules, case series confirm a shorter OS in CPS class B patients treated in the second-line or later with these agents. One case series from the University of California San Francisco with 18 CPS class B patients treated with nivolumab showed a median OS of 5.9 months [47]. Another study assessing either nivolumab or pembrolizumab in Austria/Germany with 28 CPS class B patients showed a median OS of 8.6 months [48]. Safety profiles in the prospective trials and subsequent case series were all similar and acceptable, especially considering AEs related to liver dysfunction were as low as 3% in the largest case series and 8.2% in CheckMate 040 [46, 48].

The observation that advanced HCC is one of the most statistically significant negative prognostic factors in patients with decompensated cirrhosis was noted in studies before the advent of effective anticancer therapy [9]. It can be hypothesized that in some cases treating the advanced HCC may improve liver function and improve survival. Now, with newer combinations that have higher response rates, it might be reasonable to consider these options for the treatment of select patients [49, 50]. Still, safety in those patients with CPS class B is still needed. With that in mind, it is also important to discern whether hepatic decompensation is due to progression of HCC, drug toxicity, or other provoking events. In one experience, sorafenib treatment in CPS class A patients with main portal vein invasion and high AFP was associated with hepatic decompensation [51]. In a retrospective study of patients from the SOFIA registry who permanently discontinued sorafenib, post–sorafenib survival was analyzed by tumor progression ($n = 123$), unacceptable side effects ($n = 77$), and hepatic decompensation ($n = 60$), which was defined as jaundice, ascites, gastrointestinal hemorrhage, or encephalopathy [52]. The tumor progression group included patients with a simultaneously detected progression of disease and hepatic decompensation, whereas patients in the decompensated group did not have tumor progression noted. The post–sorafenib OS in the tumor progression group was 4.6 months (3.3–5.7 months) compared to 1.8 months (1.6–2.4 months) in the decompensation group ($P < 0.001$), again emphasizing the need for careful selection of patients before treatment to avoid worsening prognosis. An exception can be made for isolated hyperbilirubinemia after initiation of sorafenib or lenvatinib because both drugs inhibit uridine diphosphate-glucuronosyltransferase (UGT) 1A1 [53, 54]. Increases in indirect bilirubin may reflect unmasking of UGT1A1 polymorphisms or poor hepatic reserve by inhibiting the small UGT1A1 activity remaining, neither of which reflects toxicity. Minimizing dose interruptions and reductions in this situation is prudent in patients with HCC that occupies most of the hepatic parenchyma, as treatment is crucial to restoring compromised hepatic function.

Conclusion

"Decompensated liver disease" is a clinically heterogeneous group of patients. When developing a treatment plan for patients who also have advanced liver cancer, our ability to improve their survival with cancer treatment is limited. Certainly, those patients with CPS class C liver disease will succumb to complications of their disease in a relatively short period of time and should receive only supportive care. CPS class B patients are more heterogeneous and if their performance status is maintained and they are able to come to the clinic, may be considered for treatment, recognizing that this is based on safety data and a lack of high-level evidence that we can improve their survival with cancer treatment. In that regard, given the lack of safety data and what appears to be increased toxicity in patients with CPS class B9 liver function, we would also recommend supportive care for these patients.

A proposed treatment algorithm for frontline treatment and subsequent treatment of patients with CPS class B7 and 8 cirrhosis and HCC is shown in Fig. 27.1. Given the available data on its safety, sorafenib remains a viable option for advanced HCC patients with CPS class B7/8 patients in the first-line setting. Lenvatinib is not a preferred first-line option given the maximally tolerated dose of lenvatinib in CPS class B patients may be subtherapeutic and is associated with higher rates of decompensation [43]. Reassessing a patient's ability to tolerate further anticancer therapy after progression on sorafenib is paramount as sequential treatment is associated with improved OS in the CPS class A population and this may be true for CPS class B patients as well. Regorafenib may be appropriate in carefully selected patients who would have otherwise fit the inclusion criteria of the RESORCE

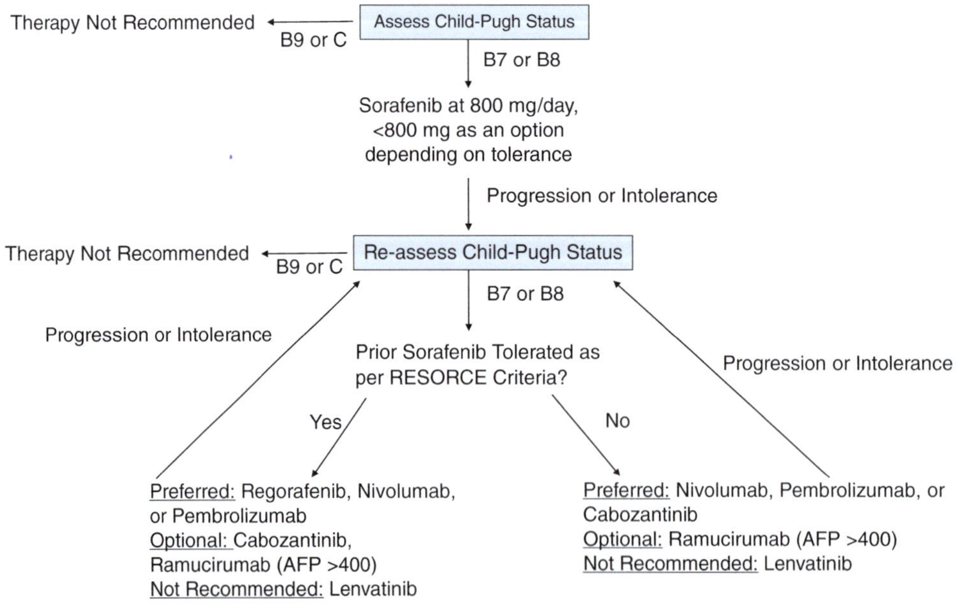

Fig. 27.1 A proposed treatment algorithm for frontline treatment and subsequent treatment of patients with Child-Pugh (CP) score of B7 and 8 cirrhosis and advanced hepatocellular carcinoma. *AFP* alpha fetoprotein

trial with the exception of CPS status. Cabozantinib may become a future option once more safety and efficacy data are available. CPS class B7/B8 patients treated with ramucirumab in the original REACH trial had high rates of AEs with a nonstatistically significant trend of benefit in patients with an AFP greater than 400 ng/mL. Given the expanding data with single-agent checkpoint inhibitors in this population, these are preferred alternatives to patients with CPS class B7/8 cirrhosis, especially if they are intolerant to first-line sorafenib, patients who were excluded from the RESORCE study. Finally, as new combinations are showing strong efficacy signals in the frontline setting, further data in the CPS class B population will be needed before applying these regimens broadly [49, 50].

References

1. Choi D, Lim HK, Rhim H, et al. Percutaneous radiofrequency ablation for early-stage hepatocellular carcinoma as a first-line treatment: long-term results and prognostic factors in a large single-institution series. Eur Radiol. 2007;17(3):684–92. https://doi.org/10.1007/s00330-006-0461-5.
2. Lencioni R, Cioni D, Crocetti L, et al. Early-stage hepatocellular carcinoma in patients with cirrhosis: long-term results of percutaneous image-guided radiofrequency ablation. Radiology. 2005;234(3):961–7. https://doi.org/10.1148/radiol.2343040350.
3. Llovet JM, Brú C, Bruix J. Prognosis of hepatocellular carcinoma: the BCLC staging classification. Semin Liver Dis. 1999;19(3):329–38. https://doi.org/10.1055/s-2007-1007122.
4. Heimbach JK, Kulik LM, Finn RS, et al. AASLD guidelines for the treatment of hepatocellular carcinoma. Hepatology. 2018;67(1):358–80. https://doi.org/10.1002/hep.29086.
5. Freeman RB, Edwards EB, Harper AM. Waiting list removal rates among patients with chronic and malignant liver diseases. Am J Transplant. 2006;6(6):1416–21. https://doi.org/10.1111/j.1600-6143.2006.01321.x.
6. Johnson PJ, Berhane S, Kagebayashi C, et al. Assessment of liver function in patients with hepatocellular carcinoma: a new evidence-based approach-the ALBI grade. J Clin Oncol. 2015;33(6):550–8. https://doi.org/10.1200/JCO.2014.57.9151.
7. Lee SK, Song MJ, Kim SH, Park M. Comparing various scoring system for predicting overall survival according to treatment modalities in hepatocellular carcinoma focused on platelet-albumin-bilirubin (PALBI) and albumin-bilirubin (ALBI) grade: a nationwide cohort study. PLoS One. 2019;14(5):e0216173. https://doi.org/10.1371/journal.pone.0216173.
8. Gustot T, Moreau R. Acute-on-chronic liver failure vs. traditional acute decompensation of cirrhosis. J Hepatol. 2018;69(6):1384–93. https://doi.org/10.1016/j.jhep.2018.08.024.
9. D'Amico G, Garcia-Tsao G, Pagliaro L. Natural history and prognostic indicators of survival in cirrhosis: a systematic review of 118 studies. J Hepatol. 2006;44(1):217–31. https://doi.org/10.1016/j.jhep.2005.10.013.
10. Sieghart W, Hucke F, Peck-Radosavljevic M. Transarterial chemoembolization: modalities, indication, and patient selection. J Hepatol. 2015;62(5):1187–95. https://doi.org/10.1016/j.jhep.2015.02.010.
11. Giorgio A, Merola MG, Montesarchio L, et al. Percutaneous radiofrequency ablation of hepatocellular carcinoma in cirrhosis: analysis of complications in a single centre over 20 years. Br J Radiol. 2017;90(1074):20160804. https://doi.org/10.1259/bjr.20160804.
12. Abou-Alfa GK. Commentary: sorafenib—the end of a long journey in search of systemic therapy for hepatocellular carcinoma, or the beginning? Oncologist. 2009;14(1):92–4. https://doi.org/10.1634/theoncologist.2008-0294.
13. Llovet JM, Ricci S, Mazzaferro V, et al. Sorafenib in advanced hepatocellular carcinoma. N Engl J Med. 2008;359(4):378–90. https://doi.org/10.1056/NEJMoa0708857.
14. Cheng AL, Kang YK, Chen Z, et al. Efficacy and safety of sorafenib in patients in the Asia-Pacific region with advanced hepatocellular carcinoma: a phase III randomised, double-blind, placebo-controlled trial. Lancet Oncol. 2009;10(1):25–34. https://doi.org/10.1016/S1470-2045(08)70285-7.
15. Pinter M, Sieghart W, Graziadei I, et al. Sorafenib in unresectable hepatocellular carcinoma from mild to advanced stage liver cirrhosis. Oncologist. 2009;14(1):70–6. https://doi.org/10.1634/theoncologist.2008-0191.
16. Labeur TA, Achterbergh R, Takkenberg B, Van Delden O, Mathôt R, Klümpen HJ. Sorafenib for patients with hepatocellular carcinoma and Child-Pugh B liver cirrhosis: lessons learned from a terminated study. Oncologist. 2020;25(9):e1274–9. https://doi.org/10.1634/theoncologist.2019-0718.
17. Guedes Leal CR, Magalhães C, Barbosa D, et al. Survival and tolerance to sorafenib in Child-Pugh B patients with hepatocellular carcinoma: a prospective study. Investig New Drugs. 2018;36(5):911–8. https://doi.org/10.1007/s10637-018-0621-x.
18. Chiu J, Tang YF, Yao TJ, et al. The use of single-agent sorafenib in the treatment of advanced hepatocellular carcinoma patients with underlying Child-Pugh B liver cirrhosis: a retrospective analysis of efficacy, safety, and survival benefits. Cancer. 2012;118(21):5293–301. https://doi.org/10.1002/cncr.27543.
19. Ogasawara S, Chiba T, Ooka Y, et al. Sorafenib treatment in Child-Pugh A and B patients with advanced hepatocellular carcinoma: safety, efficacy and prognostic factors. Investig New Drugs. 2015;33(3):729–39. https://doi.org/10.1007/s10637-015-0237-3.
20. McNamara MG, Slagter AE, Nuttall C, et al. Sorafenib as first-line therapy in patients with advanced Child-Pugh B hepatocellular carcinoma-a meta-analysis. Eur J Cancer. 2018;105:1–9. https://doi.org/10.1016/j.ejca.2018.09.031.
21. Reiss KA, Yu S, Mamtani R, et al. Starting dose of sorafenib for the treatment of hepatocellular carcinoma: a retrospective, multi-institutional study. J Clin Oncol. 2017;35(31):3575–81. https://doi.org/10.1200/JCO.2017.73.8245.
22. Lencioni R, Marrero J, Venook A, Ye SL, Kudo M. Design and rationale for the non-interventional Global Investigation of Therapeutic DEcisions in Hepatocellular Carcinoma and Of its Treatment with Sorafenib (GIDEON) study. Int J Clin Pract. 2010;64(8):1034–41. https://doi.org/10.1111/j.1742-1241.2010.02414.x.
23. Marrero JA, Kudo M, Venook AP, et al. Observational registry of sorafenib use in clinical practice across Child-Pugh subgroups: the GIDEON study. J Hepatol. 2016;65(6):1140–7. https://doi.org/10.1016/j.jhep.2016.07.020.
24. Ganten TM, Stauber RE, Schott E, et al. Sorafenib in patients with hepatocellular carcinoma—results of the observational INSIGHT study. Clin Cancer Res. 2017;23(19):5720–8. https://doi.org/10.1158/1078-0432.CCR-16-0919.
25. Iavarone M, Cabibbo G, Piscaglia F, et al. Field-practice study of sorafenib therapy for hepatocellular carcinoma: a prospective multicenter study in Italy. Hepatology. 2011;54(6):2055–63. https://doi.org/10.1002/hep.24644.
26. Suzuki E, Kaneko S, Okusaka T, et al. A multicenter phase II study of sorafenib in Japanese patients with advanced hepatocellular carcinoma and Child Pugh A and B class. Jpn J Clin Oncol. 2018;48(4):317–21. https://doi.org/10.1093/jjco/hyy010.

27. Bruix J, Qin S, Merle P, et al. Regorafenib for patients with hepatocellular carcinoma who progressed on sorafenib treatment (RESORCE): a randomised, double-blind, placebo-controlled, phase 3 trial. Lancet. 2017;389(10064):56–66. https://doi.org/10.1016/S0140-6736(16)32453-9.
28. Kudo M, Finn RS, Qin S, et al. Lenvatinib versus sorafenib in first-line treatment of patients with unresectable hepatocellular carcinoma: a randomised phase 3 non-inferiority trial. Lancet. 2018;391(10126):1163–73. https://doi.org/10.1016/S0140-6736(18)30207-1.
29. Abou-Alfa GK, Meyer T, Cheng AL, et al. Cabozantinib in patients with advanced and progressing hepatocellular carcinoma. N Engl J Med. 2018;379(1):54–63. https://doi.org/10.1056/NEJMoa1717002.
30. Zhu AX, Kang YK, Yen CJ, et al. Ramucirumab after sorafenib in patients with advanced hepatocellular carcinoma and increased alpha-fetoprotein concentrations (REACH-2): a randomised, double-blind, placebo-controlled, phase 3 trial. Lancet Oncol. 2019;20(2):282–96. https://doi.org/10.1016/S1470-2045(18)30937-9.
31. Zhu AX, Park JO, Ryoo BY, et al. Ramucirumab versus placebo as second-line treatment in patients with advanced hepatocellular carcinoma following first-line therapy with sorafenib (REACH): a randomised, double-blind, multicentre, phase 3 trial. Lancet Oncol. 2015;16(7):859–70. https://doi.org/10.1016/S1470-2045(15)00050-9.
32. El-Khoueiry AB, Sangro B, Yau T, et al. Nivolumab in patients with advanced hepatocellular carcinoma (CheckMate 040): an open-label, non-comparative, phase 1/2 dose escalation and expansion trial. Lancet. 2017;389(10088):2492–502. https://doi.org/10.1016/S0140-6736(17)31046-2.
33. Zhu AX, Finn RS, Edeline J, et al. KEYNOTE-224 investigators. Pembrolizumab in patients with advanced hepatocellular carcinoma previously treated with sorafenib (KEYNOTE-224): a non-randomised, open-label phase 2 trial. Lancet Oncol. 2018;19(7):940–52. https://doi.org/10.1016/S1470-2045(18)30351-6.
34. Lee HW, Kim HS, Kim SU, et al. Survival estimates after stopping sorafenib in patients with hepatocellular carcinoma: NEXT Score development and validation. Gut Liver. 2017;11(5):693–701. https://doi.org/10.5009/gnl16391.
35. Terashima T, Yamashita T, Sunagozaka H, et al. Analysis of the liver functional reserve of patients with advanced hepatocellular carcinoma undergoing sorafenib treatment: prospects for regorafenib therapy. Hepatol Res. 2018;48(12):956–66. https://doi.org/10.1111/hepr.13196.
36. Kim K, Jha R, Prins PA, et al. Regorafenib in advanced hepatocellular carcinoma (HCC): considerations for treatment. Cancer Chemother Pharmacol. 2017;80(5):945–54. https://doi.org/10.1007/s00280-017-3431-5.
37. Finn RS, Merle P, Granito A, et al. Outcomes of sequential treatment with sorafenib followed by regorafenib for HCC: additional analyses from the phase III RESORCE trial. J Hepatol. 2018;69(2):353–8. https://doi.org/10.1016/j.jhep.2018.04.010.
38. Ikeda K, Kudo M, Kawazoe S, et al. Phase 2 study of lenvatinib in patients with advanced hepatocellular carcinoma. J Gastroenterol. 2017;52(4):512–9. https://doi.org/10.1007/s00535-016-1263-4.
39. Hiraoka A, Kumada T, Atsukawa M, et al. Important clinical factors in sequential therapy including lenvatinib against unresectable hepatocellular carcinoma. Oncology. 2019;97(5):277–85. https://doi.org/10.1159/000501281.
40. Capdevila J, Newbold K, Licitra L, et al. Optimisation of treatment with lenvatinib in radioactive iodine-refractory differentiated thyroid cancer. Cancer Treat Rev. 2018;69:164–76. https://doi.org/10.1016/j.ctrv.2018.06.019.
41. Sasaki R, Fukushima M, Haraguchi M, et al. Response to lenvatinib is associated with optimal relative dose intensity in hepatocellular carcinoma: experience in clinical settings. Cancers (Basel). 2019;11(11):1769. https://doi.org/10.3390/cancers11111769.
42. Ikeda M, Okusaka T, Mitsunaga S, et al. Safety and pharmacokinetics of lenvatinib in patients with advanced hepatocellular carcinoma. Clin Cancer Res. 2016;22(6):1385–94. https://doi.org/10.1158/1078-0432.CCR-15-1354.
43. Hiraoka A, Kumada T, Atsukawa M, et al. Early relative change in hepatic function with lenvatinib for unresectable hepatocellular carcinoma. Oncology. 2019;97(6):334–40. https://doi.org/10.1159/000502095.
44. Zhu AX, Baron AD, Malfertheiner P, et al. Ramucirumab as second-line treatment in patients with advanced hepatocellular carcinoma: analysis of REACH trial results by Child-Pugh score. JAMA Oncol. 2017;3(2):235–43. https://doi.org/10.1001/jamaoncol.2016.4115.
45. He AR, Yau T, Hsu C, et al. Nivolumab (NIVO) + ipilimumab (IPI) combination therapy in patients (pts) with advanced hepatocellular carcinoma (aHCC): subgroup analyses from CheckMate 040. J Clin Oncol. 2020;38(4 Suppl):512. https://doi.org/10.1200/JCO.2020.38.4_suppl.512.
46. Kudo M, Matilla A, Santoro A, et al. CheckMate-040: nivolumab (NIVO) in patients (pts) with advanced hepatocellular carcinoma (aHCC) and Child-Pugh B (CPB) status. J Clin Oncol. 2019;37(4 Suppl):327. https://doi.org/10.1200/JCO.2019.37.4_suppl.327.
47. Kambhampati S, Bauer KE, Bracci PM, et al. Nivolumab in patients with advanced hepatocellular carcinoma and Child-Pugh class B cirrhosis: safety and clinical outcomes in a retrospective case series. Cancer. 2019;125(18):3234–41. https://doi.org/10.1002/cncr.32206.
48. Scheiner B, Kirstein MM, Hucke F, et al. Programmed cell death protein-1 (PD-1)-targeted immunotherapy in advanced hepatocellular carcinoma: efficacy and safety data from an international multicentre real-world cohort. Aliment Pharmacol Ther. 2019;49(10):1323–33. https://doi.org/10.1111/apt.15245.
49. Finn RS, Qin S, Ikeda M, et al. Atezolizumab plus bevacizumab in unresectable hepatocellular carcinoma. N Engl J Med. 2020;382(20):1894–905. https://doi.org/10.1056/NEJMoa1915745.
50. Llovet J, Finn RS, Ikeda M, et al (2019) A phase 1B trial of lenvatinib (LEN) plus pembrolizumab (pembro) in unresectable hepatocellular carcinoma (HCC): updated results. Paper presented at: ESMO; September 29, 2019; Barcelona, Spain.
51. Kuo YH, Wu IP, Wang JH, et al. The outcome of sorafenib monotherapy on hepatocellular carcinoma with portal vein tumor thrombosis. Investig New Drugs. 2018;36(2):307–14. https://doi.org/10.1007/s10637-017-0468-6.
52. Iavarone M, Cabibbo G, Biolato M, et al. Predictors of survival in patients with advanced hepatocellular carcinoma who permanently discontinued sorafenib. Hepatology. 2015;62(3):784–91. https://doi.org/10.1002/hep.27729.
53. Bins S, Lenting A, El Bouazzaoui S, et al. Polymorphisms in SLCO1B1 and UGT1A1 are associated with sorafenib-induced toxicity. Pharmacogenomics. 2016;17(14):1483–90. https://doi.org/10.2217/pgs-2016-0063.
54. Nagahama M, Ozeki T, Suzuki A, et al. Association of lenvatinib trough plasma concentrations with lenvatinib-induced toxicities in Japanese patients with thyroid cancer. Med Oncol. 2019;36(5):39. https://doi.org/10.1007/s12032-019-1263-3.

28 Would Effectively Downstaging Hepatocellular Carcinoma Allow All Patients to Be Considered for Liver Transplantation?

Emily Harding-Theobald and Neehar D. Parikh

Main Points

1. Hepatocellular carcinoma (HCC) tumor biology is heterogeneous, and radiographic features at presentation do not reliably correlate well with subsequent tumor behavior.
2. Strict adherence to the Milan criteria (MC) excludes patients with HCC beyond criteria who may have substantial survival benefit from transplantation.
3. Tumor downstaging uses liver-directed therapies to bring patients within the MC and potentially allow them to be eligible for LT.
4. Downstaging permits the selection of intermediate-stage HCC patients as LT candidates on the basis of two objective indicators of tumor biology: response to treatment and indolent tumor behavior over time (which requires a mandatory wait time).
5. Reducing the alpha fetoprotein (AFP) to less than 400 ng/mL before LT resulted in similar waitlist and post-LT outcomes compared to those patients who never had an AFP greater than 400 ng/mL.
6. An AFP ng/mL greater than 1000 ng/mL and a Child-Pugh score (CPS) class B or C liver function were predictors of downstaging failure.
7. Patients with HCC who are downstaged have similar outcomes compared with those who present at an earlier stage.

The MC have guided the selection of LT candidates with HCC based solely on morphologic characteristics (i.e., size and number) for the last 2 decades. Tumor downstaging, which uses liver-directed therapies to bring patients within MC, has emerged to expand the population of patients with HCC eligible for LT and further incorporates measure of tumor biology into the selection of patients for LT for HCC. The role of downstaging in LT remains controversial, and herein we detail arguments for and against routine use of downstaging for eligible patients.

E. Harding-Theobald (✉)
Division of Gastroenterology, University of Colorado, Aurora, CO, USA
e-mail: emily.harding-theobald@cuanschutz.edu

N. D. Parikh
Division of Gastroenterology and Hepatology, University of Michigan, Ann Arbor, MI, USA
e-mail: ndparikh@med.umich.edu

Point: Patients with Hepatocellular Carcinoma Who Are Downstaged Have Similar Outcomes Compared with Those Who Presented at an Earlier Stage

Access to LT is limited by the availability of organs, and thus there is an imperative to transplant those individuals most likely to derive a benefit from LT, while also maximizing post-LT outcomes. Before the introduction of MC, the post-LT 5-year survival for patients with HCC was approximately 30–40%, compared with 70–80% among patients transplanted for nonmalignant indications [1, 2]. The adoption of the MC rapidly closed this gap, resulting in equivalent post-LT survival outcomes for HCC compared to the general transplant population [3]. Strict adherence to the MC, however, excludes patients beyond criteria who may have the largest survival benefit from LT because of limitations in liver-directed and systemic therapies in patients with HCC [4]. In addition, HCC tumor biology is heterogeneous and radiographic features at presentation do not reliably correlate well with subsequent tumor behavior [5]. Downstaging protocols that use liver-directed therapies to decrease tumor burden to within the MC have been adopted by many transplant centers worldwide [6]. This permits the selection of intermediate-stage HCC patients as transplant candidates on the basis of two objective indicators of tumor biology: response to treatment and indolent tumor behavior over time. The expanded University of California San Francisco crite-

N. S. Reau, A. G. Singal (eds.), *Controversies in Hepatocellular Carcinoma*, https://doi.org/10.1007/978-3-031-90225-3_28

Table 28.1 United Network of Organ Sharing Downstaging (UNOS-DS) criteria

UNOS DS criteria
1 lesion > 5 cm and ≤ 8 cm
2 or 3 lesions that meet all of the following criteria: At least 1 > 3 cm Each lesion ≤5 cm Total diameter of all lesions ≤8 cm
4 or 5 lesions each <3 cm, and total diameter of all lesions ≤8 cm

ria, introduced in 2001, demonstrated acceptable outcomes among patients who are successfully downstaged to MC [7]. Since that time a number of downstaging criteria have been proposed, including the Pittsburgh, Hangzhou, Tokyo, and Up-to-7 criteria [8, 9]. In 2017, the United Network of Organ Sharing (UNOS) policy was updated to approve HCC exception points for patients whose disease is initially outside T2 criteria if they are successfully downstaged to within T2 (Table 28.1) [10]. The literature has thus far shown acceptable outcomes for downstaged patients compared with patients within the MC.

Posttransplant Survival After Downstaging Is Equivalent to the Milan Criteria in Prospective Studies

Any expanded HCC criteria must be carefully balanced against the requirement to maintain acceptable outcomes given limited organ availability and potential harms to other patients on the waitlist. An international consensus document from Italy concluded that a 5-year posttransplant survival of greater than 50% is the minimum acceptable cutoff for HCC patients to fulfill transplant utility, while another modeling study estimated a minimal survival of 61% to prevent harm to others on the waitlist [11, 12]. While these thresholds can be considered arbitrary, published studies of downstaged patients have largely shown posttransplant survival meeting these thresholds. A 2015 pooled analysis of 950 patients reported on 13 small retrospective studies with 1-year survival rates ranging from 87% to 100%. There was substantial heterogeneity in downstaging protocols and survival reporting, however, which limited any conclusions on longer-term posttransplant outcomes [13]. Early downstaging studies have varied substantially in their use of upper-limit exclusion criteria, downstaging target (MC vs other criteria), and duration of observation before LT. Two studies reported on downstaging to a consistent in-MC end point, each suggesting posttransplant survival comparable to patients meeting conventional criteria. Ravaioli et al. [14] reported 3-year survival of 82% both in the downstaged (n = 65) and MC groups (n = 186). Yao et al. [15] reported 5-year posttransplant survival of 78% and 81% for downstaged (n = 64) and MC (n = 122) patients, respectively. Importantly, both the Ravaioli and Yao studies used a strict upper limit on tumor burden, similar to the current UNOS inclusion criteria, and reported a high downstaging success rate between 63% and 69% [16]. A subsequent 2017 multicenter retrospective study of 109 patients transplanted after downstaging to within MC described a post-LT survival of 79.7% at 5 years, and recurrence-free probability of 87.3% [17].

While traditionally downstaging has referred to tumor burden on imaging, several analyses have focused on the effect of downstaging AFP. Previous analyses have shown consistent poor outcomes of patients with an AFP greater than 1000 ng/mL before LT and this threshold serves as an exclusion criterion to LT in the United States. An analysis of 6817 patients from the UNOS database showed that reducing the AFP to less than 400 before LT resulted in similar waitlist and posttransplant outcomes compared to those patients who never had an AFP greater than 400 [18]. Thus patients who can be successfully downstaged using surrogates of tumor biology (tumor burden on imaging, AFP) can achieve excellent outcomes. Taken together, the aggregate evidence offers a strong argument for the widespread use of downstaging as a selection tool before LT, with careful limitations on inclusion criteria and protocol.

The Influence of Locoregional Therapy and Waiting Time Before Transplant

One of the major criticisms put forward against downstaging has been the concern that the observed outcomes may not be generalizable outside the narrow context in which downstaging has been studied. Downstaging outcomes are particularly influenced by three specific characteristics: careful patient selection, mandatory waiting times, and the quality of locoregional therapy (LRT). Multicenter data of 187 patients showed that an AFP greater than 1000 ng/mL and CPS class B or C liver function were predictors of downstaging failure due to poor tumor biology and limited hepatic reserve.

The quality and response to LRTs also are highly predictive of outcomes as tumor burden on explant and risk of posttransplant recurrence are highly correlated; however, the literature shows wide variation on explant pathological features. For example, an analysis of explant pathology of downstaged patients showed the presence of microvascular invasion in 7 (6.4%) of 109 patients, and poorly differentiated grade in a single patient, [17] whereas an analysis of 32 downstaged patients from Italy showed microvascular invasion in 56% of patients and poorly differentiated grade in 75% of patients [14]. This difference is likely driven by several factors, including variation in initial tumor burden before LRT, variations in LRT practice, and wait time before LT between transplant centers, but it underscores the importance of evaluating downstaging outcomes in the context of explant pathologic tumor characteristics, which ultimately correlates with posttransplant recurrence rates [17]. This does not necessarily detract from the strong evidence that equivalent outcomes to within the MC can be achieved through downstaging; rather, it suggests that effective downstaging may be limited to patients who demonstrate a response to LRT and stability over time. In one large study of 501 patients with HCC, complete response to LRT predicted a posttransplant recurrence risk of 2% compared with 18% for those who did not achieve complete response. Importantly, initial tumor size did not correlate with complete response to treatment [19].

In addition to LRT practices, variation in mandatory waiting time also influences downstaging success. Shorter (< 6 months) waiting times are strongly associated with worse posttransplant outcomes, presumably because of the inclusion of patients with unfavorable tumor biology [20, 21]. The granular downstaging literature from the United States is largely from region 5, where median wait times before LT are 11–13 months [17]. By contrast, in short waiting time regions, an estimated 90% of patients are transplanted in a shorter time frame [22]. When UNOS formalized downstaging in 2017, it included a 6-month mandatory waiting period after successful downstaging to within the MC [10]. This maximizes the selective pressure on patients and allows sufficient time to observe stable treatment response. A 2019 study compared UNOS downstaging criteria, all comers who underwent downstaging, and those patients always within MC, with the aim of validating the UNOS-downstaging criteria. The authors showed equivalent 3-year survival between patients within the MC and those who met UNOS-downstaging criteria, whereas all comers had inferior survival. Notably, survival was statistically significantly worse among patients in short (2.6 months) and medium (6.5 months) waiting time regions, compared to longer wait time regions (12.8 months) [23]. Thus, until better surrogates of tumor biology are available, a mandatory waiting period of observation for patients who are downstaged is a rational policy to ensure low post-LT recurrence [24–26].

Downstaging Protocols Will Continue to Evolve and Improve Over Time

The practice of downstaging remains in the early stages, with emerging risk stratification methods to measure tumor biology. The incorporation of circulating and radiographic biomarkers into downstaging algorithms may improve patient selection and posttransplant outcomes following downstaging for HCC and may obviate the need for strict adherence to reducing radiographically measured burden of HCC. In 2018, Toso and colleagues [27] proposed a downstaging target that combined morphologic and biomarker data: total tumor volume and serum AFP level. The authors demonstrated similar disease-free survival among patients downstaged to this benchmark when compared to the MC [27]. The Modified Toronto criteria, based on biopsy-derived tumor pathology, AFP, and the absence of cancer-related symptoms, have also been shown to achieve acceptable survival and recurrence rates among patients outside the MC [28]. As our understanding and ability to measure HCC tumor biology improves, we will be better able to identify those patients who will derive the most benefit from transplant for HCC with increasing accuracy.

Counterpoint: Patients with Hepatocellular Carcinoma Who Are Downstaged Have Inferior Outcomes Compared with Those Who Presented at an Earlier Stage

While downstaging has resulted in the ability of patients with intermediate-stage HCC to achieve a curable option, the evidence behind downstaging is limited and prone to reporting bias. There is inadequate description of harms to patients who attempt downstaging but ultimately do not receive an LT. Importantly, while short-term post-LT outcomes for patients who have undergone downstaging appear to be similar to those patients within the MC, we lack robust long-term data showing similar outcomes.

Recurrence Rates After Downstaging May Increase in Longer-Term Follow-Up

Published evidence supporting downstaging for HCC has shown post-LT survival can reach near equivalence with patients transplanted within the MC [17, 29, 30]. While overall survival has been the primary end point of most downstaging studies, downstaged patients do have a slightly higher recurrence rate when compared to MC patients [13]. For example, in a study of 90 patients who underwent downstaging in the United States, there was a 20% rate of recurrence compared with 5% for those patients originally within the MC [31]. An analysis of national UNOS data showed a HCC recurrence rate of 13% among UNOS downstaging compared with 7% within the MC, with the caveat that there can be incomplete reporting of HCC recurrence in the data set [22]. Furthermore, most studies have also observed relatively short follow-up periods of approximately 2–3 years, which risks missing late recurrences. In 2019, Ravaioli et al. [32] updated their cohort of downstaged patients after a follow-up of 5 years. They reported a 5-year recurrence rate after downstaging of 20.9% compared with 7.6% for patients within the MC. The authors also showed that while survival had previously been equivalent between the MC and downstaged patients at 3 years (~80%), by 5 years the downstaged group's survival had decreased to 63%, compared with 77% within the MC [32]. Prospective studies with adequate power and long-term follow-up will be an essential step toward understanding the true recurrence rate and survival after downstaging.

Expansion Downstaging May Negatively Influence Outcomes Because of Variation in Practice

The expansion of downstaging to less experienced centers will inevitably influence the observed outcomes following LT for HCC. The largest studies to date have relied on careful selection of candidates with clear upper limits on tumor burden and careful monitoring of downstaging success. Variations in patient selection, quality of LRT, and imaging quality may affect downstaging outcomes in negative ways.

In an analysis of the early national experience of downstaging, Mehta and colleagues [23] reported 32.5% of patients who were intended to undergo downstaging to the MC were understaged before LT, based on explant pathology. Understaging can occur because of limitations in imaging quality, limitation in imaging interpretation, or tumor progression between the time between imaging and LT. Understaging is a risk factor for HCC recurrence posttransplant, and thus downstaged patients are at risk of development of poor posttransplant outcomes because of this variation.

The Pretransplant Harms of Downstaging Are Minimally Studied

A majority of the work studying downstaged patients focuses on outcomes following LT. Comparably little research has focused on waitlist and prelisting outcomes. Patients undergoing downstaging who do not receive an LT may experience substantial harms related to the receipt of LRT and lack of application of increasingly effective systemic therapies for HCC [33]. Only a small minority of T3 patients ever make it to transplant because of tumor biology, transplant selection criteria, and organ availability. Aggregate data suggest that approximately 30–50% of patients are never successfully downstaged to the MC; however, this is highly dependent on selection criteria for downstaging [13, 14, 34]. An additional 24–34% of the successfully downstaged patients will then dropout before LT, usually because of tumor progression or death [35]. This is compared with approximately 20–26% 1- and 2-year cumulative probability of dropout for patients within the MC [16]. Statistically significant predictors of dropout from downstaging protocols include an AFP greater than 1000 (hazard ratio [HR]: 2.42; 95% CI, 1.16–5.05) and CPS class B and C cirrhosis (HR: 2.19; 95% CI, 1.04–4.64), and thus downstaging could be generally avoided in this high-risk group [15].

Pretreatment mortality associated with LRT is estimated to be 2.5%, and patients can undergo 2 or 3 LRTs in the downstaging process [36, 37]. In patients with underlying liver decompensation, acute liver failure leading to death has been reported in up to 17% of patients [38]. Aggressive LRT of patients beyond the MC could result in substantial cumulative harm, while the benefit of LT remains available to relatively few. To the degree that downstaging replaces or delays the use of systemic chemotherapy, patients may also experience additional loss of palliative options and survival time. Therefore, while downstaging protocols have been expanded to many transplant centers worldwide, there are concerns that there may be inferior outcomes for those patients both before and after LT.

Conclusion

Downstaging HCC offers a curative option to patients who were previously incurable. Transplant for downstaged patients may be a better use of scarce organs than other patients at extremely low risk of waitlist dropout (i.e., unifocal HCC < 3 cm with preserved liver function) [39].

Survival data have shown largely equivalent outcomes to those patients within the MC with several important caveats. Most of the data supporting downstaging come from a few expert centers with established prospective protocols. The generalizability of downstaging has yet to be robustly evaluated. National analyses using UNOS data are limited by a lack of granularity and inconsistent reporting within the UNOS database; however, early data appear to support the recently adopted UNOS downstaging criteria. Ongoing validation of these criteria with long-term outcomes is needed to determine which patients should meet inclusion for downstaging. As better proxies for tumor biology are developed, including blood and radiographic biomarkers, the need for strict adherence to the MC will diminish; however, response to LRT and observation time are two practical methods to estimate tumor biology that are currently used in downstaged patients. Multicenter prospective cohorts are needed to evaluate the long-term outcomes in downstaged patients and the harms of downstaging, and to develop LRT treatment algorithms. In the meantime, the current evidence supports patients can achieve considerable benefits with downstaging, but caution must be taken in patient selection and follow-up to avoid pre-LT and post-LT harms.

References

1. Haug CE, Jenkins RL, Rohrer RJ, et al. Liver transplantation for primary hepatic cancer. Transplantation. 1992;53(2):376–82. https://doi.org/10.1097/00007890-199202010-00021.
2. Matas AJ, Smith JM, Skeans MA, et al. OPTN/SRTR 2013 annual data report: kidney. Am J Transplant. 2015;15(Suppl 2):1–34. https://doi.org/10.1111/ajt.13195.
3. Mazzaferro V, Regalia E, Doci R, et al. Liver transplantation for the treatment of small hepatocellular carcinomas in patients with cirrhosis. N Engl J Med. 1996;334(11):693–9. https://doi.org/10.1056/NEJM199603143341104.
4. Yarchoan M, Agarwal P, Villanueva A, et al. Recent developments and therapeutic strategies against hepatocellular carcinoma. Cancer Res. 2019;79(17):4326–30. https://doi.org/10.1158/0008-5472.CAN-19-0803.
5. Losic B, Craig AJ, Villacorta-Martin C, et al. Intratumoral heterogeneity and clonal evolution in liver cancer. Nat Commun. 2020;11:291. https://doi.org/10.1038/s41467-019-14050-z.
6. Majno PE, Adam R, Bismuth H, et al. Influence of preoperative transarterial lipiodol chemoembolization on resection and transplantation for hepatocellular carcinoma in patients with cirrhosis. Ann Surg. 1997;226(6):688–701.; discussion 701–703. https://doi.org/10.1097/00000658-199712000-00006.
7. Yao FY, Ferrell L, Bass NM, et al. Liver transplantation for hepatocellular carcinoma: expansion of the tumor size limits does not adversely impact survival. Hepatology. 2001;33(6):1394–403. https://doi.org/10.1053/jhep.2001.24563.
8. Marsh JW, Dvorchik I. Liver organ allocation for hepatocellular carcinoma: are we sure? Liver Transpl. 2003;9(7):693–6. https://doi.org/10.1053/jlts.2003.50086.
9. Zheng SS, Xu X, Wu J, et al. Liver transplantation for hepatocellular carcinoma: Hangzhou experiences. Transplantation. 2008;85(12):1726–32. https://doi.org/10.1097/TP.0b013e31816b67e4.
10. Prentice M. Changes to HCC criteria for auto approval. OPTN/UNOS Liver and Intestinal Organ Transplantation Committee 2016. Accessed 11 Nov 2019.
11. Cillo U, Burra P, Mazzaferro V, et al. I-BELT (Italian board of experts in the field of liver transplantation). A multistep, consensus-based approach to organ allocation in liver transplantation: toward a "blended principle model". Am J Transplant. 2015;15(10):2552–61. https://doi.org/10.1111/ajt.13408.
12. Volk ML, Vijan S, Marrero JA. A novel model measuring the harm of transplanting hepatocellular carcinoma exceeding Milan criteria. Am J Transplant. 2008;8(4):839–46. https://doi.org/10.1111/j.1600-6143.2007.02138.x.
13. Parikh ND, Waljee AK, Singal AG. Downstaging hepatocellular carcinoma: a systematic review and pooled analysis. Liver Transpl. 2015;21(9):1142–52. https://doi.org/10.1002/lt.24169.
14. Ravaioli M, Grazi GL, Piscaglia F, et al. Liver transplantation for hepatocellular carcinoma: results of down-staging in patients initially outside the Milan selection criteria. Am J Transplant. 2008;8(12):2547–57. https://doi.org/10.1111/j.1600-6143.2008.02409.x.
15. Yao FY, Mehta N, Flemming J, et al. Downstaging of hepatocellular cancer before liver transplant: long-term outcome compared to tumors within Milan criteria. Hepatology. 2015;61(6):1968–77. https://doi.org/10.1002/hep.27752.
16. Yao FY, Fidelman N. Reassessing the boundaries of liver transplantation for hepatocellular carcinoma: where do we stand with tumor down-staging? Hepatology. 2016;63(3):1014–25. https://doi.org/10.1002/hep.28139.
17. Mehta N, Guy J, Frenette CT, et al. Excellent outcomes of liver transplantation following down-staging of hepatocellular carcinoma to within Milan criteria: a multicenter study. Clin Gastroenterol Hepatol. 2018;16(6):955–64. https://doi.org/10.1016/j.cgh.2017.11.037.
18. Merani S, Majno P, Kneteman NM, et al. The impact of waiting list alpha-fetoprotein changes on the outcome of liver transplant for hepatocellular carcinoma. J Hepatol. 2011;55(4):814–9. https://doi.org/10.1016/j.jhep.2010.12.040.
19. Agopian VG, Morshedi MM, McWilliams J, et al. Complete pathologic response to pretransplant locoregional therapy for hepatocellular carcinoma defines cancer cure after liver transplantation: analysis of 501 consecutively treated patients. Ann Surg. 2015;262(3):536–45. https://doi.org/10.1097/SLA.0000000000001384.
20. Halazun KJ, Patzer RE, Rana AA, et al. Standing the test of time: outcomes of a decade of prioritizing patients with hepatocellular carcinoma, results of the UNOS natural geographic experiment. Hepatology. 2014;60(6):1957–62. https://doi.org/10.1002/hep.27272.
21. Parikh ND, Singal AG. Model for end-stage liver disease exception points for treatment-responsive hepatocellular carcinoma. Clin Liver Dis (Hoboken). 2016;7(5):97–100. https://doi.org/10.1002/cld.545.
22. Mehta N, Heimbach J, Lee D, et al. Wait time of less than 6 and greater than 18 months predicts hepatocellular carcinoma recurrence after liver transplantation: proposing a wait time "sweet spot". Transplantation. 2017;101(9):2071–8. https://doi.org/10.1097/TP.0000000000001752.
23. Mehta N, Dodge JL, Grab JD, Yao FY. National experience on down-staging of hepatocellular carcinoma before liver transplant: influence of tumor burden, alpha-fetoprotein, and wait time. Hepatology. 2020;71(3):943–54. https://doi.org/10.1002/hep.30879.
24. El-Serag HB, Siegel AB, Davila JA, et al. Treatment and outcomes of treating of hepatocellular carcinoma among Medicare recipients in the United States: a population-based study. J Hepatol. 2006;44(1):158–66. https://doi.org/10.1016/j.jhep.2005.10.002.
25. Berry K, Ioannou GN. Comparison of liver transplant-related survival benefit in patients with versus without hepatocellular carcinoma in the United States. Gastroenterology. 2015;149(3):669–80.; quiz e15-16. https://doi.org/10.1053/j.gastro.2015.05.025.
26. Volk ML. Unfair priority for HCC: a problem whose ideal solution remains unsolved. Am J Transplant. 2010;10(7):1507–8. https://doi.org/10.1111/j.1600-6143.2010.03154.x.
27. Toso C, Meeberg G, Andres A, et al. Downstaging prior to liver transplantation for hepatocellular carcinoma: advisable but at the price of an increased risk of cancer recurrence – a retrospective study. Transpl Int. 2019;32(2):163–72. https://doi.org/10.1111/tri.13337.
28. Sapisochin G, Goldaracena N, Laurence JM, et al. The extended Toronto criteria for liver transplantation in patients with hepatocellular carcinoma: a prospective validation study. Hepatology. 2016;64(6):2077–88. https://doi.org/10.1002/hep.28643.
29. De Luna W, Sze DY, Ahmed A, et al. Transarterial chemoinfusion for hepatocellular carcinoma as downstaging therapy and a bridge toward liver transplantation. Am J Transplant. 2009;9(5):1158–68. https://doi.org/10.1111/j.1600-6143.2009.02576.x.
30. Yao FY, Xiao L, Bass NM, Kerlan R, Ascher NL, Roberts JP. Liver transplantation for hepatocellular carcinoma: validation of the UCSF-expanded criteria based on preoperative imaging. Am J Transplant. 2007;7(11):2587–96. https://doi.org/10.1111/j.1600-6143.2007.01965.x.
31. Murali AR, Romero-Marrero C, Miller C, et al. Predictors of successful downstaging of hepatocellular carcinoma outside Milan criteria. Transplantation. 2016;100(11):2391–7. https://doi.org/10.1097/TP.0000000000001402.

32. Ravaioli M, Odaldi F, Cucchetti A, et al. Long term results of down-staging and liver transplantation for patients with hepatocellular carcinoma beyond the conventional criteria. Sci Rep. 2019;9:3781.
33. Likhitsup A, Razumilava N, Parikh ND. Treatment for advanced hepatocellular carcinoma: current standard and the future. Clin Liver Dis (Hoboken). 2019;13(1):13–9. https://doi.org/10.1002/cld.782.
34. Bryce K, Tsochatzis EA. Downstaging for hepatocellular cancer: harm or benefit? Transl Gastroenterol Hepatol. 2017;2:106. https://doi.org/10.21037/tgh.2017.11.1.
35. Yao FY, Breitenstein S, Broelsch CE, Dufour JF, Sherman M. Does a patient qualify for liver transplantation after the down-staging of hepatocellular carcinoma? Liver Transpl. 2011;17(Suppl 2):S109–16. https://doi.org/10.1002/lt.22335.
36. Chapman WC, Majella Doyle MB, Stuart JE, et al. Outcomes of neoadjuvant transarterial chemoembolization to downstage hepatocellular carcinoma before liver transplantation. Ann Surg. 2008;248(4):617–25. https://doi.org/10.1097/SLA.0b013e31818a07d4.
37. Marelli L, Stigliano R, Triantos C, et al. Transarterial therapy for hepatocellular carcinoma: which technique is more effective? A systematic review of cohort and randomized studies. Cardiovasc Intervent Radiol. 2007;30(1):6–25. https://doi.org/10.1007/s00270-006-0062-3.
38. Hsin IF, Hsu CY, Huang HC, et al. Liver failure after transarterial chemoembolization for patients with hepatocellular carcinoma and ascites: incidence, risk factors, and prognostic prediction. J Clin Gastroenterol. 2011;45(6):556–62. https://doi.org/10.1097/MCG.0b013e318210ff17.
39. Mehta N, Dodge JL, Hirose R, Roberts JP, Yao FY. Predictors of low risk for dropout from the liver transplant waiting list for hepatocellular carcinoma in long wait time regions: implications for organ allocation. Am J Transplant. 2019;19(8):2210–8. https://doi.org/10.1111/ajt.15353.

Should Surgery Be Considered for Hepatocellular Carcinoma If Unable to Be Performed at Presentation?

29

Michelle R. Ju and Adam C. Yopp

In a carefully selected group of patients, surgery can be safely performed after additional interventions even if resection was unable to be performed at presentation.

Main Points

1. The goal of hepatic resection for hepatocellular carcinoma (HCC) is achieving an appropriate oncological margin while maintaining a functional liver remnant (FLR).
2. Tumor size of a solitary HCC tumor should not preclude hepatic resection if patient-related factors including liver function, liver volume, and performance status are acceptable.
3. Portal hypertension is a contraindication to hepatic resection. Clinical parameters including splenomegaly greater than 12 cm, clinical signs of collateralization, and thrombocytopenia of less than 100/nL are surrogate markers for clinically significant portal hypertension.
4. Macrovascular invasion (MVI) in the form of portal venous tumor thrombus (PVTT) (Vp1-3) can be safely considered for liver resection following a multidisciplinary discussion in high-volume centers.
5. In patients with inadequate or borderline FLR of less than 20%, selective portal vein embolization (PVE) can increase remnant size.
6. Transarterial chemoembolization (TACE) plus PVE before right hepatectomy experienced higher mean increase in FLR and improved overall and disease-free survival compared with PVE alone.

M. R. Ju · A. C. Yopp (✉)
Division of Surgical Oncology, Department of Surgery, UT Southwestern Medical Center, Dallas, TX, USA
e-mail: Michelle.Ju@UTSouthwestern.edu; Adam.Yopp@UTSouthwestern.edu

HCC is the sixth most common cancer and the second leading cause of cancer-related deaths worldwide [1, 2]. In 2017, there were 953,000 incident cases of liver cancer globally and 819,000 deaths [3]. Most commonly occurring in a background of chronic liver disease, HCC represents the leading cause of death in patients diagnosed with cirrhosis [4]. Factors determining outcome following HCC diagnosis are complex and heterogeneous in nature, dependent not only on tumor burden and biology, but on underlying liver function and patient performance status as well. These complexities require maintaining a tenuous balance between tumor-related and patient-related factors, and decisions regarding treatment strategies are best made in the context of a multidisciplinary team approach [5].

Surgical resection is one of the mainstays of curative HCC treatment and has been associated with a median overall survival (OS) of greater than 60 months with 5-year OS rates approaching 60% [6–8]. Clinical practice guidelines have recommended the use of surgical resection in early-stage HCC [9]. However, owing to heterogeneity of the patient population and low use of HCC screening, in the past only 10–37% of patients were candidates for surgical resection at the time of initial HCC diagnosis [10–12]. Recent implementation of HCC screening programs has resulted in earlier tumor diagnosis, thus increasing the number of patients that might be candidates for curative surgical resection [13]. The aim of this chapter is to review the indications for curative HCC surgical resection as it pertains to underlying tumor- and patient-related factors and explore interventions that allow for surgical options when upfront resection is unable to be performed.

Indications for Surgical Resection

The decision to proceed with surgical resection is multifaceted and requires careful consideration of tumor biology, including the number of tumor nodules, tumor size, and

N. S. Reau, A. G. Singal (eds.), *Controversies in Hepatocellular Carcinoma*, https://doi.org/10.1007/978-3-031-90225-3_29

presence of vascular involvement, in addition to underlying liver dysfunction and overall patient performance status. A multidisciplinary approach with input from a medical oncologist, transplant hepatologist, transplant surgeon, surgical oncologist, radiation oncologist, interventional radiologist, and diagnostic radiologist should be used and has been associated with improved patient outcomes following HCC diagnosis [5]. The existence of a multitude of HCC staging systems exemplifies the complexity of evaluating a newly diagnosed HCC patient for curative surgical resection. According to the Barcelona Clinic Liver Cancer (BCLC) system endorsed by the American Association for the Study of Liver Diseases (AASLD) and the European Association for the Study of the Liver (EASL), early-stage (0/A) disease is recommended for consideration of surgical resection. BCLC stage 0 constitutes patients with a single nodule less than or equal to 2 cm in size with preserved liver function, while BCLC stage A represents a solitary lesion or up to 3 nodules less than or equal to 3 cm in size with preserved liver function and good patient performance status [14, 15]. The ultimate goal of hepatic resection for HCC is achieving an appropriate oncological margin while maintaining a functional liver remnant.

Tumor-Related Factors

Tumor Size

Within the initially described BCLC staging system, a solitary HCC tumor greater than 5 cm was considered intermediate stage (BCLC B) and locoregional therapy consisting of intra-arterial therapy was recommended [16]. Within the updated BCLC staging guidelines, tumor size in a solitary nodule is no longer a criterion. While tumor size by itself is not an independent predictor of HCC recurrence following surgery, increasing tumor size is associated with increasing incidence of microvascular invasion and distant metastases, factors portending increased recurrence and worse OS [17].

Despite the increased preponderance of risk factors associated with the likelihood of worse prognostic features, surgical resection of large HCC tumors (>5 cm) is associated both with similar surgical complication rates and long-term oncological outcomes compared to resections of smaller tumors. With improvements in patient selection and perioperative management, morbidity rates range from 30% to 40% with postoperative mortality rates ranging from 3% to 5% in patients undergoing surgical resections for tumors larger than 5 cm [18, 19]. Similarly, oncological outcomes following resection of solitary HCC tumors larger than 5 cm support the role of resection vs noncurative therapies in this patient cohort, with 5-year OS rates ranging from 27% to 53% [18, 20–24].

Intra-arterial therapy for large HCC tumors, including TACE or transarterial radioembolization (TARE), are the noncurative treatment options recommended by the updated BCLC guidelines [17]. A recent propensity score analysis comparing outcomes of patients with solitary HCC tumors larger than 5 cm undergoing hepatic resection vs TACE demonstrate improved outcomes with resection. Five-year OS in the resection group was 41.3% vs 18.5% in the TACE group ($P = 0.007$) [25]. Given the safety and efficacy of surgical resection combined with the lack of effective alternative curative treatment options, tumor size of a solitary HCC tumor should not preclude hepatic resection given that patient-related factors including liver function, liver volume, and performance status are considered acceptable.

Multifocality of Tumors

Indications for hepatic resection in the face of multifocal HCC tumors is controversial and generally limited to patients who lack liver transplantation (LT) options, as multifocality is an independent risk factor associated with tumor recurrence following resection [26, 27]. Several published studies have demonstrated that surgical resection in patients with multifocal HCC, but still within the Milan criteria ($\leq$ 3 tumors $\leq$ 3 cm), has 5-year OS rates ranging from 46% to 69% [19, 22, 28]. Outcome measures in patients undergoing hepatic resection for multifocal HCC tumors outside the Milan criteria are less favorable, with 5-year OS rates ranging from 12% to 24% [24, 29]. The major limitation of these surgical series is their retrospective nature with multifocality determined on pathology results from surgical resections rather than a priori based on preoperative imaging studies. Given the lack of conclusive data demonstrating an oncological benefit in multifocal HCC tumors, liver resection should be reserved for circumstances where LT is not available and only noncurative therapies including TACE, TARE, or systemic therapy are an option.

Presence of Macrovascular Invasion

The presence of MVI, either in the form of PVTT or hepatic venous tumor thrombus, is a poor prognostic factor following HCC diagnosis and is seen as a harbinger of systemic metastatic spread [30]. Patients presenting with HCC tumors demonstrating MVI are classified as BCLC C stage (advanced) and are most commonly treated with systemic therapy with median OS times ranging from 6.5 months to 13.6 months in the first-line setting [31–33].

Although 5-year OS rates following hepatic resection for HCC in the presence of MVI are dismal, ranging from 10% to 40% in retrospective series, careful patient selection based on extent of MVI and subsequent extent of hepatectomy have slightly improved outcome measures [21, 22, 34–38]. Based on a classification system developed by the Liver Cancer Study Group of Japan, [39] PVTT can be divided into five grades: Vp0, no tumor thrombus in the portal vein; Vp1, presence of a tumor thrombus in the second-order branches of the portal vein; Vp3, presence of a tumor thrombus in the first-order branches of the portal vein; and Vp4, presence of a tumor thrombus in the main trunk of the portal vein or a portal vein branch contralateral to the primarily involved lobe (or both). A large retrospective study in Japan using propensity score matching demonstrated that liver resection in the presence of PVTT had statistically significantly prolonged median survival times compared to non–liver resection therapy (locoregional or systemic therapy; 2.5 years vs 1.6 years, $P < 0.001$). In the liver resection cohort, this survival benefit was demonstrated only in patients with Vp1-3 and not Vp4 PVTT. Ninety-day mortality following liver resection was also statistically significantly higher in the Vp4 vs Vp1-3 cohort [40]. Given the lack of relative efficacious therapy compared to hepatic resection, patients presenting with MVI in the form of PVTT (Vp1-3) can be safely considered for liver resection following a multidisciplinary discussion in high-volume centers. There is no current published literature supporting hepatic resection in patients presenting with Vp4 PVTT or hepatic venous tumor thrombus.

Patient-Related Factors

Although perioperative mortality following hepatic resection for HCC has decreased over the past three decades, [41, 42] as indications for resection continue to expand, careful preoperative assessment of the degree of functional impairment of the liver is crucial to ensuring that oncological benefit outweighs the risks of post–hepatectomy liver failure (PHLF). PHLF, defined by the International Study Group of Liver Surgery as an increased prothrombin time and concomitant hyperbilirubinemia on or after postoperative day 5, is associated with perioperative mortality rates of more than 50% [43].

Preoperative Assessment: Clinical and Blood Tests

Multiple validated tools exist to stratify patients based on preresection liver function in determining both feasibility of resection and the extent of resection tolerability. In the West, the 3 most used prognostic tools are the Child-Pugh score (CPS) classification, the Model for End-Stage Liver Disease (MELD) score, and the albumin-bilirubin (ALBI) score.

In 1964, Child and Turcotte developed a classification based on serum albumin, total bilirubin, and prothrombin time, as well as the presence and grade of hepatic encephalopathy and ascites, to predict short-term mortality following portacaval shunt surgery [44]. Although widely used to stratify patients for hepatic resection for HCC, neither the original CPS nor the Pugh modification was designed specifically for this purpose [45]. Nevertheless, CPS continues to be widely used in surgical decision-making for liver resection. In most Western centers, liver resection is limited to CPS class A patients, with CPS class C status universally accepted as a contraindication to resection.

A critique of the CPS classification is that it relies on the use of subjective variables (severity of ascites and encephalopathy). In response to these concerns, Johnson et al. [46] proposed a grading system based only on the ALBI as an alternative system. Similar to CPS classification, the ALBI score was not intended to serve as a predictive biomarker for PHLF following liver resection for HCC. However, investigators have reported the ALBI grade more accurately predicts PHLF than CPS [47]. Currently, the AASLD/EASL guidelines recommend that liver resection for HCC be performed only in patients with a serum total bilirubin of less than or equal to 1 mg/dL. In some Asian centers, a cutoff of less than or equal to 2 mg/dL is widely used [48]. Given that liver resection for HCC is usually limited to patients with normal bilirubin levels, the usefulness of the ALBI grade in surgical candidates may be based simply on the albumin level alone.

The MELD score, based on serum total bilirubin, international normalized ratio, and creatinine, was first reported to predict early mortality following elective transjugular intrahepatic portosystemic shunt placement, but is now used primarily for allocation of organs for LT [49]. Multiple studies have suggested MELD score might be associated with PHLF following hepatectomy for HCC in cirrhotic patients. Teh and colleagues [50] demonstrated that cirrhotic patients with cirrhosis and a MELD score of less than 9 have generally low morbidity of around 8% and negligible mortality from PHLF. Patients with a MELD score greater than or equal to 9 have been shown to have a greater risk of postoperative liver failure and perioperative mortality [51, 52]. Originally designed as a continuous score in patients with poor underlying liver function, the use of MELD as a discrete variable with an a priori cutoff point to determine PHLF might limit its usefulness.

Preoperative Assessment: Portal Hypertension

The presence of portal hypertension, defined as a hepatic venous pressure gradient (HVPG) greater than or equal to10 mm Hg, has been specified both by AASLD and EASL guidelines as a contraindication to hepatic resection for HCC [14, 17]. These recommendations were based on a study by Bruix and colleagues [53] in 1996 in 29 patients undergoing liver resection for HCC. In this study, elevated HVPG and thrombocytopenia was associated with decompensation following hepatic resection in CPS class A cirrhotic patients [53]. Twenty-three of the 29 study patients underwent at least a major hepatectomy (sectionectomy or greater), calling into question the applicability of their findings in operations involving less liver parenchyma resected. In contrast, several other studies have demonstrated that resections for HCC can be safely performed in patients with portal hypertension with resultant low perioperative mortality rates and clear oncologic benefits [28, 54, 55].

Because HVPG is an invasive procedure requiring institutional expertise, clinical parameters including splenomegaly greater than 12 cm, clinical signs of collateralization such as a recanalized periumblical vein, and thrombocytopenia of less than 100/nL are surrogate markers for clinically significant portal hypertension. The decision to perform hepatic resection in a patient with clinically significant portal hypertension should be weighed in the context of possible alternative curative therapies, including LT and ablation.

Preoperative Assessment: Functional Imaging

Quantitative assessment by indocyanine green (ICG) clearance is the most commonly used preoperative test in the East. ICG is a water-soluble fluorescent cyanine dye that is exclusively excreted by the liver into the bile without metabolism or enterohepatic circulation and its retention, measured as percentage serum retention at 15 min (R15), is an indirect assessment of functional hepatic blood flow [56]. A surgical decision algorithm based on R15 was first reported by Makuuchi et al. [57] in 1993 and is now widely used in many Eastern centers.

A recent multicenter Japanese study developed the ALICE grading system, a system using serum albumin and ICG R15 evaluation [58]. Like ALBI, ALICE has predictive power comparable with CPS classification, but allows further stratification of CPS class A patients. These results were validated in a retrospective European cohort [59]. However, the use of ICG clearance over other functional tests is still debated and is rarely used in the United States (Table 29.1).

Assessing Future Liver Remnant Volume

Knowledge of standard liver volume proportion is necessary to determine if the extent of surgical resection will result in an appreciable decline in remnant functional

Table 29.1 Classification systems for evaluating liver function

Classification system	Components				Score
Child-Pugh		**Points**			Class A = 5–6 points (mild) Class B = 7–9 points (moderate) Class C = 10–15 points (severe)
	Variable	**1**	**2**	**3**	
	Encephalopathy	None	Mild or moderate (grade 1–2)	Severe (grade 3–4)	
	Ascites	None	Mild or moderate (diuretic responsive)	Severe (diuretic refractory)	
	Bilirubin, mg/dL	< 2	2–3	> 3	
	Albumin, g/dL	> 3.5	2.8–3.5	< 2.8	
	PT/INR	< 1.7	1.7–2.3	> 2.3	
ALBI	Bilirubin, μmol/L. Albumin, g/L. Calculated as ALBI = ($\log_{10}$ bilirubin × 0.66) + (albumin × −0.085)				Grade 1 = ≤ -2.60 Grade 2 = -2.60 to ≤ -1.39 Grade 3 = > -1.39
MELD	Bilirubin, mg/dL. INR Serum creatinine, mg/dL. Did patient receive dialysis at least twice in past wk., or received 24 h of CVVHD within prior wk.? (yes/no). Serum sodium, mmol/L[a]. Calculated as MELD = 3.8 × $\log_e$(serum bilirubin) + 11.2 × $\log_e$(INR) + 9.6 × $\log_e$(serum creatinine) + 6.4				6–40, with higher numbers correlating to more severe liver dysfunction
ALICE	ICG R15, %. Albumin, g/L. Calculated as ALICE = 0.663 × $\log_{10}$ICGR15–0.0718 × albumin				Grade 1 = ≤ 2.20 Grade 2a = -2.20 to ≤ -1.88 Grade 2b = > -1.88 to ≤ -1.39 Grade 3 = > -1.39

ALBI albumin-bilirubin score, *CVVHD* continuous venovenous hemodialysis, *ICG* indocyanine green, *INR* international normalized ratio, *MELD* Model for End-Stage Liver Disease, *PT* prothrombin time, *R15* percentage serum retention at 15 min

[a]In January 2016, the MELD score was updated to include serum sodium as a factor in calculating the MELD score (known as MELD-Na) for patients with a MELD greater than or equal to 12. Calculated as MELD-Na = MELD + 1.32 × (137-Na) – [0.033 × MELD × (137-Na)]

liver volume. On average, the right liver (segments V–VIII) contributes two-thirds of the total liver volume, while the left liver (segments II–IV) contributes one-third of the liver volume [60]. The method used to calculate the FLR is heavily debated and relies on formulas involving body surface area and radiographic imaging [61, 62]. Computed tomography with three-dimensional reconstruction or volumetric magnetic resonance imaging traces the hepatic segmental contours and multiplies the surface area by slice thickness to calculate the total liver volume [63]. To calculate the FLR, the following formula is used: (resected volume–tumor volume)/(total liver volume–tumor volume) [64, 65]. The minimum FLR considered to be safe following liver resection is based on the function and underlying disease status of the liver. Patients with HCC and cirrhosis necessitate a greater FLR than noncirrhotic patients, with liver remnants of 40% needed. Small liver remnant volume is associated with higher rates of PHLF and other perioperative complications following hepatic resection [66, 67].

Techniques for Increasing Future Liver Remnant Size

In situations where surgery cannot be performed at presentation because of inadequate FLR, there are several techniques available to induce hypertrophy of the residual liver in an attempt to allow for surgical options.

Portal Vein Embolization

In patients with inadequate or borderline FLR, selective PVE has been successfully used to increase remnant size. PVE is an image-guided procedure that induces hypertrophy of the FLR by redirecting portal blood flow away from the liver segments to be resected toward the non–tumor-bearing liver. When appropriately used, PVE has been shown to reduce postoperative morbidity and increase the number of HCC patients eligible for curative intent resection. In patients with normal liver function, PVE is indicated when FLR is less than 20%, which is often the case when extended right hepatectomy is required [68]. In cirrhotic patients with HCC, PVE is indicated when standardized FLR is 40%. Using ICG criteria, PVE has been advocated for FLR of less than or equal to 40% when ICG R15 is less than or equal to 10% and for FLR less than or equal to 50% when ICG R15 is between 10% and 20% [68].

PVE is absolutely contraindicated in cases with extensive ipsilateral tumor thrombus or clinically evident portal hypertension. When extensive ipsilateral tumor thrombus exists, PVE is contraindicated because most of the portal blood flow has already been diverted, and safe delivery of an embolic agent is difficult [69]. As clinically evident portal hypertension is a contraindication to hepatectomy, PVE is not indicated in this setting.

The goal of PVE is complete portal occlusion of targeted liver segments. Embolizing the entire portal tree, including distal branches, is critical to prevent portoportal shunts, as well as to maximize hypertrophy of the FLR and prevent hypertrophy of segments planned for resection. Multiple embolic agents have been described for PVE including n-butyl cyanoacrylate (NBCA), fibrin glue, ethanol, ethiodized oil, and microparticles such as trisacryl gelatin or polyvinyl alcohol. No randomized trials have been conducted comparing the various agents, but retrospective studies show similar efficacy [69].

The degree of hypertrophy after PVE depends on the presence/absence and severity of underlying liver disease. Patients with normal livers can be expected to regenerate at 2 weeks post procedure at rates of 12–21 cm^3/day vs 9 cm^3/day in patients with cirrhosis. In noncirrhotic patients, sufficient hypertrophy usually occurs within 2–4 weeks, whereas sufficient hypertrophy in cirrhotic patients can take up to 4 weeks or more [70]. Additionally, 10–20% of cirrhotic patients do not achieve adequate contralateral hypertrophy after undergoing PVE because of diminished liver regenerative capacity [71].

Combination Arterial and Portal Embolization

Interest has developed in applying TACE sequentially with PVE before performing hepatectomy for HCC. There are several theoretical benefits of combination TACE + PVE. First, the liver necrosis that typically occurs after TACE may lead to increased rates of regeneration. Second, after PVE, hepatic arterial flow within the embolized segment increases, which can lead to increased tumor growth as HCC tumors preferentially derive blood supply from the hepatic artery. TACE therefore might provide local control of tumors during the interval between PVE and hepatectomy. Last, HCC is associated with the formation of arterioportal shunts, which can diminish the effects of PVE, which is typically performed upstream of these shunts. TACE targets these arterioportal shunts and may render PVE more effective.

In a French series of 36 patients with HCC and cirrhosis treated between 1998 and 2004, half (n = 18) underwent TACE + PVE while the remaining half underwent PVE alone before right hepatectomy. The mean increase in percentage FLR volume was statistically significantly higher in the TACE + PVE group than in the PVE alone group. Operative blood loss, liver failure, and mortality were comparable between the groups. The TACE + PVE group had a statistically significantly higher incidence of complete tumor necrosis and higher 5-year disease-free survival (37% vs 19%) [72]. A more recent Korean study published in 2011 of 135 patients undergoing TACE + PVE or PVE alone before right hepatectomy found similar results, with patients receiving combination therapy experiencing a higher mean increase in FLR and improved OS and disease-free survival compared with PVE alone [73]. While there is growing experience with TACE + PVE, including a randomized controlled trial currently ongoing in China, this procedure is still considered experimental and is not widely used in the United States.

Associated Liver Partition with Portal Vein Ligation for Staged Hepatectomy

Associated liver partition with portal vein ligation for staged hepatectomy (ALPPS) is a novel, alternate method to PVE that has been used for FLR hypertrophy in patients with extensive colorectal liver metastasis (CRLM) undergoing extended right hepatectomy. In this technique, portal vein ligation and in situ splitting of the liver along the falciform ligament are performed to induce rapid hypertrophy of the left lateral section. After a short interval (median, 9 days), patients undergo completion hepatectomy. ALPPS has recently been attempted in patients with HCC. A retrospective review of 35 patients in the ALPPS registry from 2010 to 2015 with HCC found rapid and extensive FLR hypertrophy; however, this was statistically significantly lower than for CRLM patients (47% vs 76%) [74]. The degree of hypertrophy was negatively correlated with the severity of liver fibrosis. The 90-day perioperative mortality was high at 31% (compared to 7% for CRLM) and the long-term oncologic outcomes are unknown. This procedure is still experimental, and PVE remains the preferred technique to induce FLR hypertrophy before hepatectomy in HCC patients. There may be a limited role for ALPPS in younger, healthy patients without cirrhosis who are not candidates for LT and for whom the operative risks are deemed acceptable.

Stem Cell Infusion

A growing interest has developed in infusing stem cells into the FLR during PVE to improve hepatic regeneration. Small phase 2 trials have shown promising initial results, with stem cell–treated patients experiencing statistically significant increases in hepatic growth volume compared with PVE-only patients, as well as improvements in underlying liver function [75, 76]. Further studies are needed to determine whether stem cell infusion has an effect on surgical outcomes and survival.

Conclusion

Most HCC patients are not candidates for resection at diagnosis, often because of underlying liver dysfunction, inadequate FLR, or tumor characteristics. The perioperative risks of hepatic resection in cirrhotic patients with HCC must be balanced against the oncological benefits. Tumor characteristics including large tumor size, vascular invasion, and multifocality are not absolute contraindications to hepatic resection in well-compensated patients with a lack of alternative curative options. Underlying liver dysfunction must be carefully assessed, accounting for the degree of dysfunction, extent of planned liver resection, and alternative curative therapies, including LT and ablation.

In situations where small FLR renders a patient ineligible for surgery at initial presentation, adjuncts such as PVE or ALPPS can be used to induce hypertrophy of the residual liver. These techniques may allow an increased role for hepatic resection in HCC patients who were not surgical candidates upfront because of inadequate FLR size. In a carefully selected group of patients, surgery can be safely performed after additional interventions even if resection was unable to be performed at presentation.

References

1. Forner A, Llovet JM, Bruix J. Hepatocellular carcinoma. Lancet. 2012;379(9822):1245–55. https://doi.org/10.1016/S0140-6736(11)61347-0.
2. Rahib L, Smith BD, Aizenberg R, Rosenzweig AB, Fleshman JM, Matrisian LM. Projecting cancer incidence and deaths to 2030: the unexpected burden of thyroid, liver, and pancreas cancers in the United States. Cancer Res. 2014;74(11):2913–21. https://doi.org/10.1158/0008-5472.CAN-14-0155.
3. Global Burden of Disease Cancer Collaboration, Fitzmaurice C, Abate D, Abbasi N, et al. Global, regional, and national cancer incidence, mortality, years of life lost, years lived with disability, and disability-adjusted life-years for 29 cancer groups, 1990 to 2017: a systematic analysis for the global burden of disease study. JAMA Oncol. 2019;5(12):1749–68. https://doi.org/10.1001/jamaoncol.2019.2996.
4. Ioannou GN, Weiss NS, Kowdley KV. Relationship between transferrin-iron saturation, alcohol consumption, and the incidence of cirrhosis and liver cancer. Clin Gastroenterol Hepatol. 2007;5(5):624–9. https://doi.org/10.1016/j.cgh.2007.01.008.
5. Yopp AC, Mansour JC, Beg MS, et al. Establishment of a multidisciplinary hepatocellular carcinoma clinic is associated with improved clinical outcome. Ann Surg Oncol. 2014;21(4):1287–95. https://doi.org/10.1245/s10434-013-3413-8.
6. Pang TC, Lam VW. Surgical management of hepatocellular carcinoma. World J Hepatol. 2015;7(2):245–52. https://doi.org/10.4254/wjh.v7.i2.245.
7. Maluccio MA, Zang Y, Pi W, et al. Survival in patients with hepatocellular carcinoma (HCC): a report of 1444 patients treated within a multidisciplinary program. J Clin Oncol. 2017;35(15 Suppl):e15652.
8. Liu W, Wang K, Bao Q, Sun Y, Xing BC. Hepatic resection provided long-term survival for patients with intermediate and advanced-stage resectable hepatocellular carcinoma. World J Surg Oncol. 2016;14:62. https://doi.org/10.1186/s12957-016-0811-y.
9. Marrero JA, Kulik LM, Sirlin CB, et al. Diagnosis, staging, and management of hepatocellular carcinoma: 2018 practice guidance by the American Association for the Study of Liver Diseases. Hepatology. 2018;68(2):723–50. https://doi.org/10.1002/hep.29913.
10. Fong Y, Sun RL, Jarnagin W, Blumgart LH. An analysis of 412 cases of hepatocellular carcinoma at a Western center. Ann

Surg. 1999;229(6):790–9.; discussion 799-800. https://doi.org/10.1097/00000658-199906000-00005.
11. Fan ST, Lo CM, Liu CL, et al. Hepatectomy for hepatocellular carcinoma: toward zero hospital deaths. Ann Surg. 1999;229(3):322–30. https://doi.org/10.1097/00000658-199903000-00004.
12. Colella G, Bottelli R, De Carlis L, et al. Hepatocellular carcinoma: comparison between liver transplantation, resective surgery, ethanol injection, and chemoembolization. Transpl Int. 1998;11(Suppl 1):S193–6. https://doi.org/10.1007/s001470050459.
13. Singal AG, Lampertico P, Nahon P. Epidemiology and surveillance for hepatocellular carcinoma: new trends. J Hepatol. 2020;72(2):250–61. https://doi.org/10.1016/j.jhep.2019.08.025.
14. European Association for the Study of the Liver. EASL clinical practice guidelines for the management of patients with decompensated cirrhosis. J Hepatol. 2018;69(2):406–60. https://doi.org/10.1016/j.jhep.2018.03.024.
15. Heimbach JK, Kulik LM, Finn RS, et al. AASLD guidelines for the treatment of hepatocellular carcinoma. Hepatology. 2018;67(1):358–80. https://doi.org/10.1002/hep.29086.
16. Llovet JM, Brú C, Bruix J. Prognosis of hepatocellular carcinoma: the BCLC staging classification. Semin Liver Dis. 1999;19(3):329–38. https://doi.org/10.1055/s-2007-1007122.
17. Bruix J, Sherman M, American Association for the Study of Liver Diseases. Management of hepatocellular carcinoma: an update. Hepatology. 2011;53(3):1020–2. https://doi.org/10.1002/hep.24199.
18. Pandey D, Lee KH, Wai CT, Wagholikar G, Tan KC. Long term outcome and prognostic factors for large hepatocellular carcinoma (10 cm or more) after surgical resection. Ann Surg Oncol. 2007;14(10):2817–23. https://doi.org/10.1245/s10434-007-9518-1.
19. Poon RTP, Fan ST, Wong J. Selection criteria for hepatic resection in patients with large hepatocellular carcinoma larger than 10 cm in diameter. J Am Coll Surg. 2002;194(5):592–602. https://doi.org/10.1016/s1072-7515(02)01163-8.
20. Cho YB, Lee KU, Lee HW, et al. Outcomes of hepatic resection for a single large hepatocellular carcinoma. World J Surg. 2007;31(4):795–801. https://doi.org/10.1007/s00268-006-0359-z.
21. Pawlik TM, Poon RT, Abdalla EK, International Cooperative Study Group on Hepatocellular Carcinoma, et al. Critical appraisal of the clinical and pathologic predictors of survival after resection of large hepatocellular carcinoma. Arch Surg. 2005;140(5):450–7.; discussion 457–458. https://doi.org/10.1001/archsurg.140.5.450.
22. Ruzzenente A, Capra F, Pachera S, et al. Is liver resection justified in advanced hepatocellular carcinoma? Results of an observational study in 464 patients. J Gastrointest Surg. 2009;13(7):1313–20. https://doi.org/10.1007/s11605-009-0903-x.
23. Yamashita Y, Taketomi A, Shirabe K, et al. Outcomes of hepatic resection for huge hepatocellular carcinoma (≥ 10 cm in diameter). J Surg Oncol. 2011;104(3):292–8. https://doi.org/10.1002/jso.21931.
24. Zhong JH, Xiang BD, Gong WF, et al. Comparison of long-term survival of patients with BCLC stage B hepatocellular carcinoma after liver resection or transarterial chemoembolization. PLoS One. 2013;8(7):e68193. https://doi.org/10.1371/journal.pone.0068193.
25. Zhu SL, Ke Y, Peng YC, et al. Comparison of long-term survival of patients with solitary large hepatocellular carcinoma of BCLC stage A after liver resection or transarterial chemoembolization: a propensity score analysis. PLoS One. 2014;9(12):e115834. https://doi.org/10.1371/journal.pone.0115834.
26. Chang WT, Kao WY, Chau GY, et al. Hepatic resection can provide long-term survival of patients with non-early-stage hepatocellular carcinoma: extending the indication for resection? Surgery. 2012;152(5):809–20. https://doi.org/10.1016/j.surg.2012.03.024.
27. Shah SA, Cleary SP, Wei AC, et al. Recurrence after liver resection for hepatocellular carcinoma: risk factors, treatment, and outcomes. Surgery. 2007;141(3):330–9. https://doi.org/10.1016/j.surg.2006.06.028.
28. Ishizawa T, Hasegawa K, Aoki T, et al. Neither multiple tumors nor portal hypertension are surgical contraindications for hepatocellular carcinoma. Gastroenterology. 2008;134(7):1908–16. https://doi.org/10.1053/j.gastro.2008.02.091.
29. Torzilli G, Belghiti J, Kokudo N, et al. A snapshot of the effective indications and results of surgery for hepatocellular carcinoma in tertiary referral centers: is it adherent to the EASL/AASLD recommendations? An observational study of the HCC East-West study group. Ann Surg. 2013;257(5):929–37. https://doi.org/10.1097/SLA.0b013e31828329b8.
30. Mokdad AA, Singal AG, Marrero JA, Zhu H, Yopp AC. Vascular invasion and metastasis is predictive of outcome in Barcelona Clinic Liver Cancer stage C hepatocellular carcinoma. J Natl Compr Cancer Netw. 2017;15(2):197–204. https://doi.org/10.6004/jnccn.2017.0020.
31. Cheng AL, Kang YK, Chen Z, et al. Efficacy and safety of sorafenib in patients in the Asia-Pacific region with advanced hepatocellular carcinoma: a phase III randomised, double-blind, placebo-controlled trial. Lancet Oncol. 2009;10(1):25–34. https://doi.org/10.1016/S1470-2045(08)70285-7.
32. Kudo M, Finn RS, Qin S, et al. Lenvatinib versus sorafenib in first-line treatment of patients with unresectable hepatocellular carcinoma: a randomised phase 3 non-inferiority trial. Lancet. 2018;391(10126):1163–73. https://doi.org/10.1016/S0140-6736(18)30207-1.
33. Llovet JM, Ricci S, Mazzaferro V, SHARP Investigators Study Group, et al. Sorafenib in advanced hepatocellular carcinoma. N Engl J Med. 2008;359(4):378–90. https://doi.org/10.1056/NEJMoa0708857.
34. Ban D, Shimada K, Yamamoto Y, et al. Efficacy of a hepatectomy and a tumor thrombectomy for hepatocellular carcinoma with tumor thrombus extending to the main portal vein. J Gastrointest Surg. 2009;13(11):1921–8. https://doi.org/10.1007/s11605-009-0998-0.
35. Le Treut YP, Hardwigsen J, Ananian P, Saisse J, Gregoire E, Richa H, et al. Resection of hepatocellular carcinoma with tumor thrombus in the major vasculature. A European case-control series. J Gastrointest Surg. 2006;10(6):855–62. https://doi.org/10.1016/j.gassur.2005.12.011.
36. Inoue Y, Hasegawa K, Ishizawa T, et al. Is there any difference in survival according to the portal tumor thrombectomy method in patients with hepatocellular carcinoma? Surgery. 2009;145(1):9–19. https://doi.org/10.1016/j.surg.2008.09.005.
37. Wang Y, Yuan L, Ge RL, Sun Y, Wei G. Survival benefit of surgical treatment for hepatocellular carcinoma with inferior vena cava/right atrium tumor thrombus: results of a retrospective cohort study. Ann Surg Oncol. 2013;20(3):914–22. https://doi.org/10.1245/s10434-012-2646-2.
38. Chok KS, Cheung TT, Chan SC, Poon RT, Fan ST, Lo CM. Surgical outcomes in hepatocellular carcinoma patients with portal vein tumor thrombosis. World J Surg. 2014;38(2):490–6. https://doi.org/10.1007/s00268-013-2290-4.
39. Kudo M, Izumi N, Kokudo N, HCC Expert Panel of Japan Society of Hepatology, et al. Management of hepatocellular carcinoma in Japan: consensus-based clinical practice guidelines proposed by the Japan Society of Hepatology (JSH) 2010 updated version. Dig Dis. 2011;29(3):339–64. https://doi.org/10.1159/000327577.
40. Kokudo T, Hasegawa K, Matsuyama Y, Liver Cancer Study Group of Japan, et al. Survival benefit of liver resection for hepatocellular carcinoma associated with portal vein invasion. J Hepatol. 2016;65(5):938–43. https://doi.org/10.1016/j.jhep.2016.05.044.
41. Dokmak S, Ftériche FS, Borscheid R, Cauchy F, Farges O, Belghiti J. 2012 Liver resections in the 21st century: we are far from zero mortality. HPB (Oxford). 2013;15(11):908–15. https://doi.org/10.1111/hpb.12069.

42. Kenjo A, Miyata H, Gotoh M, et al. Risk stratification of 7,732 hepatectomy cases in 2011 from the National Clinical Database for Japan. J Am Coll Surg. 2014;218(3):412–22. https://doi.org/10.1016/j.jamcollsurg.2013.11.007.
43. Balzan S, Belghiti J, Farges O, et al. The "50-50 criteria" on postoperative day 5: an accurate predictor of liver failure and death after hepatectomy. Ann Surg. 2005;242(6):824–8.; discussion 828–829. https://doi.org/10.1097/01.sla.0000189131.90876.9e.
44. Child CG, Turcotte JG. Surgery and portal hypertension. Major Probl Clin Surg. 1964;1:1–85.
45. Pugh RN, Murray-Lyon IM, Dawson JL, Pietroni MC, Williams R. Transection of the oesophagus for bleeding oesophageal varices. Br J Surg. 1973;60(8):646–9. https://doi.org/10.1002/bjs.1800600817.
46. Johnson PJ, Berhane S, Kagebayashi C, et al. Assessment of liver function in patients with hepatocellular carcinoma: a new evidence-based approach – the ALBI grade. J Clin Oncol. 2015;33(6):550–8. https://doi.org/10.1200/JCO.2014.57.9151.
47. Wang YY, Zhong JH, Su ZY, et al. Albumin-bilirubin versus Child-Pugh score as a predictor of outcome after liver resection for hepatocellular carcinoma. Br J Surg. 2016;103(6):725–34. https://doi.org/10.1002/bjs.10095.
48. Imamura H, Seyama Y, Kokudo N, et al. One thousand fifty-six hepatectomies without mortality in 8 years. Arch Surg. 2003;138(11):1198–206. https://doi.org/10.1001/archsurg.138.11.1198.
49. Malinchoc M, Kamath PS, Gordon FD, Peine CJ, Rank J, ter Borg PC. A model to predict poor survival in patients undergoing transjugular intrahepatic portosystemic shunts. Hepatology. 2000;31(4):864–71. https://doi.org/10.1053/he.2000.5852.
50. Teh SH, Christein J, Donohue J, et al. Hepatic resection of hepatocellular carcinoma in patients with cirrhosis: Model of End-Stage Liver Disease (MELD) score predicts perioperative mortality. J Gastrointest Surg. 2005;9(9):1207–15. https://doi.org/10.1016/j.gassur.2005.09.008.
51. Cucchetti A, Ercolani G, Vivarelli M, et al. Impact of model for end-stage liver disease (MELD) score on prognosis after hepatectomy for hepatocellular carcinoma on cirrhosis. Liver Transpl. 2006;12(6):966–71. https://doi.org/10.1002/lt.20761.
52. Delis SG, Bakoyiannis A, Dervenis C, Tassopoulos N. Perioperative risk assessment for hepatocellular carcinoma by using the MELD score. J Gastrointest Surg. 2009;13(12):2268–75. https://doi.org/10.1007/s11605-009-0977-5.
53. Bruix J, Castells A, Bosch J, et al. Surgical resection of hepatocellular carcinoma in cirrhotic patients: prognostic value of preoperative portal pressure. Gastroenterology. 1996;111(4):1018–22. https://doi.org/10.1016/s0016-5085(96)70070-7.
54. Cucchetti A, Ercolani G, Vivarelli M, et al. Is portal hypertension a contraindication to hepatic resection? Ann Surg. 2009;250(6):922–8. https://doi.org/10.1097/SLA.0b013e3181b977a5.
55. Zhong JH, Ke Y, Gong WF, et al. Hepatic resection associated with good survival for selected patients with intermediate and advanced-stage hepatocellular carcinoma. Ann Surg. 2014;260(2):329–40. https://doi.org/10.1097/SLA.0000000000000236.
56. Teranaka M, Schenk WG. Hepatic blood flow measurement. A comparison of the indocyanine green and electromagnetic techniques in normal and abnormal flow states in the dog. Ann Surg. 1977;185(1):58–63. https://doi.org/10.1097/00000658-197701000-00009.
57. Makuuchi M, Kosuge T, Takayama T, et al. Surgery for small liver cancers. Semin Surg Oncol. 1993;9(4):298–304. https://doi.org/10.1002/ssu.2980090404.
58. Kokudo T, Hasegawa K, Amikura K, et al. Assessment of preoperative liver function in patients with hepatocellular carcinoma – the albumin-Indocyanine Green Evaluation (ALICE) grade. PLoS One. 2016;11(7):e0159530. https://doi.org/10.1371/journal.pone.0159530.
59. Russolillo N, Forchino F, Conci S, et al. Validation of the albumin-indocyanine green evaluation model in patients with resected hepatocellular carcinoma and comparison with the albumin-bilirubin score. J Hepatobiliary Pancreat Sci. 2019;26(1):51–7. https://doi.org/10.1002/jhbp.597.
60. Abdalla EK, Denys A, Chevalier P, Nemr RA, Vauthey JN. Total and segmental liver volume variations: implications for liver surgery. Surgery. 2004;135(4):404–10. https://doi.org/10.1016/j.surg.2003.08.024.
61. Ribero D, Chun YS, Vauthey JN. Standardized liver volumetry for portal vein embolization. Semin Intervent Radiol. 2008;25(2):104–99. https://doi.org/10.1055/s-2008-1076681.
62. Urata K, Kawasaki S, Matsunami H, et al. Calculation of child and adult standard liver volume for liver transplantation. Hepatology. 1995;21(5):1317–21.
63. Madoff DC, Abdalla EK, Gupta S, et al. Transhepatic ipsilateral right portal vein embolization extended to segment IV: improving hypertrophy and resection outcomes with spherical particles and coils. J Vasc Interv Radiol. 2005;16(2 Pt 1):215–25. https://doi.org/10.1097/01.RVI.0000147067.79223.85.
64. Kubota K, Makuuchi M, Kusaka K, et al. Measurement of liver volume and hepatic functional reserve as a guide to decision-making in resectional surgery for hepatic tumors. Hepatology. 1997;26(5):1176–81. https://doi.org/10.1053/jhep.1997.v26.pm0009362359.
65. Ogasawara K, Une Y, Nakajima Y, Uchino J. The significance of measuring liver volume using computed tomographic images before and after hepatectomy. Surg Today. 1995;25(1):43–8. https://doi.org/10.1007/BF00309384.
66. Farges O, Malassagne B, Flejou JF, Balzan S, Sauvanet A, Belghiti J. Risk of major liver resection in patients with underlying chronic liver disease: a reappraisal. Ann Surg. 1999;229(2):210–5. https://doi.org/10.1097/00000658-199902000-00008.
67. Vauthey JN, Chaoui A, Do KA, et al. Standardized measurement of the future liver remnant prior to extended liver resection: methodology and clinical associations. Surgery. 2000;127(5):512–9. https://doi.org/10.1067/msy.2000.105294.
68. Kishi Y, Abdalla EK, Chun YS, et al. Three hundred and one consecutive extended right hepatectomies: evaluation of outcome based on systematic liver volumetry. Ann Surg. 2009;250(4):540–8. https://doi.org/10.1097/SLA.0b013e3181b674df.
69. Madoff DC, Abdalla EK, Vauthey JN. Portal vein embolization in preparation for major hepatic resection: evolution of a new standard of care. J Vasc Interv Radiol. 2005;16(6):779–90. https://doi.org/10.1097/01.RVI.0000159543.28222.73.
70. Madoff DC, Hicks ME, Vauthey JN, et al. Transhepatic portal vein embolization: anatomy, indications, and technical considerations. Radiographics. 2002;22(5):1063–76. https://doi.org/10.1148/radiographics.22.5.g02se161063.
71. Kianmanesh R, Regimbeau JM, Belghiti J. Selective approach to major hepatic resection for hepatocellular carcinoma in chronic liver disease. Surg Oncol Clin N Am. 2003;12(1):51–63. https://doi.org/10.1016/s1055-3207(02)00090-x.
72. Ogata S, Belghiti J, Farges O, Varma D, Sibert A, Vilgrain V. Sequential arterial and portal vein embolizations before right hepatectomy in patients with cirrhosis and hepatocellular carcinoma. Br J Surg. 2006;93(9):1091–8. https://doi.org/10.1002/bjs.5341.
73. Yoo H, Kim JH, Ko GY, et al. Sequential transcatheter arterial chemoembolization and portal vein embolization versus portal vein embolization only before major hepatectomy for patients with

hepatocellular carcinoma. Ann Surg Oncol. 2011;18(5):1251–7. https://doi.org/10.1245/s10434-010-1423-3.
74. D'Haese JG, Neumann J, Weniger M, et al. Should ALPPS be used for liver resection in intermediate-stage HCC? Ann Surg Oncol. 2016;23(4):1335–43. https://doi.org/10.1245/s10434-015-5007-0.
75. Schulte am Esch J, Schmelzle M, Fürst G, et al. Infusion of CD133+ bone marrow-derived stem cells after selective portal vein embolization enhances functional hepatic reserves after extended right hepatectomy: a retrospective single-center study. Ann Surg. 2012;255(1):79–85. https://doi.org/10.1097/SLA.0b013e31823d7d08.
76. Han HS, Ahn KS, Cho JY, et al. Autologous stem cell transplantation for expansion of remnant liver volume with extensive hepatectomy. Hepato-Gastroenterology. 2014;61(129):156–61.

Part VI

Post Liver Transplantation

Should Rapamune (Sirolimus)-Based Immunomodulation Be Given Post Transplant for Those Transplanted for or Found to Have Incidental Hepatocellular Carcinoma?

30

Arvind R. Murali and K. V. Narayanan Menon

Sirolimus and mammalian target of rapamycin inhibitors (mTORis) do not have a clear advantage for transplanted patients with hepatocellular carcinoma (HCC) but may be considered because they may provide a dual advantage over calcineurin inhibitors (CNIs') renal protective effect as well as an antineoplastic effect.

Main Points

1. The backbone of current immunosuppressive strategies post liver transplantation (LT) are the CNIs, which have oncogenic potential.
2. The mTOR is a serine/threonine protein kinase. mTORis have antineoplastic effects.
3. mTORis have also been shown to decrease the risk of de novo malignancies after solid-organ transplant.
4. Meta-analysis of retrospective data suggested increased recurrence-free survival for LT recipients who received mTORi compared to CNI at 1 and 3 years but not at 5 years.
5. A prospective multicenter trial showed no difference in HCC recurrence-free survival or overall survival (OS) after 8 years in those who received sirolimus compared to those who did not, though there was an advantage in the first 4 years. There was no benefit in those transplanted with HCC outside the Milan criteria (MC).

A. R. Murali
Orlando Health Liver Center, Orlando Health, University of Iowa, Iowa City, IA, USA

K. V. N. Menon (✉)
Medical City Dallas Transplant Institute, Dallas, TX, USA
e-mail: Kozhikode.Menon@hcahealthcare.com

HCC is the most common primary malignancy of the liver. It is the fifth most common cancer worldwide and the third most common cause of cancer-related mortality. The incidence of HCC is on the rise in the United States and is expected to continue to rise until 2030. HCC is one of the most common indications for LT and with the rise in the incidence of HCC, the use of LT for HCC is bound to increase.

Surgical resection and LT are the 2 most commonly performed and recommended curative treatments for HCC. Surgical resection is usually performed for solitary lesions in the absence of portal hypertension. LT is the treatment of choice for patients with cirrhosis and portal hypertension and who satisfy the MC (single lesion < 5 cm, up to 3 HCC lesions with the largest < 3 cm, without vascular invasion or distant metastasis) [1]. Patients outside the MC may be successfully downstaged to within the MC with locoregional therapy such as transarterial chemoembolization or radioembolization and then considered for LT after a wait time of 6 months to ensure stability of the treated liver lesion(s). Patients successfully downstaged have been shown to have similar posttransplantation outcomes compared to patients who were originally within the MC [2].

Hepatocellular Carcinoma Recurrence After Liver Transplantation

Patients who underwent LT for HCC are unfortunately at risk for recurrence of HCC post LT. The risk of recurrence varies from 5% to 20% [3, 4]. Several pretransplant and posttransplant factors have been shown to influence the risk of recurrence of HCC. Patients within the MC are at low risk for HCC recurrence compared to patients with HCC outside the MC at the time of transplant [1]. Pretransplant alpha fetoprotein above 400 ng/mL and total tumor volume have also been shown to increase HCC recurrence after LT [3, 5]. Histological factors on explant that increase the risk

N. S. Reau, A. G. Singal (eds.), *Controversies in Hepatocellular Carcinoma*, https://doi.org/10.1007/978-3-031-90225-3_30

of recurrence of HCC after LT include the presence of microvascular or macrovascular invasion, poor differentiation of HCC, and viable tumor on explant that exceeds the MC [6, 7]. While pre-LT risk factors of HCC recurrence can be controlled for, explant characteristics such as the presence or absence of microvascular or macrovascular invasion, HCC differentiation, and number and size of viable tumor on explant may not be accurately determined before LT. It is thus imperative to have an active surveillance protocol for HCC post LT to identify and treat HCC recurrence, if any, at the earliest. Unfortunately, even with periodic surveillance, a substantial percentage of patients with recurrence present with extrahepatic HCC and have limited curative options, resulting in poor outcomes [4]. Preventive strategies to decrease the risk of recurrence of HCC after LT are thus of importance in improving post-LT outcomes of patients with HCC.

Immunosuppression After Liver Transplant

The backbone of current immunosuppressive strategies post-LT are the CNIs—tacrolimus, and to a lesser extent, cyclosporine. However, CNIs have several side effects, including diabetes mellitus, hypertension, hypercholesterolemia, cardiovascular complications, and nephrotoxicity, among others. CNI-induced nephrotoxicity is the most frequent cause of renal dysfunction after LT and up to 20–30% of patients develop chronic kidney disease stage 3 or higher by 5 years after LT [8, 9]. CNIs also have oncogenic potential; they can enhance the survival and proliferation of cancer cells. CNIs have been shown to increase the risk of skin cancers, solid organ cancers, and hematologic malignancies after LT [10]. Thus, there has been a need to identify alternative immunosuppressive regimens for post-LT patients.

Rapamycin Inhibitors

mTOR is a serine/threonine protein kinase that belongs to the family of phosphatidylinositol-3 protein–related kinases. mTOR plays an important role in the signal induction pathway through 2 multiprotein complexes, mTORC1 and mTORC2. These complexes integrate signals from multiple tyrosine kinase pathways and affect cellular growth, proliferation, and survival. The mechanisms for the immunosuppressive effect of mTORis are different from that of the CNIs. CNIs inhibit the first phase of T-cell activation, from G0 to G1 phase. CNIs, by blocking calcineurin, interrupt signals from T-cell receptors, thereby inhibiting production of interleukin-2 and preventing the stimulation of T cells. In contrast, mTORis affect the second phase of T-cell activation from the G1 to the S phase, by blocking the signal transduction pathway required for the interleukin-stimulated proliferation of T cells [11].

Sirolimus, the prototypic mTORi, and its analog everolimus have been used as immunosuppressive agents in solid-organ transplant recipients. Sirolimus has a longer half-life of 60 h and reaches steady state in 6 days, whereas everolimus has a shorter half-life of 30 hours and reaches steady state in 4 days. The use of mTORis gained interest mainly because of their protective effect on renal function. CNI nephrotoxicity is associated with a four- to fivefold increased risk of death. Randomized controlled trials (RCTs) have demonstrated that patients on mTORi-based immunosuppressive regimens, especially everolimus, have sustained preservation of renal function at 1 and 3 years after LT compared to patients on a CNI-based immunosuppressive regimen [12–14]. Importantly, patients on mTORi have comparable OS as well as graft loss compared to patients on CNIs [13, 15]. Based on these studies, many centers introduce mTORis 4–6 weeks after LT with a simultaneous dose decrease/discontinuation of CNI. In addition to their renoprotective effect, mTORis also have an antineoplastic effect as discussed next that may further warrant their use in patients with HCC.

Rapamycin Inhibitors and Malignancy

Dysregulation in the mTOR signaling pathway has been shown to be an important mechanism in the development of several malignancies. mTORis have antiproliferative and antiangiogenesis effects. Thus, mTORs have emerged as therapeutic targets in the treatment for cancer. mTORis prevents angiogenesis by reducing levels of vascular endothelial growth factor and inhibiting the growth of endothelial cells and fibroblasts. In vitro and in vivo studies have shown that mTORis suppress the growth of cancer cells as well as inhibit metastatic tumor growth [16, 17]. Studies have shown that drugs that inhibit the mTOR pathway are effective in solid organ malignancies [18–21]. Temsirolimus, a prodrug of sirolimus, and everolimus have been demonstrated to be effective in the treatment of renal cell carcinoma [18, 19]. mTORis have also been shown to decrease the risk of de novo malignancies after solid-organ transplant. An analysis of more than 30,000 kidney transplant recipients reported to the Organ Procurement and Transplantation Network database showed a statistically significant decreased incidence of de novo cancers in the sirolimus/everolimus group compared to the CNI-only group [22]. A similar effect was also seen for de novo solid cancers [22]. Thus, there is increasing interest in the use of mTORis as the primary immunosuppressive agent in patients who have received LT for HCC with an intent to decrease the risk of recurrence of HCC.

Sirolimus and Hepatocellular Carcinoma

Villanueva et al. [11] demonstrated the importance of mTOR signaling in the development of HCC. They showed in their in vivo studies that everolimus inhibited tumor growth and improved survival in experimental HCC mouse models. Several cohort studies have subsequently evaluated the efficacy of sirolimus in post-LT patients with HCC but have shown variable benefit. A meta-analysis published in 2012 showed decreased HCC recurrence, increased recurrence-free survival, and increased OS in patients who received sirolimus compared to sirolimus-free regimens [23]. A more recent meta-analysis also showed statistically significant increased recurrence-free survival for the mTORi group compared to the CNI group at 1 and 3 years but not at 5 years [24]. However, the problem with these meta-analyses was that the included studies were predominantly retrospective. In addition, there was considerable heterogeneity in the studies included in the analysis. The best data on the use of sirolimus in post-LT patients with HCC comes from the RCT conducted by Geissler and colleagues [15]. In this prospective multicenter trial, the authors randomly assigned 525 LT recipients with HCC to a sirolimus-free CNI group vs sirolimus group. Sirolimus was introduced at 4–6 weeks after LT. The primary end point was HCC recurrence-free survival. Intention-to-treat analysis performed after 8 years showed no statistically significant difference in the recurrence-free survival or OS between the two groups at the end of the study [15]. However, OS seemed to be superior for the sirolimus group compared to the sirolimus-free group up to 5 years post LT. It is possible that the lack of survival benefit for mTORis beyond 5 years after LT could be from the fact that non–liver-related conditions such as metabolic syndrome (diabetes and hypertension) and cardiopulmonary diseases are more important contributors of death beyond 5 years post LT [25]. More interestingly on subgroup analysis, patients with HCC within the MC seemed to benefit most from sirolimus compared to a CNI regimen, while there was no statistically significant difference in the recurrence-free survival between the two groups in patients outside the MC. The benefit of sirolimus on recurrence-free survival in patients within the MC was seen only in the first 4 years after LT; the survival advantage was not seen beyond 4 years [15]. While it was believed that patients at high risk for HCC recurrence such as those outside the MC may best benefit from sirolimus, the RCT by Geissler et al. [15] suggested the opposite. Further RCTs are needed to establish the clinical benefit of sirolimus in patients transplanted for HCC.

The lack of a clear, demonstrable clinical benefit of sirolimus in the prevention of HCC recurrence after LT could be explained by several hypotheses. The inhibition of mTOR by sirolimus and its prodrugs is incomplete and thus they can only inhibit cell proliferation but do not induce apoptosis [26]. Additionally, the doses of mTORis used in the treatment of solid-organ cancers such as renal cell carcinoma are much higher than the doses used for immunosuppression after LT [19]. Finally, some tumors express markers such as KRAS and BRAF, which resist response to mTORis [26]. Thus not all patients receiving LT for HCC may benefit from the use of sirolimus after LT. Identifying biomarkers that predict the response to mTORis may be challenging, but if successful would help in appropriate patient selection for use of mTORis after LT to decrease HCC recurrence.

Adverse Events of Mammalian Target of Rapamycin Inhibitors

The introduction of mTORis is not without drawbacks. Complete withdrawal of CNIs is associated with an increased risk of acute cellular rejection [27, 28]. mTORis have their own list of adverse events including hypertriglyceridemia, dyslipidemia, hypertension, insulin resistance, stomatitis, painful oral ulcers, delayed wound healing, interstitial pneumonitis, and proteinuria, among others [15, 27]. A substantial percentage of patients (~25%) require drug discontinuation because of adverse effects [15, 27]. Initial reports also raised concern for the development of hepatic artery thrombosis when used early after LT [28]. In fact, mTORis carry a black box warning for hepatic artery thrombosis when used in the post-LT setting. However, several future studies have shown no increased risk of hepatic artery thrombosis with mTORis compared to CNIs [15].

Conclusion

Sirolimus has antineoplastic properties by blocking the mTOR signal induction pathway, which has been shown to play a pivotal role in the development of HCC. In vitro and in vivo studies have demonstrated that sirolimus decreases tumor proliferation and improves survival. Though RCT studies demonstrating a clear advantage of sirolimus over CNIs in improving OS and recurrence-free survival are lacking, sirolimus may be considered for patients who have undergone LT for HCC because it may provide a dual advantage over CNIs' with its renal protective effect as well as an antineoplastic effect.

References

1. Mazzaferro V, Regalia E, Doci R, et al. Liver transplantation for the treatment of small hepatocellular carcinomas in patients with cirrhosis. N Engl J Med. 1996;334(11):693–9. https://doi.org/10.1056/NEJM199603143341104.
2. Yao FY, Hirose R, LaBerge JM, et al. A prospective study on downstaging of hepatocellular carcinoma prior to liver transplantation. Liver Transpl. 2005;11(12):1505–14. https://doi.org/10.1002/lt.20526.
3. Murali AR, Romero-Marrero C, Miller C, et al. Predictors of successful downstaging of hepatocellular carcinoma outside Milan criteria. Transplantation. 2016;100(11):2391–7. https://doi.org/10.1097/TP.0000000000001402.
4. Fernandez-Sevilla E, Allard MA, Selten J, et al. Recurrence of hepatocellular carcinoma after liver transplantation: is there a place for resection? Liver Transpl. 2017;23(4):440–7. https://doi.org/10.1002/lt.24742.
5. Toso C, Asthana S, Bigam DL, Shapiro AM, Kneteman NM. Reassessing selection criteria prior to liver transplantation for hepatocellular carcinoma utilizing the scientific registry of transplant recipients database. Hepatology. 2009;49(3):832–8. https://doi.org/10.1002/hep.22693.
6. Mazzaferro V, Llovet JM, Miceli R, Metroticket Investigator Study Group, et al. Predicting survival after liver transplantation in patients with hepatocellular carcinoma beyond the Milan criteria: a retrospective, exploratory analysis. Lancet Oncol. 2009;10(1):35–43. https://doi.org/10.1016/S1470-2045(08)70284-5.
7. Donat M, Alonso S, Pereira F, et al. Impact of histological factors of hepatocellular carcinoma on the outcome of liver transplantation. Transplant Proc. 2016;48(6):1968–77. https://doi.org/10.1016/j.transproceed.2016.04.002.
8. Gonwa TA, Mai ML, Melton LB, et al. End-stage renal disease (ESRD) after orthotopic liver transplantation (OLTX) using calcineurin-based immunotherapy: risk of development and treatment. Transplantation. 2001;72(12):1934–9. https://doi.org/10.1097/00007890-200112270-00012.
9. Ojo AO, Held PJ, Port FK, et al. Chronic renal failure after transplantation of a nonrenal organ. N Engl J Med. 2003;349(10):931–40. https://doi.org/10.1056/NEJMoa021744.
10. Carenco C, Assenat E, Faure S, et al. Tacrolimus and the risk of solid cancers after liver transplant: a dose effect relationship. Am J Transplant. 2015;15(3):678–86. https://doi.org/10.1111/ajt.13018.
11. Villanueva A, Chiang DY, Newell P, et al. Pivotal role of mTOR signaling in hepatocellular carcinoma. Gastroenterology. 2008;135(6):1972–1983, 1983.e1–11. https://doi.org/10.1053/j.gastro.2008.08.008.
12. De Simone P, Nevens F, De Carlis L, H2304 Study Group, et al. Everolimus with reduced tacrolimus improves renal function in de novo liver transplant recipients: a randomized controlled trial. Am J Transplant. 2012;12(11):3008–20. https://doi.org/10.1111/j.1600-6143.2012.04212.x.
13. Fischer L, Saliba F, Kaiser GM, H2304 Study Group, et al. Three-year outcomes in de novo liver transplant patients receiving everolimus with reduced tacrolimus: follow-up results from a randomized, multicenter study. Transplantation. 2015;99(7):1455–62. https://doi.org/10.1097/TP.0000000000000555.
14. Teperman L, Moonka D, Sebastian A, Spare-the-Nephron Trial Liver Transplantation Study Group, et al. Calcineurin inhibitor-free mycophenolate mofetil/sirolimus maintenance in liver transplantation: the randomized spare-the-nephron trial. Liver Transpl. 2013;19(7):675–89. https://doi.org/10.1002/lt.23658.
15. Geissler EK, Schnitzbauer AA, Zülke C, et al. Sirolimus use in liver transplant recipients with hepatocellular carcinoma: a randomized, multicenter, open-label phase 3 trial. Transplantation. 2016;100(1):116–25. https://doi.org/10.1097/TP.0000000000000965.
16. Guba M, von Breitenbuch P, Steinbauer M, et al. Rapamycin inhibits primary and metastatic tumor growth by antiangiogenesis: involvement of vascular endothelial growth factor. Nat Med. 2002;8(2):128–35. https://doi.org/10.1038/nm0202-128.
17. Rupertus K, Dahlem C, Menger MD, Schilling MK, Kollmar O. Rapamycin inhibits hepatectomy-induced stimulation of metastatic tumor growth by reduction of angiogenesis, microvascular blood perfusion, and tumor cell proliferation. Ann Surg Oncol. 2009;16(9):2629–37. https://doi.org/10.1245/s10434-009-0564-8.
18. Hudes G, Carducci M, Tomczak P, Global ARCC Trial, et al. Temsirolimus, interferon alfa, or both for advanced renal-cell carcinoma. N Engl J Med. 2007;356(22):2271–81. https://doi.org/10.1056/NEJMoa066838.
19. Motzer RJ, Escudier B, Oudard S, RECORD-1 Study Group, et al. Phase 3 trial of everolimus for metastatic renal cell carcinoma: final results and analysis of prognostic factors. Cancer. 2010;116(18):4256–65. https://doi.org/10.1002/cncr.25219.
20. Baselga J, Campone M, Piccart M, et al. Everolimus in postmenopausal hormone-receptor-positive advanced breast cancer. N Engl J Med. 2012;366(6):520–9. https://doi.org/10.1056/NEJMoa1109653.
21. Yao JC, Shah MH, Ito T, RAD001 in Advanced Neuroendocrine Tumors, Third Trial (RADIANT-3) Study Group, et al. Everolimus for advanced pancreatic neuroendocrine tumors. N Engl J Med. 2011;364(6):514–23. https://doi.org/10.1056/NEJMoa1009290.
22. Kauffman HM, Cherikh WS, Cheng Y, Hanto DW, Kahan BD. Maintenance immunosuppression with target-of-rapamycin inhibitors is associated with a reduced incidence of de novo malignancies. Transplantation. 2005;80(7):883–9. https://doi.org/10.1097/01.tp.0000184006.43152.8d.
23. Menon KV, Hakeem AR, Heaton ND. Meta-analysis: recurrence and survival following the use of sirolimus in liver transplantation for hepatocellular carcinoma. Aliment Pharmacol Ther. 2013;37(4):411–9. https://doi.org/10.1111/apt.12185.
24. Grigg SE, Sarri GL, Gow PJ, Yeomans ND. Systematic review with meta-analysis: sirolimus- or everolimus-based immunosuppression following liver transplantation for hepatocellular carcinoma. Aliment Pharmacol Ther. 2019;49(10):1260–73. https://doi.org/10.1111/apt.15253.
25. Watt KD, Pedersen RA, Kremers WK, Heimbach JK, Charlton MR. Evolution of causes and risk factors for mortality post-liver transplant: results of the NIDDK long-term follow-up study. Am J Transplant. 2010;10(6):1420–7. https://doi.org/10.1111/j.1600-6143.2010.03126.x.
26. Hua H, Kong Q, Zhang H, Wang J, Luo T, Jiang Y. Targeting mTOR for cancer therapy. J Hematol Oncol. 2019;12(1):71. https://doi.org/10.1186/s13045-019-0754-1.
27. Abdelmalek MF, Humar A, Stickel F, Sirolimus Liver Conversion Trial Study Group, et al. Sirolimus conversion regimen versus continued calcineurin inhibitors in liver allograft recipients: a randomized trial. Am J Transplant. 2012;12(3):694–705. https://doi.org/10.1111/j.1600-6143.2011.03919.x.
28. Asrani SK, Wiesner RH, Trotter JF, et al. De novo sirolimus and reduced-dose tacrolimus versus standard-dose tacrolimus after liver transplantation: the 2000-2003 phase II prospective randomized trial. Am J Transplant. 2014;14(2):356–66. https://doi.org/10.1111/ajt.12543.

Should Systemic Therapy Be Offered to All Patients Post Transplant for Those Transplanted for or Found to Have Incidental Hepatocellular Carcinoma?

31

Shikha Jain

Currently, there is no conclusive evidence that supports the utility of adjuvant therapy post transplant for patients with a diagnosis of hepatocellular carcinoma (HCC) undergoing treatment with curative intent.

Main Points

1. The goal of adjuvant therapy after transplant is to improve overall patient outcomes and survival and reduce the risk of recurrence.
2. In the adjuvant setting, Licartin may be an option to prevent recurrence in the posttransplant setting and prolong survival; however, more trials are needed to further evaluate its efficacy, effect, and safety profile.
3. In small studies, sorafenib use in the adjuvant setting was safe and seemed to decrease the risk of HCC recurrence in the high-risk transplant population. However, larger studies are needed.
4. There has been no definitive evidence to support the use of systemic therapy post transplant to prevent recurrence of HCC.

Treatment with curative intent of HCC is best achieved with hepatic resection, local ablation, and/or liver transplantation (LT). Of these 3, LT is the best therapy if feasible because hepatectomy removes any potential unrecognized sites of spread, as well as any opportunity for de novo HCC to arise. Transplant should be considered in patients who meet United Network for Organ Sharing (UNOS) criteria, which includes the following: a single tumor less than or equal to 5 cm in diameter or 2 or 3 tumors each less than or equal to 3 cm in diameter, and no evidence of macrovascular involvement or extrahepatic disease. At select institutions, patients with tumor characteristics slightly outside UNOS guidelines may still be considered.

A randomized controlled trial (RCT) of 200 individuals showed that partial hepatectomy was associated with a better overall and recurrence-free survival (RFS) when compared to combination transarterial chemoembolization (TACE) and radiofrequency ablation (RFA). Initial treatment with either partial hepatectomy or transplant is preferred and should be considered for patients with Child-Pugh score (CPS) class A, a lack of portal hypertension, and who fit UNOS criteria [1]. The patient must also have an acceptable performance status. The 10-year RFS rate after transplant is 50–70% compared to 22–25% after resection [2]. However, patient overall survival (OS) after resection in a well-selected population is comparable to OS achieved with transplant as many patients survive despite recurrence. Many patients post transplant also die of non–tumor-related causes. The ability to detect recurrence post locoregional therapy (LRT) has also increased with advanced imaging and other techniques, and recurrence of HCC after resection is nearly 50% in the first few years after resection. This is likely due to the presence of occult HCC in other parts of the liver [3].

LT has been proven to be an extremely effective cure for certain patients with early-stage disease; however, posttransplant management to improve outcomes is lacking. Posttransplant management requires an individualized approach as the heterogeneity of HCC burden and variance in tumor biology that present for transplant require a tailored approach.

Posttransplant recurrence is an important negative predictor of survival in the HCC transplant population. A review in 2015 looked at 61 studies and found a mean rate of HCC recurrence of 16% with a mean time from transplant to HCC recurrence of 13 months (range, 2–132 months) [4]. It is important to note that almost 51% of transplant recipients in this review were classified as beyond the Milan criteria (MC), and as more institutions transplant patients with HCC

S. Jain (✉)
Division of Hematology and Oncology, University of Illinois Cancer Center, Chicago, IL, USA
e-mail: sjain25@uic.edu

N. S. Reau, A. G. Singal (eds.), *Controversies in Hepatocellular Carcinoma*, https://doi.org/10.1007/978-3-031-90225-3_31

that are beyond the MC, it is likely the number of HCC recurrences may be larger than those estimated.

The goal of adjuvant therapy after transplant is to improve overall patient outcomes and OS and reduce the risk of recurrence. Adjuvant therapy could potentially result in improved outcomes by treating undetectable micrometastatic disease present at the time of transplant. Thus the potential benefits and risks must be assessed in concert with the possible toxicities in the posttransplant setting.

Systemic Therapy Is Effective and Should Be Offered to All Patients Post Transplant for Hepatocellular Carcinoma

The recurrence of HCC after transplant is between 8% and 20%. Owing to the chance of recurrence post transplant, the utility of systemic therapy in the adjuvant setting post transplant is often a point of discussion. To detect for recurrence, protocols include serial imaging, monitoring of tumor markers such as alpha fetoprotein (AFP), and regular clinic visits for several years post transplant. The majority of recurrences occur within the first 2 years, but they can occur later [5]. The majority are extrahepatic metastatic disease. Treatment at the time of recurrence can range from local ablation to radiation therapy, surgical resection, chemotherapy, targeted agents, or clinical trials. In cases of recurrence, the best outcomes are in tumors amenable to ablation or resection, although selection bias may affect published data because patients who are candidates for these procedures typically recur with only a single lesion. In a study from the University of Pennsylvania of 96 patients with recurrent HCC, 26 had resection and 5 had ablation. Patients with local therapy had a longer median survival (33 months for resection, 21 months for ablation) compared to 7.7 months for those who received systemic therapy [6].

In the posttransplant setting, there have been studies examining specific immunosuppression and possible negative or beneficial effects on the prevention of HCC recurrence. The class of systemic therapy known as mammalian target of rapamycin inhibitors (mTORis) have been shown to have some antitumor activity post transplantation. There have been some encouraging studies such as the results published by Geissler et al. [7] of a multicenter, multinational prospective RCT that included 525 patients who received orthotopic LT for HCC. Patients were then treated with a calcineurin inhibitor (CNI) alone or in conjunction with an mTORi. Those who received the mTORi with the CNI showed higher OS and RFS at 5 years; however, the results were not statistically significant. A further subgroup analysis showed that low-risk patients and patients younger than 60 years of age who received the mTORi had a greater survival benefit and decrease in tumor-related deaths [7]. The results from a University of Pennsylvania retrospective review also showed that the mTORi and immunosuppressant agent sirolimus was associated with lower recurrence risk [6].

The regimen a patient takes for immunosuppression is a potential factor in preventing recurrence. The immune system is a considerable defense against cancer, both in controlling viruses linked to cancer, and also by attacking dysplastic cells themselves [8].

An immune-radioconjugate containing metuximab called Licartin is an antineoplastic agent that has shown some promise in the adjuvant setting for patients with advanced HCC in the posttransplant setting. Metuximab is an antibody fragment that targets the HCC-associated antigen HAb18G/CD147 overexpressed in fibroblasts and HCC. The expression of this antigen has been found to be associated with cancer cell progression, increased invasion, adhesion, and metastasis, and metuximab is conjugated to the radioisotope iodine I [9]. In one RCT, a group of 66 HCC patients beyond the MC received Licartin 3 doses every 28 days 3 weeks after LT, and had a statistically significantly lower rate of recurrence than the control group (27% vs 57%, $P = 0.017$); however, larger studies are still needed [10]. A more recent study of 60 patients was reported from the Tianjin First Central Hospital. In the transplant-only group 30 patients died compared to 5 patients in the group that received Licartin. The median follow-up period was 30.6 months with a range of 4–48 months. The primary cause of death was tumor progression. The HCC in situ recurrence rate was not statistically significant (15.6% vs 13.3%) and the incidence of hepatocellular metastasis for the 2 groups was 22.2% and 20%, also not statistically significant. However, the median survival time of the transplant-only group was 34 months and cumulative survival rates at 6, 12, 18, 24, 36, and 48 months post transplant were 93.3%, 86.6%, 73.3%, 62.2%, 48.9%, and 33.3%. The cumulative survival rates of the group that received adjuvant Licartin at the same time intervals post transplant were 100%, 93.3%, 86.7%, 80.0%, 66.7%, and 66.7%, respectively. This difference was found to be statistically significant ($P = 0.038$) [11]. In the adjuvant setting, Licartin may be an option to prevent recurrence in the posttransplant setting and prolong survival; however, more trials are needed to further evaluate its efficacy, effect, and safety profile.

Sorafenib, an oral multikinase inhibitor of the vascular endothelial growth factor receptor (VEGFR), the platelet-derived growth factor receptor (PDGFR) and Raf, was the first systemic therapy approved in the frontline setting for metastatic HCC in those with CPS class A. Studies have been conducted to assess the utility of sorafenib as adjuvant therapy post transplant. A retrospective, case-control, matched analysis was performed to evaluate the feasibility and efficacy of sorafenib as adjuvant therapy in LT patients.

This retrospective review found that during a mean follow-up of 17.75 months, 1 of 8 patients treated with sorafenib developed HCC recurrence. During a mean follow-up of 32 months, 4 of 8 matched controls developed recurrence. The disease-free 1-year survival in the 2 groups was 85.7% and 57.1%, respectively, and 1-year survival was 87.5% and 62.5%, respectively. This retrospective review suggested some benefit to the addition of sorafenib in the adjuvant setting [12].

A prospective pilot study examined the use of sorafenib as adjuvant therapy in high-risk transplant recipients. This pilot study treated 7 patients with sorafenib compared to 12 historical controls. Two of the 7 patients who received systemic treatment had a recurrence. Of the comparison group, 9 suffered recurrence and all succumbed to the disease. The overall rate of recurrence was decreased in the group receiving adjuvant therapy when compared to the historical controls (29% vs 75%, $P = 0.07$). Disease-free 1-year survival was 100% vs 66%, respectively, and this pilot study concluded sorafenib use in the adjuvant setting was safe and seemed to decrease the risk of HCC recurrence in the high-risk transplant population [13]. However, larger studies are needed.

In the neoadjuvant setting, bridge therapy is considered in certain cases to decrease tumor progression and decrease the dropout rate from the LT waiting list. One study reported an analysis of 205 patients from a transplant center registry who had been diagnosed with HCC and showed that bridging LRT was associated with survival post transplant ($P = 0.005$). Studies have examined RFA, transarterial embolization, TACE, TACE with drug-eluting beads, transarterial radioembolization (TARE) with yttrium-90 microspheres, conformal radiation therapy, and sorafenib as potential "bridge" therapies [14]. However, a meta-analysis found that bridge therapy did not statistically significantly affect mortality, survival, and recurrence rates post transplant when compared to transplant alone [15].

A 2019 study prospectively assessed the feasibility of systemic sorafenib in the neoadjuvant setting while awaiting LT. This study also examined perfusion computed tomography (CT) parameters to assess tumor response and properties early after initiating sorafenib therapy. They found that there was no change in CT perfusion parameters as a result of sorafenib, and they also determined that sorafenib as a neoadjuvant therapy was associated with intolerability and dose reductions. Thus, tolerability of the drug before LT was inadequate and the study failed to show a correlation between changes in CT perfusion parameters during systemic therapy and mRECIST responses [16].

Many of these studies are small and retrospective, and the populations examined in these studies are heterogeneous. With the lack of RCTs examining the utility of bridge therapy, limited conclusions can be drawn to its level of effectiveness. However, many institutions do use this methodology, particularly in cases when there will likely be a long wait time for transplant [14].

Systemic Therapy Is Not Effective and Should Not Be Offered to All Patients in the Adjuvant Posttransplant Setting

There have been some promising studies looking at the utility of immunotherapy and checkpoint inhibitors post transplant. Unfortunately, severe rejection and high percentage of graft loss has been reported, thus not making these viable options at this time [17]. The mechanism of rejection has not been confirmed; however, it is thought to be due to the activation of CD8+ effector cells activation of cellular immunity and the downregulation of T-helper cells. Concerns that immunosuppression may interfere with checkpoint inhibitor efficacy have also been expressed as they rely on an intact T-cell response. It remains unclear if these agents will have a role in the adjuvant setting post transplant for patients with HCC [18].

Adjuvant therapy with sorafenib was previously considered to be promising; however, new data show that there is minimal, if any, added benefit. Many of the studies completed are limited as they are single-center studies, case series, or studies with a small number of participants. One of the largest such studies examined adjuvant sorafenib in 25 patients vs 20 patients who received standard of care in patients with advanced HCC on explant pathology. This study concluded that the addition of preemptive sorafenib did not improve OS or RFS [19]. Data examining neoadjuvant sorafenib in patients with potentially curative resection or ablation also did not have a benefit when examining chance of recurrence [20]. The postoperative use of sorafenib has proven to be inefficient as described in the STORM trial; however, in the posttransplant setting, results may vary [20]. Several studies support the fact that sorafenib use either with or without an mTORi can improve survival when used for recurrence post transplant. However, some drug-induced toxicity often led to dose reductions and even discontinuation of therapy [21–23]. The use of adjuvant sorafenib pre or post transplant will need to be further studied to determine if there is efficacy and utility to this regimen in this setting.

Several studies have evaluated the utility of doxorubin in the posttransplant setting, alone or in combination with other agents. Two randomized trials of single-agent doxorubicin during transplant did not show a statistically significant benefit [24, 25]. Epirubicin has also not shown a survival benefit in randomized trials of adjuvant therapy. This is consistent with what has been seen in patients who receive doxorubicin in the setting of advanced HCC.

There has been no definitive evidence to support the use of systemic therapy post transplant to prevent recurrence of HCC. In an RCT of 58 patients, adjuvant oxaliplatin-based chemotherapy for patients with HCC beyond the MC did show an improvement in 1- and 3-year OS; however, there was no difference in the 3-year disease-free survival in both groups [9].

Ongoing clinical trials are attempting to evaluate the effects of nivolumab immunotherapy in patients with high-risk recurrence after curative local therapy such as resection or ablation, but these do not include those who have already received an LT [26].

Recurrence After Transplant

Studies have examined optimal ways in which to assess risk of tumor recurrence post transplant. A group in France suggested that the MC may be overly restrictive and developed a predictive model of recurrence that combined tumor size and number with AFP value [27]. Another analysis showed that microvascular invasion, AFP at time of transplant, and sum of the largest diameter of viable tumor plus the number of viable tumors on explant were associated with recurrence [28].

If the patient's cancer recurs after LT, options for treatment include clinical trials, best supportive care, and systemic therapy. If there is minimal extrahepatic disease present at the time of recurrence, LRTs can be considered with the understanding that it is possible other new areas of extrahepatic lesions will appear at some point in. the future.

Sorafenib was the first systemic therapy approved in the frontline setting for metastatic HCC in patients with CPS class A. The SHARP study published in 2008 that led to its approval found an OS of 10.7 months in the sorafenib group compared to 7.9 months in the placebo group [29]. This study was conducted looking primarily at those with CPS class A, and caution is advised in patients with CPS class B and C, and those with an elevated bilirubin as there is minimal information in these groups. In 2018, lenvatinib was approved in the first-line setting for patients with unresectable HCC. Its approval was based on the international, multicenter, randomized, noninferiority trial REFLECT. This trial included 954 patients with previously untreated metastatic or unresectable HCC. Lenvatinib was found to be noninferior to sorafenib for OS in this population, and the drug was approved [30]. Limited data support the use of chemotherapy such as FOLFOX in patients with recurrence [31]. Subsequent lines of therapy include regorafenib, cabozantinib, ramucirumab if the patients' AFP is greater than or equal to 400 ng/mL, nivolumab, or pembrolizumab. For patients with HCC that is *NTRK* gene fusion positive, larotrectinib and entrectinib are options as well.

In 2017, the Food and Drug Administration granted accelerated approval to nivolumab for the treatment of HCC in patients who had previously received sorafenib. The approval for this immunotherapy was based on a subgroup of the Checkmate-040 trial, a multicenter, open-label trial that examined patients with CPS class A cirrhosis and HCC who progressed on sorafenib.

There are currently no data that specify the optimal treatment for those who progress after lenvatinib therapy, nor for the use of lenvatinib after sorafenib therapy [14].

Conclusion

The efficacy of adjuvant therapy in the posttransplant setting was initially evaluated in the 1990s. Several nonrandomized studies suggested there may be some benefit from the addition of adjuvant systemic therapy; however, most RCTs have not shown this benefit. It is also essential to remember the population of HCC patients who receive a transplant are heterogeneous, and an individualized approach must be taken when assessing the utility of posttransplant therapy. Most posttransplant adjuvant therapy studies have had a small sample size, and the majority have not been RCTs.

Currently, there is no conclusive evidence that supports the utility of adjuvant therapy post transplant for patients with a diagnosis of HCC undergoing treatment with curative intent. More RCTs will be needed to further assess the utility of this treatment paradigm. Several single-arm retrospective studies suggest there *may* be a role for adjuvant therapy in the posttransplant setting; however, the evidence is not strong enough to change treatment regimens. Controlled studies have shown mixed results. Thus adjuvant therapy in the posttransplant setting should not be recommended at this time outside the setting of a clinical trial.

References

1. Liu H, Wang ZG, Fu SY, et al. Randomized clinical trial of chemoembolization plus radiofrequency ablation versus partial hepatectomy for hepatocellular carcinoma within the Milan criteria. Br J Surg. 2016;103(4):348–56. https://doi.org/10.1002/bjs.10061.
2. Vibert E, Schwartz M, Olthoff KM. Advances in resection and transplantation for hepatocellular carcinoma. J Hepatol. 2020;72(2):262–76. https://doi.org/10.1016/j.jhep.2019.11.017.
3. Aufhauser DD Jr, Sadot E, Murken DR, et al. Incidence of occult intrahepatic metastasis in hepatocellular carcinoma treated with transplantation corresponds to early recurrence rates after partial hepatectomy. Ann Surg. 2018;267(5):922–8. https://doi.org/10.1097/SLA.0000000000002135.
4. de' Angelis N, Landi F, Carra MC, Azoulay D. Managements of recurrent hepatocellular carcinoma after liver transplantation: a systematic review. World J Gastroenterol. 2015;21(39):11185–98. https://doi.org/10.3748/wjg.v21.i39.11185.

5. Fernandez-Sevilla E, Allard MA, Selten J, et al. Recurrence of hepatocellular carcinoma after liver transplantation: is there a place for resection? Liver Transpl. 2017;23(4):440–7. https://doi.org/10.1002/lt.24742.
6. Ekpanyapong S, Philips N, Loza BL, et al. Predictors, presentation and treatment outcomes of recurrent hepatocellular carcinoma after liver transplantation: a large single center experience. J Clin Exp Hepatol. 2020;10(4):304–15. https://doi.org/10.1016/j.jceh.2019.11.003.
7. Geissler EK, Schnitzbauer AA, Zülke C, et al. Sirolimus use in liver transplant recipients with hepatocellular carcinoma: a randomized, multicenter, open-label phase 3 trial. Transplantation. 2016;100(1):116–25. https://doi.org/10.1097/TP.0000000000000965.
8. Hanahan D, Weinberg RA. Hallmarks of cancer: the next generation. Cell. 2011;144(5):646–74. https://doi.org/10.1016/j.cell.2011.02.013.
9. Zhang Q, Chen H, Li Q, et al. Combination adjuvant chemotherapy with oxaliplatin, 5-fluorouracil and leucovorin after liver transplantation for hepatocellular carcinoma: a preliminary open-label study. Investig New Drugs. 2011;29(6):1360–9. https://doi.org/10.1007/s10637-011-9726-1.
10. Chen ZN, Mi L, Xu J, et al. Targeting radioimmunotherapy of hepatocellular carcinoma with iodine (131I) metuximab injection: clinical phase I/II trials. Int J Radiat Oncol Biol Phys. 2006;65(2):435–44. https://doi.org/10.1016/j.ijrobp.2005.12.034.
11. Li J, Qin L, Ding EC, Shen J, Li J. Efficacy and safety of licartin with repeated administration in treatment of HCC patients after liver transplantation [article in Chinese]. Zhonghua Yi Xue Za Zhi. 2018;98(33):2645–9.
12. Saab S, McTigue M, Finn RS, Busuttil RW. Sorafenib as adjuvant therapy for high-risk hepatocellular carcinoma in liver transplant recipients: feasibility and efficacy. Exp Clin Transplant. 2010;8(4):307–13.
13. Shetty KI, Dash C, Laurin J. Use of adjuvant sorafenib in liver transplant recipients with high-risk hepatocellular carcinoma. J Transp Secur. 2014;2014:913634. https://doi.org/10.1155/2014/913634.
14. National Comprehensive Cancer Network Hepatobiliary Cancers (version 4.2019). https://www.nccn.org/professionals/physician_gls/pdf/hepatobiliary.pdf
15. Kulik L, Heimbach JK, Zaiem F, et al. Therapies for patients with hepatocellular carcinoma awaiting liver transplantation: a systematic review and meta-analysis. Hepatology. 2018;67(1):381–400. https://doi.org/10.1002/hep.29485.
16. Eilard MS, Andersson M, Naredi P, et al. A prospective clinical trial on sorafenib treatment of hepatocellular carcinoma before liver transplantation. BMC Cancer. 2019;19(1):568. https://doi.org/10.1186/s12885-019-5760-8.
17. Hu B, Yang XB, Sang XT. Liver graft rejection following immune checkpoint inhibitors treatment: a review. Med Oncol. 2019;36(11):94. https://doi.org/10.1007/s12032-019-1316-7.
18. Llovet JM, Montal R, Sia D, Finn RS. Molecular therapies and precision medicine for hepatocellular carcinoma. Nat Rev Clin Oncol. 2018;15(10):599–616. https://doi.org/10.1038/s41571-018-0073-4.
19. Satapathy SK, Das K, Kocak M, et al. No apparent benefit of preemptive sorafenib therapy in liver transplant recipients with advanced hepatocellular carcinoma on explant. Clin Transpl. 2018;32(5):e13246. https://doi.org/10.1111/ctr.13246.
20. Bruix J, Takayama T, Mazzaferro V, et al. STORM investigators. Adjuvant sorafenib for hepatocellular carcinoma after resection or ablation (STORM): a phase 3, randomised, double-blind, placebo-controlled trial. Lancet Oncol. 2015;16(13):1344–54. https://doi.org/10.1016/S1470-2045(15)00198-9.
21. Sposito C, Mariani L, Germini A, et al. Comparative efficacy of sorafenib versus best supportive care in recurrent hepatocellular carcinoma after liver transplantation: a case-control study. J Hepatol. 2013;59(1):59–66. https://doi.org/10.1016/j.jhep.2013.02.026.
22. Zavaglia C, Airoldi A, Mancuso A, et al. Adverse events affect sorafenib efficacy in patients with recurrent hepatocellular carcinoma after liver transplantation: experience at a single center and review of the literature. Gastroenterol Hepatol. 2013;25(2):180–6. https://doi.org/10.1097/MEG.0b013e328359e550.
23. Gomez-Martin C, Bustamante J, Castroagudin JF, et al. Efficacy and safety of sorafenib in combination with mammalian target of rapamycin inhibitors for recurrent hepatocellular carcinoma after liver transplantation. Liver Transpl. 2012;18(1):45–52. https://doi.org/10.1002/lt.22434.
24. Söderdahl G, Bäckman L, Isoniemi H, et al. A prospective, randomized, multi-Centre trial of systemic adjuvant chemotherapy versus no additional treatment in liver transplantation for hepatocellular carcinoma. Transpl Int. 2006;19(4):288–94. https://doi.org/10.1111/j.1432-2277.2006.00279.x.
25. Pokorny H, Gnant M, Rasoul-Rockenschaub S, et al. Does additional doxorubicin chemotherapy improve outcome in patients with hepatocellular carcinoma treated by liver transplantation? Am J Transpl. 2005;5(4 Pt 1):788–94. https://doi.org/10.1111/j.1600-6143.2005.00780.x.
26. Trial C. A study of nivolumab in patients with hepatocellular carcinoma who are at high risk of recurrence after curative hepatic resection or ablation (CHECKMATE 9DX).
27. Duvoux C, Roudot-Thoraval F, Decaens T, Liver Transplantation French Study Group, et al; Liver transplantation for hepatocellular carcinoma: a model including alpha-fetoprotein improves the performance of Milan citeria. Gastroenterology 2012;143(4):986–994.e3; quiz 994.e14-e15. https://doi.org/10.1053/j.gastro.2012.05.052.
28. Mehta N, Heimbach J, Harnois DM, et al. Validation of a risk estimation of tumor recurrence after transplant (RETREAT) score for hepatocellular carcinoma recurrence after liver transplant. JAMA Oncol. 2017;3(4):493–500. https://doi.org/10.1001/jamaoncol.2016.5116.
29. Llovet JM, Ricci S, Mazzaferro V, SHARP Investigators Study Group, et al. Sorafenib in advanced hepatocellular carcinoma. N Engl J Med. 2008;359:378–90. https://doi.org/10.1056/NEJMoa0708857.
30. Yamashita T, Kudo M, Ikeda K, et al. REFLECT a phase 3 trial comparing efficacy and safety of lenvatinib to sorafenib for the treatment of unresectable hepatocellular carcinoma: an analysis of Japanese subset. J Gastroenterol. 2020;55(1):113–22. https://doi.org/10.1007/s00535-019-01642-1.
31. Qin S, Bai Y, Lim HY, et al. Randomized, multicenter, open-label study of oxaliplatin plus fluorouracil/leucovorin versus doxorubicin as palliative chemotherapy in patients with advanced hepatocellular carcinoma from Asia. J Clin Oncol. 2013;31(28):3501–8. https://doi.org/10.1200/JCO.2012.44.5643.

Should Post-liver Transplantation Surveillance for Hepatocellular Carcinoma (HCC) Stop If There Is No HCC Recurrence After 5 Years?

32

Ranya Selim, Jinyu Zhang, and Reena Salgia

Main Points

1. Posttransplant HCC recurrence occurs in up to 16% of cases and is a major cause of mortality in the posttransplant setting, with a mean time to recurrence of 13 months and median survival of less than 1 year.
2. Late recurrence (2–5 years post LT) is believed to occur because of indolent tumor growth originating from latent pretransplant HCC cells.
3. There are several scoring systems that help guide surveillance for the first 5 years after transplant.
4. Risk of recurrence might be higher in those patients downstaged before transplant.
5. After 5 years, HCC recurrence is uncommon but very late recurrence did occur in 9.5% of a large post-LT cohort. These patients had longer median postrecurrence survival in contrast to that of the early-recurrence population.

HCC has increased in incidence over the past decade to become the fourth leading cause of malignancy-related mortality worldwide [1]. In Mazzaferro's seminal paper, [2] the benefit of LT for patients with unresectable, early-stage HCC was established. Specifically included were patients defined as having one lesion smaller than or equal to 5 cm or 3 lesions all smaller than or equal to 3 cm without evidence of vascular invasion or extrahepatic spread. The study reported a 75% 4-year survival, compared to prior studies demonstrating less than 35% 5-year survival. Following the promising results of this study, patients meeting the Milan criteria (MC) were offered transplantation, and ultimately given priority for listing through the Model for End-Stage Liver Disease scoring system. Today, HCC accounts for more than 27.2% of all LTs in the United States [3]. Posttransplant survival in early-stage patients is comparable to individuals transplanted without underlying HCC and provides a curative treatment option that is superior to alternatives at this stage.

R. Selim · J. Zhang · R. Salgia (✉)
Division of Gastroenterology and Hepatology, Henry Ford Hospital, Detroit, MI, USA
e-mail: Ranya.selim@uth.tmc.edu; rsalgia1@hfhs.org

Risk Factors for Post-liver Transplant Hepatocellular Carcinoma Recurrence

Despite the curative intent of LT for early-stage HCC, there remains a risk for short- and long-term HCC recurrence. Posttransplant HCC recurrence has been reported to occur in up to 16% of cases and is a major cause of mortality in the posttransplant setting, with a mean time to recurrence of 13 months (range: 2–132 months), and median survival of less than 1 year from diagnosis of recurrence [4]. The established risk factors for tumor recurrence can be categorized into three broad categories: tumor characteristics, serologic biomarkers, and response to local treatment.

Tumor burden has long been felt to be influential in assessing one's risk for post-LT HCC recurrence. Traditionally patients have been at a higher risk for HCC recurrence when the tumor burden has exceeded the MC, based on the number or size of the tumor nodules, or based on the presence of lymphovascular invasion or extrahepatic spread. Studies looking at slightly more liberal tumor volumes, such as the University of California San Francisco criteria or "Up-to-7" criteria have been noted to have survival that is just slightly inferior to that shown with the MC [5, 6]. Vascular invasion has a substantial effect on rate of recurrence. While macrovascular invasion is readily detected on preoperative imaging, microvascular invasion is by definition detected only postoperatively on assessment of the explant. In the event of microvascular invasion, risk of tumor recurrence is statistically significantly increased, with a reported relative risk of 2.42 [7]. Tumor histology, defined by glandular distortion, can affect risk of HCC recurrence. Poorly differentiated tumor grade has been associated with a three-fold risk of tumor recurrence (39.3% vs 13%) [8].

N. S. Reau, A. G. Singal (eds.), *Controversies in Hepatocellular Carcinoma*, https://doi.org/10.1007/978-3-031-90225-3_32

Unfortunately, percutaneous biopsy has a poor sensitivity in recognizing poorly differentiated tumors, often lacking accurate correlation with explant histology.

The most commonly used biomarker for HCC screening, diagnosis, and response to treatment is AFP. Many thresholds for AFP have been studied to determine values correlating with a high risk for HCC recurrence. More recently, the United Network for Organ Sharing (UNOS) has adopted a policy by which individuals with a pretransplant AFP greater than 1000 ng/mL must see a decrease to less than 500 ng/mL with locoregional therapy (LRT) to proceed with LT [9]. Regardless of which threshold is set, AFP has been shown to be a valuable marker for assessing risk of HCC recurrence post LT and is included in many prognostic models. Other biomarkers have been studied; however, their role in assessing for posttransplant HCC recurrence has not been fully defined [10]. These include des-gamma-carboxyprothrombin, AFP-L3%, and absolute AFP-L3. A peripheral neutrophil-lymphocyte ratio greater than or equal to 5 has been found to be associated with a decreased HCC recurrence-free survival [11]. This ratio has been linked with a poor prognosis with other malignancies as well.

Response to treatment has been associated with post-LT HCC recurrence. Specifically, a favorable response to LRT has been associated with favorable tumor biology. Concurrent increase in tumor burden despite LRT is a negative prognostic indicator of posttransplant outcomes [12].

While a number of donor features and graft types have been studied, there is not a clear correlation as yet suggesting that HCC risk is tied to donor characteristics.

Last, there is a balance post transplant between immunosuppression and prevention of oncogenesis that must be achieved. The optimal strategy has yet to be determined in a rigorous form. The available data suggest an overall goal of minimizing exposure to calcineurin inhibitors while relying more on antiproliferative agents such as mammalian target of rapamycin (mTOR) inhibitors. This goal of minimizing HCC recurrence post transplant must be balanced with an adequate immunosuppressive regimen to prevent graft rejection [7, 13].

Patterns of Hepatocellular Carcinoma Recurrence

HCC recurrence post transplant has been defined as early (within 2 years) vs late (2–5 years post LT). After 5 years, HCC recurrence is much less frequent. Early recurrence is felt to occur because of undetectable or underestimated tumor burden pre transplant, with circulating tumor cells that flourish post LT. This portends a more aggressive pattern of recurrence and is associated with a worse prognosis [14]. Late recurrence is believed to occur because of indolent tumor growth originating from latent pretransplant HCC cells. Survival after recurrence is influenced by the location and whether the site is amenable to curative resection or ablation. This is felt to be the case only in up to 30% of recurrent HCCs [15]. While intrahepatic recurrences can occur, studies show a high frequency of extrahepatic recurrence along with multifocal involvement at the time of initial recurrence. Most common locations of HCC metastasis include the lungs and bones, followed by soft tissue, adrenal glands, and peritoneum [16].

Hepatocellular Carcinoma Recurrence Risk Prediction Models

How best to offer HCC surveillance to patients after LT remains an area of active discussion. There are several published models that use various factors to predict HCC recurrence (Table 32.1). These models may be used to risk-stratify patients into low- and high-risk categories to guide surveillance. No clinical trial–based data are available to guide specific post-LT HCC surveillance strategies at this time; rather, expert consensus based on published data has been used to guide surveillance programs [6, 17–22]. Transplant centers vary in their approach to HCC surveillance, with some using a predictive model, while others may rely on their center-specific protocol.

The Risk Estimation of Tumor Recurrence After Transplant (RETREAT) score has been validated as a 5-year risk stratification tool to guide posttransplant surveillance. This score is calculated using AFP range at the time of transplantation, microvascular invasion, and sum of the diameter of the largest viable tumor and number of smaller viable nodules. The sum of the scores for each of these categories then determines 3-year recurrence risk, as well as a proposed surveillance regimen. For a score of 0, no screening is recommended; for a score of 1–3, screening every 6 months for 2 years; for a score of 4, screening every 6 months for 5 years; and for a score of 5 or greater, screening every 3 to 4 months for 2 years, followed by screening every 6 months for the second to fifth year. Surveillance is recommended with multiphasic cross-sectional imaging (including chest imaging, as recurrence can occur in the form of pulmonary metastases), along with concurrent AFP [18].

The "Up-to-7" model uses the sum of primary tumor size in centimeters and number of nodules. A sum greater than 7 predicts a higher risk of recurrence [6]. Chan's model [19] uses primary tumor diameter, bilobar extension, macrovascular invasion, and tumor differentiation. Patients are then stratified into low risk (score of 0), moderate risk (score 1–2), and high risk (score of ≥ 3). Decaens' model[20] uses primary tumor diameter, number of nodules, and tumor differentiation. Scores 4 or greater predict a higher 5-year

Table 32.1 Comparison of posttransplant hepatocellular carcinoma recurrence models

Model	Factors/score		Risk for HCC recurrence
Up-to-7	Primary tumor size, cm		Sum > 7, high risk
	No. of nodules		
Chan	Primary tumor diameter ≥ 4 cm	+1	0, low risk
	Bilobar extension	+2	1–2, moderate
	Macrovascular invasion	+3	risk
	Tumor differentiation (well)	−3	≥ 3, high risk
Decaens	Primary tumor diameter		≥ 4, high risk
	≤ 2	0	
	2–3	+1	
	3–5	+2	
	≥5	+5	
	No. of nodules		
	1	0	
	2–3	+1	
	>3	+2	
	Tumor differentiation		
	Well	0	
	Moderate	+1	
	Poor	+2	
Iwatsuki	Primary tumor diameter, cm		<7.5, low risk
	2–5	+4.5	≤ 11, intermediate risk
	>5	+6.7	<15, high risk
	Bilobar extension	+3.1	≥ 15, very high risk
	Vascular invasion		
	Micro	+4.4	
	Macro	+15	
MoRAL	**Pre-MoRAL:**		Combo-moral (sum of pre and post):
	Neutrophil/lymphocyte ratio ≥ 5	6	0–2, low risk
	AFP > 200 ng/mL	4	3–6, medium risk
	Largest tumor size > 3 cm	3	7–10, high risk
	Post-MoRAL:		>10, very high risk
	Grade 4 tumors	6	
	Vascular invasion	2	
	Tumor size > 3 cm	3	
	No. of lesions > 3	3	
RETREAT	AFP range		0, low risk
	0–20	0	1–3, intermediate risk
	21–99	1	4, high risk
	100–999	2	≥ 5, very high risk
	≥ 1000	3	
	Microvascular invasion	2	
	Sum of diameter of largest viable tumor and No. of smaller viable tumors		
	0	0	
	1–4.9	1	
	5–9.9	2	
	≥ 10	3	

AFP alpha fetoprotein, *HCC* hepatocellular carcinoma

recurrence risk. The Iwatsuki model [21] uses primary tumor diameter, bilobar extension, and vascular invasion. The Model of Recurrence After Liver (MoRAL) Transplant for HCC uses a combination of pre- and posttransplant factors to predict recurrence. The pre-MoRAL score uses a neutrophil/lymphocyte ratio greater than 5, AFP greater than 200 ng/mL, and largest tumor size greater than 3 cm; this score can be used pre transplant to predict recurrence with a *C* statistic of 0.82 [22]. The post-MoRAL score uses grade 4 tumors, vascular invasion, largest tumor size greater than 3 cm, and total number of lesions greater than 3; the *C* statistic associated with this score is 0.87. Combining the 2 scores yields the Combo-MoRAL score with a *C* statistic of 0.91 [22].

While these various models do not propose a posttransplant HCC surveillance algorithm, with the exception of Mehta et al., [18] one might consider stratifying patients from these models into low-, moderate-, and high-risk categories to devise a surveillance strategy. One nuance to address would be the differing variables and risk predictions between each of the models. In this case, we suggest that a transplant center should use the model that predicts the highest risk to guide patient-level surveillance. For patients in the low-risk category, one strategy would be to perform cross-sectional imaging surveillance every 6 months for 2 years. Likewise, for patients with a moderate-risk for recurrence, performing cross-sectional imaging every 6 months, however, extending the length of surveillance to 5 years. Last, for patients who fall into a high-risk category for HCC recurrence, cross-sectional imaging every 3 months for 2 years can be pursued, then every 6 months for years 2 through 5. Finally, given its relative low cost and the lack of data regarding AFP post transplant, serum AFP can be monitored at the time of imaging.

Pros and Cons: Post-liver Transplant Surveillance for Hepatocellular Carcinoma Can Stop If There Is No Recurrence After 5 Years

LT continues to have a vital role in improving survival for patients with early-stage HCC. With approximately one-quarter of LT being performed in the United States for this indication, it is especially important that we minimize post-LT HCC recurrence. Yet there remains a gap in the current literature as to the optimal surveillance strategy. Similar to other cancer surveillance strategies, there are benefits and risks to acknowledge. In addition, when is the right time to stop surveillance? To answer this, we approach the pro and con side of the statement "Post-LT surveillance for HCC can stop if there is no recurrence after 5 years."

Pro

Post-LT surveillance for HCC can stop if there is no recurrence after 5 years. Studies to date have shown that most cases of recurrence occur within the first 2 years, [23] and then between years 2 and 5; one can argue that there is low utility to continue screening past 5 years. Currently, there is no standard protocol regarding how to conduct surveillance. Some centers recommend computed tomography or magnetic resonance imaging at 3 months, 6 months, and then yearly thereafter [19]. Some centers risk-stratify patients after 5 years based on factors mentioned previously such as AFP, tumor size, and vascular invasion [17]. Although radiation risk is low with one computed tomography imaging scan, most LT patients have had a number of imaging studies performed as part of their pretransplant workup and during their hospital admissions for decompensated disease; so there is considerable cumulative radiation exposure to consider [24]. Also, in today's world of rising health care costs, cost is another concern. One study evaluated the cost-effectiveness of continued screening. The authors found that the incremental cost of screening all patients was $340,000/life-year gained but for those whose explant exceeded the MC, the incremental cost was $138,000/life-year gained [25]. Therefore, continuing imaging studies after 5 years increases cost and is not necessarily cost-effective.

Con

Post-LT surveillance for HCC should continue beyond 5 years even if there has been no recurrence at 5 years. Although most studies support that recurrence typically occurs within the first 5 years, there are described cases of HCC recurrence beyond that point [17]. In a shared decision-making world, it may be challenging to tell patients that we will no longer continue HCC surveillance at 5 years when there remains a small risk that the cancer may recur. In a study evaluating postrecurrence outcomes, patients were categorized into early recurrence ($\leq$ 1 year), late recurrence (1–5 years), and very late recurrence ($\geq$ 5 years). Although the number of cases for very late recurrence was small, 22 out of a total of 232 patients (9.5%), the authors found that the very late recurrence population had a longer median postrecurrence survival period of 37 months in contrast to that of the early recurrence population, which was 10.2 months [26]. Techniques for LRT for intrahepatic recurrence or radiation therapy for solitary metastasis continue to evolve and offer patients more treatment options. In addition, our treatment options for HCC have evolved over the past few years to include more options for first-line, second-line, and potentially even third-line systemic therapy for HCC

[27]. If we turn to other cancer etiologies, for example, breast cancer or colorectal cancer, postrecovery surveillance continues for many years to follow, and does not stop after 5 years [28, 29]. One strategy, which we use at our center, is to use ultrasound surveillance paired with AFP after year 5 for patients with a higher risk of HCC recurrence (i.e., RETREAT score of 4 or 5). Tailoring the surveillance strategy based on risk minimizes testing to those felt to be most likely to benefit from ongoing surveillance.

Our Recommendation for Approaching Post-liver Transplant Hepatocellular Carcinoma Surveillance

Currently, there is consensus that known risk factors for post-LT HCC recurrence are based on tumor burden, the presence or absence of vascular invasion, pretransplant AFP, and often with consideration of the degree of tumor differentiation [13, 17].

Although patients do get downsized before transplant, we do not have models that take this into account. A small retrospective study evaluating patients between 1997 and 2010 found that downstaging with transarterial chemoembolization can lead to decrease in AFP, and that a statistically significant drop in AFP was associated with less HCC recurrence and overall improved survival [30]. Another single-center study found that downstaging HCC to within T2 criteria was associated with good 5-year survival without statistically significant survival difference or HCC recurrence difference when compared to the T2 group [31]. A multicenter study found that the recurrence-free period at 5 years was 87% in the downsized group, and that there was no statistically significant difference among the various centers. However, the authors did find that treatment failure (partially defined as recurrence following transplant) was associated with high pretreatment AFP (> 1000 ng/mL) and Child-Pugh score class B or C status [32]. More recently, a large retrospective study with close to 4000 HCC patients specifically evaluated outcomes for patients with tumor burden meeting UNOS downstaging inclusion criteria (UNOS-DS), initial tumor burden beyond UNOS downstaging criteria (AC-DS), and within the MC. The authors found that although survival rates were similar, there was statistically more HCC recurrence at 3 years in the AC-DS group. Additionally, in the downstaged groups, an AFP greater than or equal to 100 ng/mL before transplant was an independent predictor of HCC recurrence [33]. Based on the available literature, it appears that risk of recurrence is likely higher within the first 3 years in patients who were downstaged. The common risk factor for recurrence across studies appears to be AFP, which is part of the RETREAT model. Ultimately, a concern with downstaging is that patients may have microvascular invasion and/or macrovascular invasion seen on explant that was not initially found on radiologic studies. In the large retrospective study by Mehta et al., [33] the authors found that the AC-DS group had a statistically higher rate of vascular invasion on explant although this was not noted to be an independent risk factor for recurrence. More studies are warranted in this arena. Additionally, recurrence in downsized patients did not differ among long, middle, or short wait time regions. However, with recent changes in UNOS allocation guidelines, further studies are warranted.

We recommend that based on the available literature and for consistency of application, LT centers should adopt a risk-prediction model and stratify patients into low- to moderate- vs high-risk categories for recurrence. Patients deemed to be at higher risk should continue surveillance imaging and AFP beyond 5 years; however, it is reasonable to consider limiting this to annual surveillance or ultrasound to minimize harm associated with overtesting.

Conclusion

HCC is a leading cause of malignancy-related mortality worldwide, and the incidence of transplantation for HCC is increasing [27]. It is not entirely clear what the ideal surveillance strategy should be following LT for HCC. Currently, there is reasonable evidence regarding predictors of recurrence risk. Multiple prediction models exist, though they lack long-term prospective validation. Over time, we will be better able to evaluate outcomes of different surveillance strategies and models. For now, based on the available evidence and nuance, we believe surveillance strategies should be tailored to the individual with discontinuation of surveillance at 5 years for low-moderate risk patients, and continued modified surveillance beyond 5 years for those at high risk.

References

1. Global Burden of Disease Liver Cancer Collaboration, Akinyemiju T, Abera S, Ahmed M, et al. The burden of primary liver cancer and underlying etiologies from 1990 to 2015 at the global, regional, and national level: results from the global burden of disease study 2015. JAMA Oncol. 2017;3(12):1683–91. https://doi.org/10.1001/jamaoncol.2017.3055.
2. Mazzaferro V, Regalia E, Doci R, et al. Liver transplantation for the treatment of small hepatocellular carcinomas in patients with cirrhosis. N Engl J Med. 1996;334(11):693–9. https://doi.org/10.1056/NEJM199603143341104.
3. Yang JD, Larson JJ, Watt KD, et al. Hepatocellular carcinoma is the most common indication for liver transplantation and placement on the waitlist in the United States. Clin Gastroenterol Hepatol. 2017;15(5):767–775.e3. https://doi.org/10.1016/j.cgh.2016.11.034.
4. de' Angelis N, Landi F, Carra MC, Azoulay D. Managements of recurrent hepatocellular carcinoma after liver transplantation: s

systematic review. World J Gastroenterol. 2015;21(39):11185–98. https://doi.org/10.3748/wjg.v21.i39.11185.
5. Unek T, Karademir S, Arslan NC, et al. Comparison of Milan and UCSF criteria for liver transplantation to treat hepatocellular carcinoma. World J Gastroenterol. 2011;17(37):4206–12. https://doi.org/10.3748/wjg.v17.i37.4206.
6. Lei JY, Wang WT, Yan LN. Up-to-seven criteria for hepatocellular carcinoma liver transplantation: a single center analysis. World J Gastroenterol. 2013;19(36):6077–83. https://doi.org/10.3748/wjg.v19.i36.6077.
7. Agopian VG, Harlander-Locke M, Zarrinpar A, et al. A novel prognostic nomogram accurately predicts hepatocellular carcinoma recurrence after liver transplantation: analysis of 865 consecutive liver transplant recipients. J Am Coll Surg. 2015;220(4):416–27. https://doi.org/10.1016/j.jamcollsurg.2014.12.025.
8. Donat M, Alonso S, Pereira F, et al. Impact of histological factors of hepatocellular carcinoma on the outcome of liver transplantation. Transplant Proc. 2016;48(6):1968–77. https://doi.org/10.1016/j.transproceed.2016.04.002.
9. Mehta N, Dodge JL, Roberts JP, Hirose R, Yao FY. Alpha-fetoprotein decrease from >1,000 to <500 ng/mL in patients with hepatocellular carcinoma leads to improved posttransplant outcomes. Hepatology. 2019;69(3):1193–205. https://doi.org/10.1002/hep.30413.
10. Citores MJ, Lucena JL, de la Fuente S, Cuervas-Mons V. Serum biomarkers and risk of hepatocellular carcinoma recurrence after liver transplantation. World J Hepatol. 2019;11(1):50–64. https://doi.org/10.4254/wjh.v11.i1.50.
11. Halazun KJ, Hardy MA, Rana AA, et al. Negative impact of neutrophil-lymphocyte ratio on outcome after liver transplantation for hepatocellular carcinoma. Ann Surg. 2009;250(1):141–251. https://doi.org/10.1097/SLA.0b013e3181a77e59.
12. Otto G, Schuchmann M, Hoppe-Lotichius M, et al. How to decide about liver transplantation in patients with hepatocellular carcinoma: size and number of lesions or response to TACE? J Hepatol. 2013;59(2):279–84. https://doi.org/10.1016/j.jhep.2013.04.006.
13. Filgueira NA. Hepatocellular carcinoma recurrence after liver transplantation: risk factors, screening and clinical presentation. World J Hepatol. 2019;11(3):261–72. https://doi.org/10.4254/wjh.v11.i3.261.
14. Sapisochin G, Goldaracena N, Astete S, et al. Benefit of treating hepatocellular carcinoma recurrence after liver transplantation and analysis of prognostic factors for survival in a large Euro-American series. Ann Surg Oncol. 2015;22(7):2286–94. https://doi.org/10.1245/s10434-014-4273-6.
15. Zimmerman MA, Ghobrial RM, Tong MJ, et al. Recurrence of hepatocellular carcinoma following liver transplantation: a review of preoperative and postoperative prognostic indicators. Arch Surg. 2008;143(2):182–8.
16. Kneuertz PJ, Cosgrove DP, Cameron AM, et al. Multidisciplinary management of recurrent hepatocellular carcinoma following liver transplantation. J Gastrointest Surg. 2012;16(4):874–81. https://doi.org/10.1007/s11605-011-1710-8.
17. Verna EC, Patel YA, Aggarwal A, et al. Liver transplantation for hepatocellular carcinoma: management after the transplant. Am J Transplant. 2020;20(2):333–47. https://doi.org/10.1111/ajt.15697.
18. Mehta N, Heimbach J, Harnois DM, et al. Validation of a risk estimation of tumor recurrence after transplant (RETREAT) score for hepatocellular carcinoma recurrence after liver transplant. JAMA Oncol. 2017;3(4):493–500. https://doi.org/10.1001/jamaoncol.2016.5116.
19. Chan EY, Larson AM, Fix OK, et al. Identifying risk for recurrent hepatocellular carcinoma after liver transplantation: implications for surveillance studies and new adjuvant therapies. Liver Transpl. 2008;14(7):956–65. https://doi.org/10.1002/lt.21449.
20. Decaens T, Roudot-Thoraval F, Badran H, et al. Impact of tumour differentiation to select patients before liver transplantation for hepatocellular carcinoma. Liver Int. 2011;31(6):792–801. https://doi.org/10.1111/j.1478-3231.2010.02425.x.
21. Costentin CE, Amaddeo G, Decaens T, Liver Transplantation French Study Group, et al. Prediction of hepatocellular carcinoma recurrence after liver transplantation: comparison of four explant-based prognostic models. Liver Int. 2017;37(5):717–26. https://doi.org/10.1111/liv.13388.
22. Halazun KJ, Najjar M, Abdelmessih RM, et al. Recurrence after liver transplantation for hepatocellular carcinoma: a new MORAL to the story. Ann Surg. 2017;265(3):557–64. https://doi.org/10.1097/SLA.0000000000001966.
23. Parfitt JR, Marotta P, Alghamdi M, et al. Recurrent hepatocellular carcinoma after transplantation: use of a pathological score on explanted livers to predict recurrence. Liver Transpl. 2007;13(4):543–51. https://doi.org/10.1002/lt.21078.
24. Hendee WR, O'Connor MK. Radiation risks of medical imaging: separating fact from fantasy. Radiology. 2012;264(2):312–21. https://doi.org/10.1148/radiol.12112678.
25. Ladabaum U, Cheng SL, Yao FY, Roberts JP. Cost-effectiveness of screening for recurrent hepatocellular carcinoma after liver transplantation. Clin Transpl. 2011;25(2):283–91. https://doi.org/10.1111/j.1399-0012.2010.01212.x.
26. Alshahrani AA, Ha SM, Hwang S, et al. Clinical features and surveillance of very late hepatocellular carcinoma recurrence after liver transplantation. Ann Transplant. 2018;23:659–65. https://doi.org/10.12659/AOT.910598.
27. Couri T, Pillai A. Goals and targets for personalized therapy for HCC. Hepatol Int. 2019;13(2):125–37. https://doi.org/10.1007/s12072-018-9919-1.
28. Smith TJ. Breast cancer surveillance guidelines. J Oncol Pract. 2013;9(1):65–7. https://doi.org/10.1200/JOP.2012.000787.
29. Kahi CJ, Boland CR, Dominitz JA, United States Multi-Society Task Force on Colorectal Cancer, et al. Colonoscopy surveillance after colorectal cancer resection: recommendations of the US Multi-Society Task Force on Colorectal Cancer. Gastroenterology. 2016;150(3):758–768.e11. https://doi.org/10.1053/j.gastro.2016.01.001.
30. Bhat M, Hassanain M, Simoneau E, et al. Magnitude of change in alpha-fetoprotein in response to transarterial chemoembolization predicts survival in patients undergoing liver transplantation for hepatocellular carcinoma. Curr Oncol. 2013;20(5):265–72. https://doi.org/10.3747/co.20.1270.
31. Yao FY, Mehta N, Flemming J, et al. Downstaging of hepatocellular cancer before liver transplant: long-term outcome compared to tumors within Milan criteria. Hepatology. 2015;61(6):1968–77. https://doi.org/10.1002/hep.27752.
32. Mehta N, Guy J, Frenette CT, et al. Excellent outcomes of liver transplantation following down-staging of hepatocellular carcinoma to within Milan criteria: a multicenter study. Clin Gastroenterol Hepatol. 2018;16(6):955–64. https://doi.org/10.1016/j.cgh.2017.11.037.
33. Mehta N, Dodge JL, Grab JD, Yao FY. National experience on down-staging of hepatocellular carcinoma before liver transplant: influence of tumor burden, alpha-fetoprotein, and wait time. Hepatology. 2020;71(3):943–54. https://doi.org/10.1002/hep.30879.

Index

N. S. Reau, A. G. Singal (eds.), *Controversies in Hepatocellular Carcinoma*, https://doi.org/10.1007/978-3-031-90225-3

I

MIX
Papier aus verantwortungsvollen Quellen
Paper from responsible sources
FSC® C105338

If you have any concerns about our products,
you can contact us on
ProductSafety@springernature.com

In case Publisher is established outside the EU,
the EU authorized representative is:
Springer Nature Customer Service Center GmbH
Europaplatz 3, 69115 Heidelberg, Germany

Printed by Libri Plureos GmbH
in Hamburg, Germany